GEMS IN MYTH LEGEND AND LORE

Revised Edition

Bruce G. Knuth

Jewelers Press

Parachute, Colorado

This work is a compilation of research conducted by the author with the assistance of those mentioned in the acknowledgments. The opinions and remedies expressed may be conflicting. These are not necessarily the opinions held by the author or the publisher. The information and advice contained in this text are in no way to be considered a substitute for consultation with duly licensed medical practitioners.

Copyright© 2007 Bruce G. Knuth
Jewelers Press
Parachute, Colorado, U.S.A.

by: Bruce G. Knuth

book design and illustrations: Bruce G. Knuth
except illustrations on pages 315-317 & 319
by permission of Letterpress Graphics

All rights reserved. No part of this publication may be reproduced or transmitted in any form or by any means, electronic or mechanical, including photocopying, recording or any storage and retrieval system now known or to be invented, except by a reviewer who wishes to quote brief passages in connection with a review written for inclusion in a magazine, newspaper or broadcast.

Information and ordering:
Jewelers Press
652 Meadow Creek Drive
Parachute, Colorado 81635

Printed in the United States of America

Library of Congress Catalog Card Number: 99-94023

ISBN 978-0-9643550-4-0

10 9 8 7 6 5 4 3 2

Preface

The literature regarding precious stones has a tradition spanning two thousand years. The vast majority of these writings have dealt with the mystical, metaphysical and medicinal qualities these natural materials are supposed to possess. These texts contain the earliest studies of medicine and the physical sciences, the religious traditions of various cultures, and attempts to explain the curious coincidences of life that seem to relate to precious objects. The term most often applied to these works is *lapidary*, a word that means referring to precious stones or working with precious stones."

The purpose of this text is to collect and organize the writings of these lapidaries and other major works that speak of gems. These include the Holy Bible, the Koran, the Vedic Texts of India, the writings of leaders in the Roman Catholic Church, the opinions of ancient philosophers, and the works of numerous authors in European literature.

Besides the sources commonly consulted for most writings about gems, the author perused the complete works of Chaucer, Milton, Shakespeare, Sir Walter Scott, Emerson, and Sir Thomas Moore to find as many references to gems as possible. A new and complete translation of the writings of Chevalier Jean de Mandeville from the original French was also commissioned for this publication. An English translation of the writings of Marbode, Bishop of Rennes, is included to allow a first hand reading of this most influential work.

The citations included in this text have been derived from numerous sources. When feasible, these were gathered from original sources or generally accepted translations. A comparison by the author of original texts and later translations has shown many discrepancies in wording, context, and basic facts. Upon reading the extended quotations in this work, it will become clear that standards of grammar and punctuation have varied throughout the centuries. Readers should be assured that the often awkward writings of some scholars have been retained in an attempt to remain faithful to history and tradition. Many obvious errors found in successive transcriptions and translations have resulted in misinformation passing as truth.

A close study of the attributes listed for some gems will reveal the sources of our contemporary beliefs regarding precious stones. It will also show that the virtues given for many gems have changed over time, sometimes to the opposite of early beliefs. Noting these contradictions helps explain why so much confusion and misinformation exists concerning the claimed powers of gems.

The physical and optical characteristics of each gem are given in their opening descriptions along with the common-practice names used in the jewelry trade. Terms and gem names that are technically correct but are now considered archaic or have lost their original meaning in the jewelry trade are included and defined. Examples are the gem names sard, onyx, and sardonyx. Modern jewelers rarely distinguish between the lighter and more orange stone carnelian and the darker-brownish gem called sard. The name onyx is commonly used to refer to dyed solid-colored chalcedony. Onyx is properly used to describe chalcedony composed of straight, parallel bands. Today it is almost impossible to find sardonyx, sard alternating with bands of black or white, in the inventory of gem dealers. Hopefully, this book will serve to bring clarity and consisitency to the discussion of gems, gem lore and the terminology used in the gem trade.

A renewed interest in gems, minerals, metaphysics, and the use of precious stones in healing has shown the need for information and understanding. It is the hope of the author that this text will help bring clarity to this fascinating field of study.

The illustrations that accompany the text in this book were either drawn by the author specifically for this work or are redrawings of early woodcuts. A complete collection of woodcut illustrations by Johannis de Cuba, originally produced for his *Ortus Sanitatis* of 1483, has been included. Each of the originals have been re-drawn by the author using Adobe Illustrator® software to allow clear reproduction.

Acknowledgments

To everyone who contributed to this book, the author expresses his thanks. This includes the research and interlibrary loan staff of the Denver Public Library, the library archives department of the Colorado School of Mines for allowing use of their original copies of Agricola's *De Natura Fossilium* and the *Musaeum Metallicum* by Aldrovandi, and the University of Oklahoma for their loan of an original English translation of the *Speculum Lapidum* of 1502. Thanks also go to Kate Blanas for her arduous translation of Jean de Mandeville's fourteenth century French manuscripts.

The critical reading and comments of my teaching colleague Marta Hedde were also extremely helpful in improving clarity, style, and accuracy.

Finally, a special thank you to my wife Susan and my daughters Melanie and Lisa for being a loving and supportive family throughout the preparation of this work.

B. Knuth

Table of Contents

Introduction 1

Crystal Shapes 8

A History of Writings 11

A Modern Lapidarium 22

Gems in Literature 250

Birthstones 291

The Power of Shape 328

Appendices 333

Endnotes 345

Index 369

*And they wer set as thikke of nouchis (a setting for jewels)
Fulle of the fynest stones faire, That men rede in the Lapidaire*
Chaucer, The Hous of Fame[1]

Introduction

Gems rank as the first precious objects of early man. From the time the first colorful pebble was recovered on the banks of a stream, to the current opinion that a diamond will seal the bonds of love, people have held the belief that gems possess magical properties. These metaphysical properties became the subject of oral traditions, numerous pieces of literature, and a wealth of contemporary notions.

The value placed on gems is due to their unique properties. Their colors are the richest and purest found in nature. The durability of many is unrivaled and it is a gem that ranks as the hardest substance known. Is it any wonder that man in his temporary existence should revere that which seems everlasting? Many ancient cultures believed that rocks and minerals were the skeleton of Mother Earth and the repository of her power. The most blessed of all rocks, those of greatest beauty, must therefore hold powers beyond our comprehension.

The oldest manmade objects known are those of stone. One can imagine a prehistoric cave dweller feeling that a special flint or jasper spearpoint would never fail to conquer the most fearsome beast. Gems have been found in prehistoric burial grounds, the monuments of the Pharaohs, and in the ruins of Sumerian cities. The mummy of an ancient Egyptian queen was adorned with a gold and turquois bracelet made 7500 years ago. Cylinder seals, signet rings and beads of agate were fashioned by the ancient lapidaries of Mesopotamia in 3000 B.C. and the earliest man made objects of Northern Europe are thought to be amber carvings dating from the Mesolithic period, about 7000 B.C. These animal carvings are thought to be primitive hunting amulets giving the owner an affinity with, and power over, his prey.

Gems as Talismans and Amulets

It is accepted by many scholars that the wearing of gems originated as a talismanic practice and not for mere adornment. This, however, cannot be either proved or disproved by documentation. Even the oldest texts of a civilization do not represent it in its primitive state. Any record of the use of gems as talismans or as bodily ornaments was certainly long predated by the practice. By the time a tradition of literature has risen in a culture; religious beliefs, superstitions, modes of dress and personal adornment have been established.

Practices of fetishism in all forms depend on a particular mind set defining life and soul. Will and consciousness must be assigned to what is considered by modern western man as inanimate. Inanimate literally means to be "without a spirit" or, based on its Latin root, "unable to breath." The concept of assigning life to objects is common to most early civilizations. It often persists in some form throughout a civilization's history. The origin of the practice is demonstrated by animals and infants. In their innocence, they assume that any moving object is alive.

The animation of precious stones, since they do not move, must have originated for another reason. They are not large and imposing in their form and they rarely duplicate a human or animal shape. Their brilliancy and rich color are more likely the reasons they were first selected. Many animals are attracted by rich color, their propagation often depends on it. Babies are often attracted to small brilliant things and will reach for them without fear. Precious stones possess no threatening characteristics in their appearance and nothing to interfere with the pleasure associated with color. Numerous studies have demonstrated the natural attractive powers of color, its ability to soothe or excite, and capacity to change mood. Certainly anything that naturally exhibited these attributes was probably assumed to contain many others.

The first objects to be worn were those that could be easily strung or gathered together. Shells, seeds, and soft stones are found in many old burial sites. The lack of tools to work harder gems made them impossible to drill. These bright objects were probably collected and kept as pretty things, waiting for the appropriate technology. This may have been a knowledge of harder materials that could wear away a gem or the ability to work soft metals, allowing the wrapping or setting of stones. Once the ability to fashion gems arrived, their use as personal talismans and objects of adornment followed. When stones were worn, there was an inclination to connect occurrences in ones' life with these unique things. This grew into a belief the gems contained powers and probably the force of some spirit. To wear the gem would mean the spirit and its powers could be contained, controlled, and kept close at hand.

An examination of the etymology of the term "amulet" reveals two opinions as to its origin. The common belief is that it derives from the Arabic word *hamalât*, meaning to "suspend something when worn." A more studied opinion is the Latin derivation of the term. Pliny the Elder uses the word *amuletum*, but not always to speak of a worn object. An etymology by Varro (118-29 B.C.) states the source of amuletum to be the verb *amoliri*, to "remove" or "drive away." This follows the belief that one function of an amulet is to keep away danger or remove sickness and harm. The word "talisman" was not used in classical literature, but undoubtedly comes from the Arabic *tilsam*. This word was derived from the Greek *telesma(telesma)*, meaning an "initiation" or "incantation." In modern use, the words amulet and talisman have become interchangeable, although some prefer to reserve the term talisman for engraved or sculpted objects.

The choice of particular gems as amulets may be due to various factors. The first is the fact that some gems occur in forms that resemble other objects. The staurolite is an example of a stone that is found in the form of a sacred symbol. Its twinned crystals take the shape of a cross. Many specimens are so symmetrical and perfectly formed, they look as though they were carved by some mystic hand. Pearls resemble drops of liquid, some perfectly round and others in the form of a tear. They appear to have been placed in their homely environment rather than forming as a natural product of a lowly mollusc. Amber is another example of a gem which appears almost magical in appearance. Many pieces are found containing fossilized insects, an example of a living being encased in what appears to be stone. To many observers in the past some of these delicate winged fossils seemed to be spirits, angels, or even miniature human forms.

A second explanation is expressed in the "doctrine of sympathy and antipathy." Simply stated, it is the belief that something that has an attribute similar to something else, like color, will have an effect on that thing. The long held belief that red stones have a natural connection to blood is one example, that blue stones, such as lapis-lazuli and turquois, have a correlation to water or the sky is another. An additional expression of this doctrine is the view that the substance of some stones is susceptible to a change in the health or thoughts of the wearer. Stones were thought to dim or lose luster in the presence of illness, impending death, perjury, or unfaithfulness.

Certain stones are known to be changed by wear or contact with the skin. Turquois will darken or turn green in the presence of body oils, pearls experience a loss of luster due to wear, and opals lose their "fire" when exposed to heat, solvents, or acids, all chemicals present in perspiration. The opinion that more stable gems such as diamond or ruby are also capable of such changes may be accounted for by the natural abrasion of inferior stones mistakenly labeled as more durable stones. Clear quartz was often mistaken for diamonds and red garnets were at one time classified as rubies. Both of these gems are readily abraded. An example of a stone that may improve in appearance with wear is the emerald. Soaking this gem in oil will enhance its color and hide small fractures. Conversely, an oiled stone will suffer if subjected to heat or emulsifying solvents. Wearing an emerald can cause it to be exposed to body oils, body heat, and cleaning solutions. As a result, an emerald's appearance could fluctuate radically over time.

A third reason gems were chosen is their emission of natural energies. Beyond the readily evident visible energy we call light is a wide range of energy in the electro-magnetic spectrum. Their existence in gems has been speculated since ancient times. The visible portion of this spectrum contains the brilliant colors which first attracted early man to many stones. Beyond visible light exists many other forms of energy;

A woodcut from the Ortus Sanitatis *of Johannis de Cuba, a fifteenth century medical text, depicts a jeweler preparing rings for sale. The use of rings and other forms of jewelry set with stones was a common medical practice in treating and preventing disease.*

ultraviolet, infra-red, x-rays, gamma rays, radar, and radio waves. Gems and minerals generate or transmit all these energies.

The mineral radium was the first recognized to emit x-rays and the radioactivity of uranium is commonly understood. Quartz was proved to possess the property of piezo-electricity in 1921. A plate of quartz cut parallel to the direction of its prism faces produces an electrical charge when compressed or stressed. When vibrated by the introduction of an electrical current, quartz stabilizes the frequency of radio transmissions. "Crystal set" radios, popular in the nineteen thirties and forties, used galena crystals for their similar ability to modulate radio frequencies. Quartz is also pyro-electric and will generate a static charge when heated or rubbed vigorously. Tourmaline and amber have also been demonstrated to be pyro-electric. Amber's ability to attract small particles was noted by the ancient Greeks. And finally, the naturally magnetic crystal lodestone was observed to attract iron by the ancient Romans. These energies, and others possibly undiscovered, make connection with and may alter the energies which naturally exist in each person.

Magic, Miracles, and Science

Man's basic needs to understand and explain gave rise to all manner of beliefs, superstitions, and religions. The belief that forces outside our understanding influence man and nature has been with us since the first human sought the reason for his existence. The attributing of magical forces to gems is a understandable expansion of such beliefs. In ancient cultures, when gods and goddesses were believed to interact daily with mortals, the power of gems was taken for granted. Medieval man continued to believe and accept unseen forces beyond his understanding. The belief that gems contained mystic powers was reinforced by the scholars of the church. These patriarchs and their writings were strong controlling forces in each person's life. As stated by G. F. Kunz, "When the existence of miracles is acknowledged, there will always be a tendency to regard every singular and unaccountable happening as a miracle."[2]

The Renaissance brought a spirit of investigation and a search for reason. It is interesting to find little doubt expressed in this period regarding the power of gems. The resources of science were devoted to finding a plausible explanation for the virtues assumed to exist in stones. In this "Age of Reason," the lore of their effects on health, mental condition, and personal fortune was expanded and not refuted. It was not until the "Age of Skepticism" of the late nineteenth and the twentieth century that the metaphysical capabilities of gems was brought into question by the general population. The abilities of science to explain many things often accepted as miracles brought a sense of doubt regarding the attributes of gems. Despite the belief that the power of mysticism was false, the use of gems as talismans continued. The late twentieth century has seen a renewal of interest in metaphysics, mysticism, and magic. The wearing of birthstones and crystal amulets has never been more popular than it is today, and the practice of metaphysical arts has increased as the new millennium approaches. The next century may find some of the claims of the past regarding the virtues of gems substantiated. The operation of some unknown force of nature may await discovery.

A Basis in Fact

The study of past beliefs and the application of modern science reveals truths behind many ancient claims. As with some superstitions and myths, the healing powers of some gems is based in scientific fact. Some of the applications have, however, become exaggerated and confused. One example is the use of powdered pearls as a cure for digestive distress. Pearls are approximately ninety percent calcium carbonate. Similar chemical compounds are found in modern antacids and other stomach remedies. Early physicians, without the benefit of modern chemistry, found a natural source for this remedy.

The use of stones to stop bleeding is mentioned in many lapidaries. The gems recommended are most often red, an expression of the "sympathy of color" theory. It is also suggested in many references that these stones should be cooled and applied directly to the wound. The credit was given to the gem, the treatment is what is now known as a cold compress. A cooled gem would certainly retain its temperature and act as an effective compress. Hematite, known as the "stone which bleeds" was also recommended for the treatment of wounds. This dense iron oxide exhibits a blood red streak on a touch stone. It also has strong astringent and styptic properties. Egyptian physicians were aware of these qualities over three thousand years ago. Hematite powder will act to quickly coagulate blood and remove other fluids from a wound.

Egyptian physicians also used green crystals to treat eye diseases. They used copper oxides mixed with boric acid as an eye wash. Known as *lapis armenus*, this astringent has been shown by modern medical research to be effective and is still prescribed. The prescribing of powdered emeralds in eye washes may be traced to the Romans. The copper oxides used are green or blue in color. The Romans had a practice of referring to all green gems as *smaragdos*. A word which latter became commonly associated with emeralds. It is likely the Latin physicians, when translating Egyptian prescriptions, mistakenly substituted emeralds for the effective copper oxide crystals. The same explanation may apply to the use of sapphire and lapis-lazuli as eye remedies. Other Egyptian texts describe the copper oxides as blue crystals. The name lapis-lazuli translates literally from its Latin roots as "blue stone." The same problems of transcription may be the reason many blue stones are recommended for eye treatment.

Amber, and its fumes when burned, have served as an aromatic remedy for centuries. The scent serves the same purpose as camphor or menthol inhalants. A mixture known as oil-of-amber consists of powdered amber and alcohol or mineral spirits. This concoction has maintained its reputation as a cure for gout, whooping-cough and bronchitis, and as an antispasmodic for asthma. As recently as 1935, official publications of the United States government suggested oil-of-amber as a viable pharmaceutical product.

The use of magnets to relieve pain has become a common practice in the past few years.[3] Magnets now serve as external anesthetics in the training rooms of many professional sports teams. The Roman warrior, Alexander the Great, may have known of this effect when he issued loadstone, a naturally occurring magnetic crystal, to his troops. Alexander believed the gem would protect his men from pain as well as the spells of *jinns*, or evil spirits.

As one studies the collected gem-lore of the past, other modern medical treatments and their resemblance to ancient practices may be found. Further studies, additional scientific research, and the discovery of yet unknown forces may reveal additional foundations of numerous ancient gemological practices and beliefs.

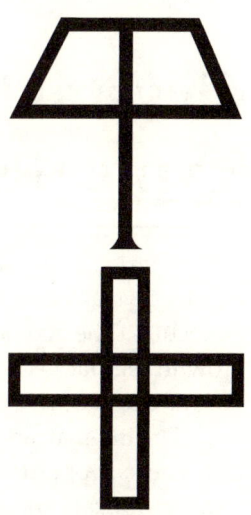

These were signs for lapis or stone in alchemy and early chemistry.

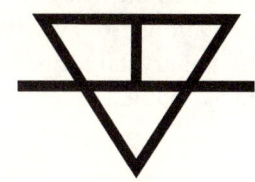

The symbol shown below is found in eighteenth-century chemistry texts meaning lapis or stone.

> *Give me an amulet*
> *That keeps intelligence with you,—*
> *Red when you love, and rosier red,*
> *And when you love not, pale and blue.*
> Emerson, The Amulet[4]

Crystal Systems and Typical Forms

Nearly all gemstones are minerals, and nearly all minerals grow in regular crystal forms. Crystals possess a regular lattice of atoms, ions, and molecules. The geometric arrangement of these parts causes minerals to grow in regular geometric shapes. These crystal shapes are bounded by flat surfaces known as faces.

The inner arrangement, or lattice, sets the crystal's physical properties. These properties include outer shape, hardness, cleavage, specific gravity, optical properties, and type of cleavage.

Crystals are divided into systems based on their lattice structure. Each system has different optical axes and angles at which the axes intersect.. Irregularities in rate of growth in a particular direction or along a particular face causes most crystals to be irregular in shape. The angles at which the faces intersect and the optical axis remain constant despite deviations from the typical shapes. Many minerals occur in a combination of crystal forms. The arrangement prefered by a mineral is referred to as "habit."

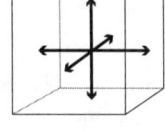 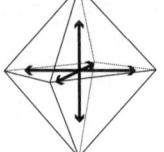 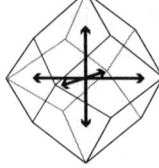

Isometric or cubic system

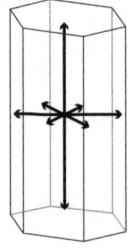 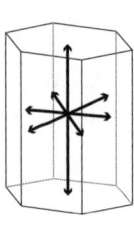 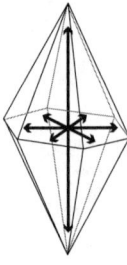

Hexagonal system

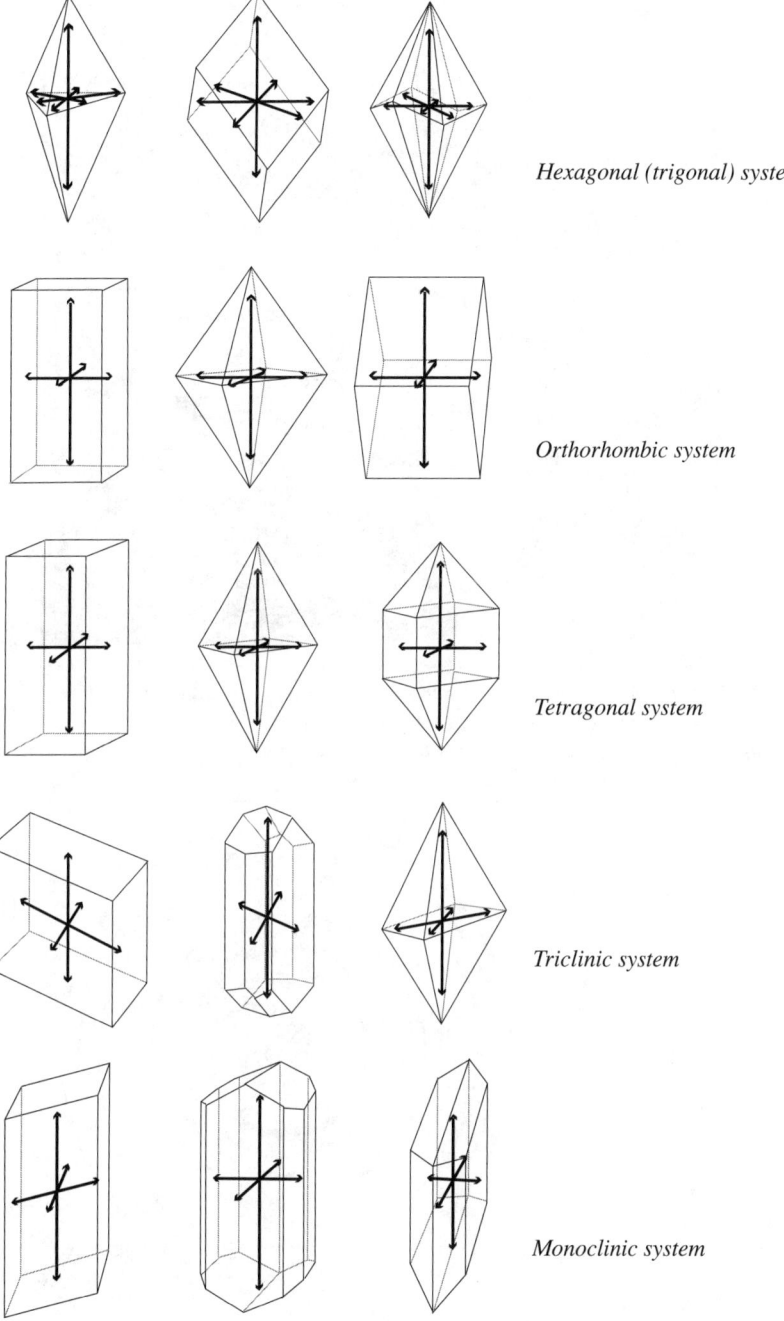

Hexagonal (trigonal) system

Orthorhombic system

Tetragonal system

Triclinic system

Monoclinic system

Crystal Systems

A nobleman looks for gems and minerals with a divining rod, an old practice alluding to the energies emitted by gems.
Illustration from De Re Mettallica *by the Renaissance scholar Georgius Agricola.*

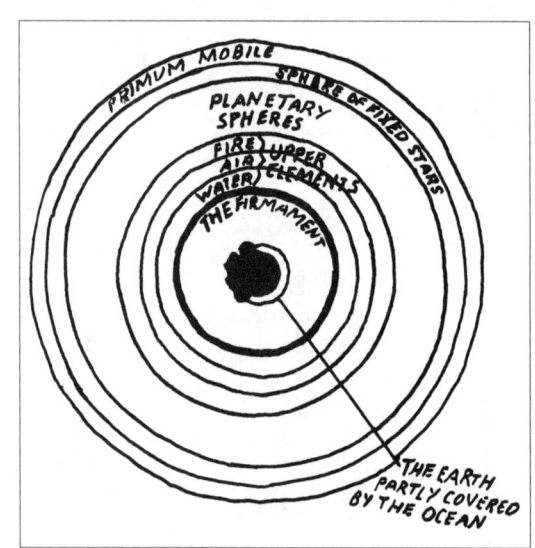

The Universe of Aristotle as conceived by an anonymous medieval writer. The outer rings represent aether, Aristotle's fifth element.

> *The universe is full of magical things*
> *patiently waiting for our wits to grow sharper.*
> Eden Phillpotts

A History of Writings

References to gemstones appear in many ancient writings. They may be mentioned as decoration, a means of barter, a symbol of wealth, the source of power and magic or the objects of adoration. In these earliest records the gem was auxiliary to the primary topic and not a subject in and of itself. It was not until the writings of Aristotle (384-322 B.C.) that the physical world and its constituent parts became worthy topics of study and documen-tation. Aristotle's division of philosophy into separate branches— logic, metaphysics, physics, ethics, politics, and art led to the systematic study of nature. The natural world also was divided into parts or elements. These elements were earth, water, air, fire, and ether. Gems could then be classified and described in these terms and their properties recorded. The study of gems and minerals began with the recording of observations, documenting of sources, and the cataloging of legends and folklore.

The first writings were attempts to compile all that was known at the time. Along the way, a mix of science and pseudo-science filled the works produced. As the understanding of the natural sciences grew, so did the sciences of mineralogy and gemology. The following is a brief chronology of these milestones in gem literature.

Origins

Theophrastus (ca. 372-287 B.C.), a student of Plato and successor to Aristotle as head of the Peripatetic school of philosophy, is credited with writing the first known treatise dedicated to gems. Of his *Peri Lithon (Of Stones)* only a fragment of the original work is preserved. This Aristotelian mix of physics and metaphysics is credited as the source for much of the lore passed along in latter works. The first English translation of *Peri Lithon* by John Hill is superseded by two modern translations, the first of these by E. R. Caley and J. F. C. Richards and the second by D. E. Eichholz.

The Roman author and historian Gaius Plinius Secundus, a.k.a. Pliny the Elder (23-79 A.D.) compiled the *Historia Naturalis (Natural History)* from a reported two thousand sources. This thirty-seven volume work, published in 77A.D., embraces the domain of natural history in the broadest sense. It includes astronomy, geography and meteorology. Beyond what is today considered natural history, it also includes fine arts, inventions, and human relations. It stands as the chief source from which future historians have obtained informative descriptions of life in the first century.

Much of Pliny's information on gems was not written from direct observation. His studies were often a retelling from second-hand sources, now lost in antiquity. Many stones are lumped together by color, mining location, or attributes passed on in legends. His recounting of legends and myths, and assumptions drawn from these, are the basis of much of the writing which appeared into the Middle

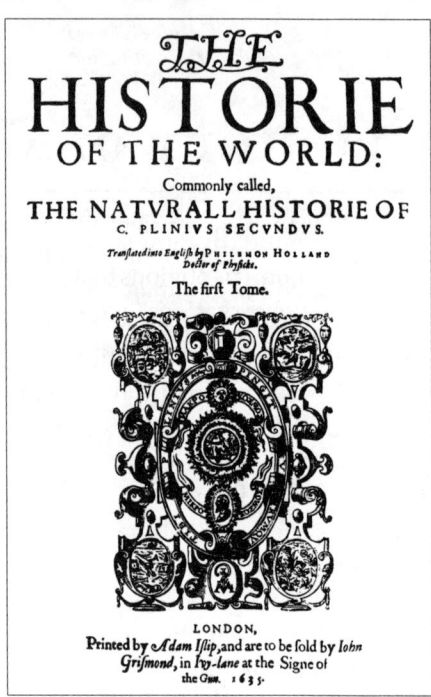

Title page of a 1635 printing of Philemon Holland's translation of Pliny's Natural History, *first published in 1601. This was the first English translation of this classic text.*

Ages. He became the most often quoted writer on the subject, and his information was relied upon and repeated without question in book after book. In the first thirteen centuries of European history, little scientific in-formation was added to the limited knowledge of gems garnered from Pliny's accounts. However, many later authors embellished the writings and added their own speculations and unwarranted conclusions.

Pliny is credited with his attempts to classify gems according to color and observable external characteristics. This led to the practice of assigning the attributes of one stone to all stones of similar color or appearance. The green stones named *smaragdus* in Latin are prime examples. Pliny classified most green stones as varieties of emerald. In all, twelve varieties are listed; green sapphire, turquois, smithsonite, malachite, jasper, and even glass are identifiable by his descriptions. He also is credited with making some astute classifications. It is thought that he was the first to recognize the connection between beryls and emeralds. "Beryls are the same nature as emeralds, or at least very similar."[1] Some of the stones grouped by Pliny share little in common. It is obvious that some of his "observations" were gathered through other's eyes. Errors in translation or reliance on oral tradition may have led to the inclusion of misinformation about stones that Pliny was not able to observe in nature.

Since the original Latin version, over two hundred and fifty editions of his classic have been published in numerous languages. The first complete edition of *Natural History* was published in Venice in 1469, and the first critical edition was offered by Hardouin in 1685. The most detailed study of Pliny's gemstone accounts is a work by S. H. Ball.

Title page of Conrad Gesner's publication of St. Epiphanus' treatise on the Gems of the

Christian Lapidaries

Saint Epiphanius (ca. 315-420 A.D.), Bishop of Constantia in Cyprus, wrote the first treatise dealing with the gemstones mentioned in Biblical writings. This work served as a model for the treatment of Biblical gems in future lapidaries. It is also the first reference to the possible composition and significance of the twelve stones in the Breastplate of Aaron. Epiphanius is known to have traveled throughout the Eastern Mediterranean and studied in Egypt and Palestine. Lynn Thorndike writes that these early opinions concerning the stones of Aaron's Breastplate, "perhaps gives an excuse and sets the fashion for the Christian medieval Lapidaries."[2] Dorothy Wyckoff states in her studies of lapidaries that even though the Church had banned the practice of heathen worship and superstitious practices:

Even devout Christians could not entirely shake off the old belief that precious stones possess some sort of supernatural powers or significance.... this interest was to some extent legitimized by focusing attention on the stones mentioned in the Bible, especially the two (different) lists of the "twelve stones"—those in the breastplate... and those in the foundations of the New Jerusalem.³

The title page of Marbod's extended poem on precious stones published in Vienna in 1531.

The entire text of Epiphanius' Biblical Lapidary was first placed in print as part of *De Omne Rerum Fossilium* by Conrad Gesner of Zurich in 1565. Noted for the fervor in his writings, Epiphanius had been quoted and referenced by numerous authors centuries prior to this first complete publication.

Isidore, Bishop of Seville (ca.560-636), was the leading clergyman of his day and exerted a great deal of influence over the thought and literature of the Middle Ages. His *Etymologiarum sive Originum Libri XX* (*Etymologies*) was an encyclopedic work containing more than one thousand manuscripts dealing with the collected knowledge of the day. The *Etymologies* was intended to serve as a dictionary rather than an encyclopedia or instructional manual. His descriptions, classifications, and attributes given to many gems made his writings the standard reference for scholars into the twelfth century and beyond. This work is often referred to by later writers who explore the subject of gems and minerals.

The eleventh century Bishop of Rennes is known by a variety of names; Marbod, Marbodei, or Marbodus followed by the names Redonensis or Andecavensis. Commonly referred to as Marbod, this medieval scholar lived from about 1035-1123. His *Libellus de lapidibus*, a.k.a. *Lapidaire en Vers*, is considered the most important and influential work of its kind. The chronicler of Marbod, Lynn Thorndike, calls it "the classic on the subject of the marvelous properties of stones."[4] These poetic works were so popular that more than one hundred and sixty manuscripts were prepared in Latin and several translations were written in the vernaculars of the day.

The text consists of seven hundred and thirty-four hexameters concerning sixty kinds of stones. It is prefaced by twenty-three lines describing the source of information as Damigeron or Evax, king of Arabs. F. D. Adams states that it is "the earliest lapidary of the Middles Ages, and also the one which is quoted most widely."[5] The first published text appeared in Vienna in 1511.

The entire work may be divided into five sections according to Adams. The first deals with twenty-six mythical stones and is so obscure that little connection can be made with known minerals. The second section describes stones of animal or organic origin. The third category includes four stones which may be distinguished as separate recognizable minerals. Descriptions of fourteen quartz varieties characterize the fourth section, and fifteen other recognizable minerals make up the final listing. This last grouping contains little in the way of physical characteristics, but is filled with legend and lore of these gems. Marbod attributes his knowledge of the magical properties of stones to Evax.

Despite its standing as a primary reference, Marbod's writings rely on information from other earlier writers. Little material is included that had not appeared in the works by Solinus (third century Roman grammarian and author of geographic compilations) and Isidore. The geographic information listed by Solinus is attributed to his studies of Pliny.

Albertus Magnus (1193-1280) Bishop of Ratisbon, teacher and lecturer is noted for his application of science and the scientific method to theology. He served as a teacher at the University of Paris and counted Thomas Aquinas as one of his students. Magnus is known as one of the founders of the school of Scholasticism and believer in reason as a supplement to faith. He made Aristotelian thought intelligible to Latin scholars.

His *De Mineralibus* is characterized by Dorothy Wyckoff as "an impressive attempt to organize the science of mineralogy."[6] The primary criticism has been that it still included a great deal of superstition and speculation. The first printed version appeared in Padua in 1476. Adams states, "(*De Mineralibus* is) one of the best and most comprehensive of the western medieval lapidaries . . . enlarges at length on their mystical and wonder-working powers and virtues."[7] This lapidary has been lauded for its attempt to explain the formation of minerals and gemstones, the causes of their color, and other physical properties.

The great writer and thinker of the thirteenth century, Alfonso X the Learned (1221-1284 A.D.), King of Castile and Leon, composed the first major work on astrology and gemstones. His principle object was to demonstrate the connections of gemstones and other minerals to celestial bodies. The astrological tables he devised were based on Arabic translations and were calculated for the Toledo Meridian. These tables, however, became the standard for all of Europe for centuries. The text was extensively illuminated and illustrated.

Alfonso classed gems by color and placed each under one of the twelve zodiac signs. Remarks on properties, uses, and medicinal virtues for each stone were included with attention paid to how these virtues are influenced by the planets and stars. His writings serve as a model for the connections between gems and astrological signs to the present day.

Title page of a 1705 printing of The Mirror of Stones, *first English translation of Camillus Leonardus'* Speculum Lapidum, *published in Latin in 1502.*

Introduction of the Scientific Method

Camillus Leonardus, physician of Pesaro, Italy, gathered information from previously published works, but used scientific methods to distinguish gems. His findings were published under the title *Speculum Lapidum* in 1502 and republished in England under the title *Mirror of Stones* in 1750. Leonardus treated physical properties, hardness, specific gravity, compactness, color, form, and geographic origins more thoroughly than previous works. He held that the virtues, medicinal use, and properties of gems was imparted or enhanced by having their surfaces engraved with suitable symbols and signs. He further stated that some powers attributed to gems existed only by such engraving.

The Science of Mineralogy

Georgius Agricola (1494-1555) has been called the father of mineralogy. His *De Natura Fossilium* (1546) is considered one of the most important mineralogical/gemological works of all time. Unlike his predecessors, he provided comments of considerable accuracy on the nature, properties, and treatment of gemstones. The descriptions indicate a first hand knowledge of many of the gems listed. It is evident that he relied on the writings of Pliny to describe stones of which he had no personal knowledge. He also "asserted that minerals and certain rocks originated from petrified juices."[8] He makes reference in his text to scratch tests for hardness and includes descriptions of physical characteristics and characteristic flaws of many gems. These physical descriptions have proved to be quite precise, even under today's standards.

De Omne Rerum Fossilium (1565) is a compilation by Conrad Gesner (1516-1565) of eight works on gemstones. It contains Epiphanius's treatise on Biblical stones, as well as Fanciscus Rueus' *De Gemmis Aliquot*, a general treatise on stones published in 1547. It also contains an original writing by Gesner, *Lapidum et Gemmarum*, which classified stones by their external characteristics into fifteen categories. Its publication marked the first time a purely mineralogical work was illustrated with woodcuts of minerals, crystals, fossils, and cut gems.

The title page of Gesner's essay on gemstones, a portion of a larger work published in 1565.

The work perpetuates the classification and naming of stones by color and the tradition of differentiating within classes by place names. In his system true emerald is called "Occidental emerald", tourmaline is called "Brazilian emerald", green sapphire is "Oriental emerald", and yellow or orange sapphires are called "Oriental topaz". The tradition of using these misnomers has continued in the jewelry trade to the present day.

Gemmarum et Lapidum Historia (1609) by Anselmus Boëtius de Boodt (ca. 1550-1632) is one of the landmarks in lapidaries. It deals with gems from the New World and compares them to those available in Europe and Asia. De Boodt lists five degrees of hardness in stones and speculates concerning the existence of a distinct atomic structure in minerals. His treatise gathers together information which demonstrates the best scientific knowledge of the day. He cites material he borrowed from other writers and adds new observations. His position as physician to the royal court of Rudolph II of Prague gave him the opportunity to study numerous gems first-hand. He also expresses skepticism over the mystical and medicinal virtues of gems and minerals. His writings demonstrate advances in the science of gemology and show a greater understanding than writers who preceded him. Adams calls de Boodt's writings, "in many respects the most important lapidary of the seventeenth century."[9]

Musaeum Metallicum in Libros IV Distributum by Ulyssis Aldrovandi (1522-1605) was published posthumously in 1648. It is an illustrated text on the mineral kingdom. Little or no new information is included, but the term geoligia of geology appears for the first time in the context it is used today. With this work, the field of gemology split into two distinct disciplines; the science of mineralogy and the romance and lore of gemstones.

The gemological historian, F. D. Adams, closed his chapter on medieval mineralogy with a discussion of Aldrovandi's work by stating:

> Medieval mineralogy in fact is not a science . . . . not a solid tower of learning . . . but a fairy castle, the insubstantial fabric of a dream, often quaint and even beautiful, but destined to crumble away because it had no foundation in reality . . . it was now to be succeeded by a true science of mineralogy built upon the basis of close observation and diligent study of the materials of the earth's crust.[10]

Nearly three centuries of gemology have followed this tradition.

The first major work published in English is by Thomas Nicols and carries the curious title *A Lapidary: or, the History of Pretious Stones,* with *Cautions for the Undeceiving of All Those That Deal with Pretious Stones* (1652). Nicols describes in detail gems and their imitations, glass replicas, and the use of foil as a backing agent. The essays list supernatural char-acteristics of gems and gives the causes and methods of detection for these phenomena. Nicols cited de Boodt as a primary source, but his research did add some new information.

Robert Boyle (1627-1691), English philosopher, chemist, and physicist, wrote *An Essay About the Origine and Virtues of Gems* in 1672. It was considered the most scientific work possible in its time. It covers the formation of minerals and their crystals. Important new work was included on crystal formation and crystallization from solutions as reported by direct observation. Minor remarks were made on the impossibility of the medicinal uses of gems due to their insolubility. Boyle doubted the ability of gems to have a remedial affect, but admitted that some soluble minerals may be of benefit. He also mentioned the use of hydrostatic weighing (specific gravity) to ascertain a gem's density and the measuring of refractive properties as means of separation and identification.

London physician Dr. Robert Pitt continued observations on the medicinal value of gems in his 1703 book, *The Craft and Frauds of Physick Expos'd*. This second work on the theme of medication was:

> written to show the small cost of the really useful drugs, the worthlessness of some

The first English work on gemology was published in Cambridge by Thomas Nicols in 1652

Title page of Robert Boyle's scientific study of gems and minerals.

expensive ones, and the folly of taking too much physic. The book gives a clear exposition of the therapeutics of that day, and is full of shrewd observation.[11]

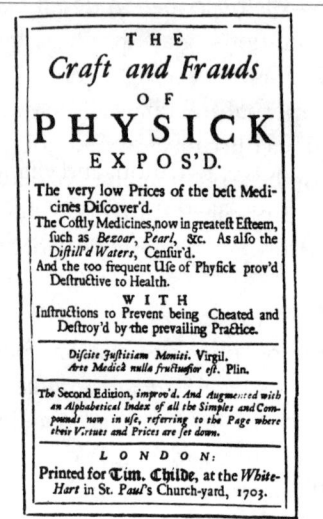

Title page of Dr. Robert Pitt's treatise on medical fraud

Pitt reinterated the insolubility of gems and the fact that this would make it impossible for them to affect the body. He gave as proof the example of stones ingested by birds. He observed that they are passed without any change in color or shape.

Modern Gemology

The early eighteenth century saw the fields of mineralogy and geology adopted as pure sciences. The introduction of better record keeping and specialized tools led to more accurate and reliable information. Writings in these areas concentrated on careful observation, chemical and physical experimentation, and the documentation of the results. From the eighteenth to the twentieth century a vast amount of technical data was gathered.

Gemology, however, relied upon the retail jewelry industry for any research or study. Since legend, lore, and the romance of gems was perceived to be "good for business," research was in this specialized area. Published information did little to expand on limited knowledge which had existed for centuries.

The research and writings of Dr. George Frederick Kunz in the late nineteenth and early twentieth century revived interest in the study of gems. His work cataloging the gems of North America, as consultant to Tiffany's, and as founder and curator of the mineral collections at the American Museum of Natural History, brought gemology to a new level. His technical writings gained little public notice, but his compilations of ancient pseudoscience—*The Curious Lore of Precious Stones* (1913) and *The Magic of Jewels and Charms* (1915)—have remained popular works.

Robert M. Shipley's founding of the Gemological Institute of America in 1931 and the American Gem Society in 1934 brought professionalism to gemology and the jewelry industry. These educational and trade organizations continue to bring standardized and reliable information to jewelers and consumers. The GIA quarterly publication, *Gems and Gemology,* is a primary source of the most current research on gems and their characteristics.

Over two thousand years of speculation, research, and writings have been collected by these scholars. The sciences of gemology and mineralogy have grown from infancy and become more and more sophisticated. Physics, chemistry, biology, and medicine have all experienced great changes over two millennia. The study of gems and their properties has contributed to all of these sciences.

The approach of the third millennium has seen a renewed interest in the powers and virtues of gems. The use of gems and minerals as healing agents, personal protectors, and conduits to other realms of con-sciousness has gained great popularity. Mysticism and metaphysics remain strong influences in the lives of many, but misinformation and confusion abounds. Modern authorities profess great insights regarding gems and their uses, but a great deal of the information being disseminated does not recognize the knowledge of the past and the foundations of gem study. As one explores the collected wisdom of these great men, it is evident that there is much yet to be understood.

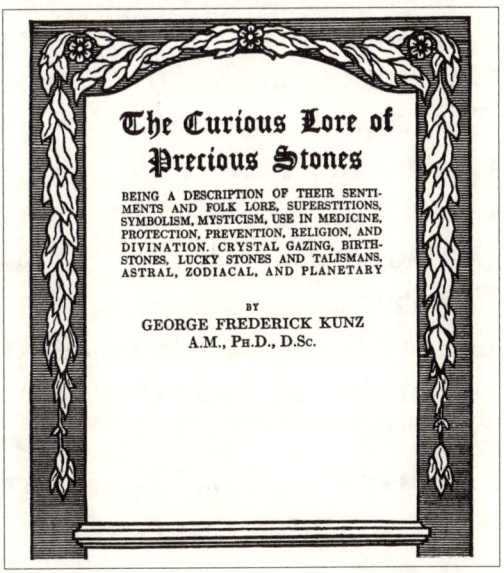

Title page of the classic text by G. F. Kunz, The Curious Lore of Precious Stones, *published in 1913.*

The most obviously beautiful things in the world of Nature are birds and flowers and the stones we call precious.
                                                      Havelock Ellis

A Modern Lapidarium

Agate 23	Jasper 132
Alexandrite 31	Jet 136
Amber 32	Lapis-lazuli 138
Amethyst 42	Lodestone 142
Aquamarine 46	Malachite 145
Beryl 47	Moonstone 146
Bloodstone 52	Obsidian 148
Carbuncle 56	Onyx 151
Carnelian 57	Opal 153
Cat's-eye 60	Pearl 160
Chalcedony 62	Peridot 168
Chiastolite 64	Quartz 171
Chrysoberyl 66	Ruby 182
Chrysolite 67	Sapphire 189
Chrysoprase 67	Sard 197
Citrine 69	Sardonyx 199
Coral 71	Serpentine 201
Diamond 75	Spinel 202
Emerald 92	Staurolite 204
Garnet Group 108	Topaz 206
Hematite 119	Tourmaline 210
Jacinth 121	Turquois 214
Jade Group 122	Zircon 229

Mythical Gems 234

His heart like an agate with your print impress'd
Love's Labour's Lost,
Act 2, Scene 1, line 236

Crystal System

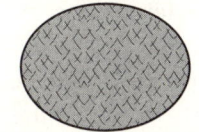

Agate

Microcrystalline aggregates

Color: Various, banded or layered **Mohs' Hardness:** $6\frac{1}{2} - 7$ **Cleavage:** None **Refractive Index:** 1.544 - 1.553 **Dispersion:** None	**Streak:** White **Specific Gravity:** 2.60 - 2.65 **Fracture:** Uneven, Concoidial **Birefringence:** None **Pleochroism:** None

Crystal System: Hexagonal (trigonal) microcrystalline aggregates
Chemical Composition: SiO_2 silicon dioxide
Transparency: Translucent to opaque
Fluorescence: Varies w/ band; partly strong, yellow, blue-white

Agate is a term correctly applied to chalcedony which is characterized by color distributed in curved bands or layers. The term is also loosely used with a prefix to described non-banded, patterned material; moss agate or mocha agate (milky-white chalcedony with green, black, or brown inclusions in dendritic patterns), landscape agate, lace agate, or fortification agate (straight bands which intersect).

Agate is believed to have been among man's earliest possessions. Stone Age trinkets, along with spear and arrow tips, have been found in prehistoric grave sights. Egyptians were mining agate as early as 3500 B.C., and early Sumerian inhabitants of Mesopotamia fashioned cylinder seals, signet rings, beads, and other ornamental objects from this versatile stone. Artifacts from 3000-2300 B.C. were created by these earliest lapidaries. The Sumerians were believed to be the first to distinguish precious gems from common minerals and may have been the first to give specific gems extraordinary attributes. The agate was believed to give the wearer special favor with the gods. Although it is evident that the stone was used abundantly, the source of Mesopotamia's agate has never been determined.

The first reference to agate in western literature is also the source of the name used in modern times. The Greek writer Theophrastus (372-287 B.C.) referred to beautiful stones which sold at a high price. He further stated that this stone was to be found in the river Achates (or Gagates) in Sicily. In 77 A.D. Pliny the Elder repeats this tale regarding the source of agate in his *Natural History* and goes on to detail all knowledge of the stone to that date. Pliny states, "The achates was in older times highly valued but now it is cheap." He continues by cataloging the varieties distinguished and gives the names of each.

> It occurs in large masses and in various colors, hence its numerous names: iaspisachates [agate-jasper], cerachates [chalcedony], smaragdachates [green agate], haemachates [red agate or agate with red], leucachates [chalcedony], and dendrachates [moss agate]. As to the variety called autachates [possibly amber], as it is burned it gives off a smell like that of myrrh.[1]

Pliny's writings are the source of many of the attributes which are given to agate. He recounts in his writings the beliefs which existed in first century Rome; looking at an agate rests the eyes, and an agate held in the mouth will quench the most desperate thirst. The coralloachates (most likely lapis-lazuli) was described as "spotted all over, like sapphirus, with drops of gold, was commonly found in Crete where it was also known as 'sacred agate'."[2] This particular color of agate was said to be capable of curing the sting of poisonous spiders and scorpions. Orpheus writes of another type:

> If thou wear a piece of the Tree-Agate upon thy hand,
> the immortal Gods shall be well pleased with thee.
> If the same be tied to the horns of thine oxen when
> ploughing, or about the ploughman's sturdy arm,
> wheat-crowned Ceres shall descend from heaven
> with full lap upon thy furrows.[3]

Magicians were warned to avoid agates which exhibited a pattern like a hyena's hide, those stones were said to bring domestic turmoil. Stones which resembled a lion's skin were said to be particularly potent against poisons, and were even more potent if worn on a cord made of the hair of a lion's mane. The second century B.C. writer Damigeron states, "The Agate stone has great powers, and if it has a color like that of a lion's skin it is powerful against scorpion bites, for if it is tied on, or rubbed on with water it immediately takes away all the pain and it cures the bite of a viper."[4] A red-yellow type, also called "lion agate," was said to be favored by Roman gladiators as a source of strength and courage.

The association of agate with the lion is repeated in Hebrew tradition. A hair plucked from a lion's mane was a source of courage for the Jews. The agate was considered a suitable substitute, and in time of need, possibly easier to find. The stone and the lion were said to be intertwined in their power. This is thought to be a source for the connection of agate with the zodiac sign Leo. The gem was said to belong to the tribe Nephtali, a name which translates as "my wrestling." The agate would give the strength of a lion to those "wrestling" with both earthy and spiritual problems.

Roman tradition states that a wrestler who wears an agate is invincible, but only if the stone is of true color. Pliny's writings provide a method used to test a stone to be certain its color is true. The tester is directed to place the stone in a pot of hot oil. After boiling for two hours, the color of the stone will tint the oil if it is genuine. Modern gemologists may note that this test for color is the reverse of a primitive method used to dye chalcedony.

Pliny wrote of the use of agate in the preparation of medicines. He specifically names two varieties as having curative powers. Powdered agate from India, described as a green stone, is to be administered as a remedy for eye diseases. Agate from Egypt or Crete, red in color, is best when a cure for spider or scorpion bites is required. No prescription was provided, however, as to the preparation or application of the stones. He does state that it is the cooling effect the gem has on the body that cures venomous bites. This ancient writer also confirms that mortars and pestles of agate were used by Roman physicians to grind their medicines.

Another portion of Pliny's *History* describes the use of agate for other purposes. This stone was used to make cups for serving and consuming wine. These cups were said to cool the beverage. He tells of men who "search amid the regions of the clouds for vessels with which to cool our draughts and to excavate rocks, towering to the very heavens, in order that

> *Who comes with summer to this earth,*
> *And owes to June her day of birth,*
> *With ring of Agate on her hand,*
> *Can health, wealth and peace command.*[5]
> *A popular rhyme prior to the reordering of birthstones by The National Association of Jewelers in 1912*

we may have the satisfaction of drinking from ice."[6] Pliny often referrs to cups made of *myrrha*. The term is used without further definition, but many authors have translated it as a reference to amber. Pliny's description of autachates as an aromatic stone (like that of myrrh) adds to this speculation. The fact that Pliny states the stone was gathered from towering rocks makes any confusion with amber unlikely. The term myrrha seems to serve as a generic term for "hard stone" or any stone resembling quartz. He does relate a tale that seems to refer to a form of agate is if it were amber. He wrote that Persians were said to use the pungent odor of burning agate to avert storms and to subdue violent rivers.

An additional attribute of agate is found in the work of a fourth century Greek author, Physiologus. This early writer states that agate is useful in searching for pearls.[7] The agate was lowered on a cord to near the bottom of the sea. As it descended, the suspended stone would turn toward the location of a pearl. A diver who follows the cord would be rewarded with a pearl bearing oyster. This method may not have proved very reliable, as this is the only known reference to this unique method of pearl collection.

Stories concerning the virtues of agate have passed to us from numerous ancient cultures. The Egyptians held that each color of agate, white, gray, dull red, blue, yellow, and brown, possesses distinct and separate virtues. They made images of their gods in each of these colors. The pharoh's physicians prescribed the milky gray type to prevent a stiff neck and to ease symptoms of colic. Eye agates, called *Aleppo*, are still treasured in Arabia. These stones contain rings of brown and white or blue and white and are used to combat the "evil eye." Their appearance led to their use as eyes in ancient images of the gods and as life-like eyes in mummified kings and priests. The Aleppo was also prescribed as a cure for boils. The stone was to be applied directly to the sore to relieve suffering.

Persians, Armenians, and Arabs did not draw a clear destinction between agate, carnelian and other chalcedony. All of these stones were considered to be of common origin. These cultures did divide stones by the place they were found. The most esteemed of the agate family were those called *yamani*. The yamani were the most attractive of the stones brought from Yemen. Wearing a yamani ring guarded against the collapse of a wall or an entire house. African tribes use black and white banded agate as a *hoodoo* against their enemies.

Travelers to the Near East returned to Europe with tales of stones and their reported virtues. The belief that red agates stop the flow of blood, and that a white carnelian, known as "milk stone," increases the production of milk in lactating women are both given Islamic origin. From Tavernier's early cataloging we have the name *agates arborisées*. This was Tavernier's translation of the Persian word for tree-stone, the stone we know as moss agate. This variety is called *Mocha* stone in Hindustan and is at times the object of worship. The name is taken from the Arabian seaport where these stones are exported to the West and the Near East. The Islamic beliefs became mixed with the traditions passed down from Greek and Roman writers.

Travels furthur to the East brought knowledge of additional beliefs. Chinese herbalists and Taoist doctors classify agate as a special variety of stone called *ma-nao*. Literally translated the word means "horse's brain," due to the undulating patterns found on the surface of many rough specimens. When written in Chinese characters, the term ma-nao is preceded by the symbols for jewel or precious stone to eliminate confusion. A popular notion still exists in China that agates are created when they are spit from a horse's mouth. The Chinese rank agate as neither common stone nor jade. It is ranked as inferior to jade but next to it in importance. The story of crystal quartz being petrified ice is paralleled in Chinese folklore. It is held that agate is the frozen blood of departed ancestors. A test for genuine agate is to rub it with a stick — if it is heated, it is false — if it remains cool, it is genuine.

Writers of the Middle Ages expanded upon the beliefs recorded by the ancients. The eleventh century Lapidary of Marbod states, "The wearer of an agate shall be made agreeable and persuasive to man, and have the favor of God."[8] This single passage is said to be the source of the tradition that agate is the proper talisman of successful businessmen. The fourteenth century *Lapidaire* of Chevalier Jean de Mandeville gives an extensive list of attributes for the gem. Many of these were obviously drawn from earlier authorities, but Mandeville expands upon the stones virtues when he writes:

> This stone destroys all venom and protects against snake bites and other venomous animals, and when one puts it in one's mouth, it works against thirst, it comforts the sight, and protects the body's health: it makes one a beautiful speaker and gracious with words, he who wears it is pleasing to God and to people, it gives one color in the face and helps one

acquire intelligence and sense, it pulls one away from bad deeds and when an agate is put on a woman who is in a difficult labor, it makes her deliver, be the baby alive or dead, and if one rubs it against fire, the agate will give off a great odor; it will also give victory. It is said to be because of this stone that Anchise, an emperor, has won many victories and dodged many perils.[9]

Cardano reinforced the belief the stone is of value in debate and negotiations when he wrote in 1585 that he tested agates and found wearing them made him, "temperate, continent, and cautious; therefore, they are all useful for acquiring riches."[10] The scholar Camillo Leonardo also claims in his *Speculum Lapidum* of 1502 that the stone gives victory and strength to its owner. The wearer is protected from all dangers, able to defeat terrestrial obstacles and endowed with a bold heart. Its ability to avert storms and lightning is an additional attribute.

Mandeville includes an inventory of the various agates identified by Medieval scholars. Mandeville states;

Agates are a stone found in many parts of the earth. There are many kinds: some are a black color with white veins, others bring to mind the color of coral, with red or golden veins, others are a crystal color, with saffron or red veins, others are a wax color with red and white spots. Another kind comes from India and is found in diverse colors and has pictures in different forms, some in the form of a man's head, others shaped like trees, animals and birds.[11]

Leonardo lists seven types of agate based on where they where thought to originate—Sicilian, Cretian, Indian. Egyptian, Arabian, Cyprian, and Persian. His descriptions, and those of Mandeville, include other gems, such as a stone that is probably amber. "The Persian being heated, smells like Myrrh, as some say." Leonardo adds, "all of them agree in this, to make men solicitous."[12] Two colors of agate have been popular as new years gifts, especially on the last day of the year, St. Sylvester's Day. White agate is used to symbolize the old year and black to represent the unknown of the new year.

Agate was said to cure insomnia and insure the bearer pleasant dreams. The *Dream Book* states that "To dream of agates will bring pleasant dreams to those born in June." The text continues with a warning to those not born in this summer month, "but misfortune will befall those who wear the agate, or dream of this stone,"[13] Additional attributes passed from the ancients include treatments for numerous ailments. Powdered in liquid, it could be ingested as a cure for insomnia or to give immunity from snake bite. This same tonic could be applied to the skin to add softness and suppleness. Agate was also prescribed as an eye stone. As recently as the early Twentieth Century, druggists would "keep tiny eye stones (of agate) to chase irritating cinders from the eye."[14] The second century B.C. historian Damigeron suggests, "Ground and sprinkled on the wound and taken in drink, it cures."[15] Still other writers had recommended holding the stone in the mouth as a cure for thirst, fevers and inflammatory diseases. One belief states that eagles carry agates to their nests to protect their young from the bites of venomous creatures.

Modern practitioners of metaphysics list numerous attributes for agate. Most of the qualities deal with enhanced communication abilities. Agate is said to increase the ability to make connections between the physical and spiritual worlds. The agate is used to stimulate analytical capabilities and precision thought. For these abilities, it continues as a popular talisman for business persons. It is also used to enhance meditative states and as a focal point in contemplation. The stone is also said to bring good luck, good health, wealth, and happiness.

Other qualities are connected to the many forms of agate. Those most often mentioned in contemporary literature and their associated qualities are listed below.

Blue Agate— strengthens the will and the ability to communicate effectively.

Blue Lace Agate— opens and balances the throat; promotes positive expression, optimism, and truth; opens the senses on all levels of awareness.

Black Agate— used as a "grounding" stone.

Botswana Grey Agate— Aids a smoker's recovery, eases lung disorders, and promotes tissue regeneration.

> *Adorned with this, thou woman's heart shall gain,*
> *And by persuasion thy desire obtain;*
> *And if of men thou aught demand, shalt come*
> *With all thy wish fulfilled rejoicing home.*
>   *Attributed to the author of Lithica in*
> King's *Natural History of Precious Stones, 1865*

Botswana Pink Agate- opens expression of feminine energy; promotes sensuality, passion, and artistic ability.

Brazilian Agate— calms nervousness; promotes happiness and peace.

Dendritic Agate— aids memory development, visualization abilities, and the ability to recall past lives.

Fire Agate— stimulates the whole physical system, especially the endocrine systems; aids tissue regeneration and cell memory; balances spiritual insight; brings harmony and aids in the development of compassionate detachments. It is also used to balance sexual and emotional issues.

Mexican Lace Agate— inspires happiness, the appreciation of beauty, and aids in creativity.

Moss Agate— good for the circulatory system; balances left brain activity and the emotional body. It is also associated with nature, rain, prosperity, and fertility.

Prairie Agate— a "grounding" stone, it promotes the ability to work in the material and physical plane.

Snakeskin Agate— aids development of spiritual insight at higher levels, opens intuitive abilities, and helps bring the spirit to mind; used in the healing of old issues and promotes spiritual rebirth.

Tree Agate— said to aid in the development of the inner strength, character, courage, and self fulfillment.

White Agate— grounds and balances the spiritual body and the higher mind, develops inner peace and tranquility, and aids the spirit in body balancing.

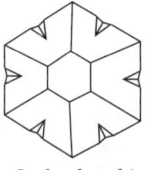

Alexandrite

Crystal System

Orthorhombic

Color: Green in daylight, light red in artificial light **Mohs' Hardness:** $8\,1/2$ **Cleavage:** Imperfect **Refractive Index:** 1.746 - 1.755 **Dispersion:** 0.009	**Streak:** White **Specific Gravity:** 3.70 - 3.74 **Fracture:** Weak **Birefringence:** 0.009 **Trichroism:** Strong; reddish-yellow, yellow, light- green, green

Crystal System: Orthorhombic; twinned, intergrown triplets
Chemical Composition: $Al_2(BeO_4)$ beryllium aluminum oxide
Transparency: Transparent
Fluorescence: Usually none, Green; weak, dark red

Alexandrite is a form of chrysoberyl which shows a distinct color change depending on the source of light used for viewing. The finest stones change from purplish-red in artificial light to blue-green in the sun. Both colors may be tinged with a subtle brown. The color change is due to the absorption of certain wavelengths of light.

Distinguished in the nineteenth century, alexandrite does not share the ancient history and lore of most other gems. Some authors claim that the stone has been known for centuries in the Orient and valued as an omen of good fortune. The stone was named by the mineralogist Nordenskjöld for Czar Alexander II of Russia. Legend says it was discovered April 29, 1839 on the twenty-first birthday of the then Czar apparent. The color change from red to green was said to honor the royal family by displaying the national colors of Imperial Russia. The stone was looked upon as a good omen in Russia until the demise of the Romanov family during the Bolshevik Revolution of 1917. Recent nationalistic fervor has begun to restore the desire for this gem in its mother-land.

Contemporary practitioners of crystal healing claim it can be used in the treatment of disorders of the spleen, testicles, and pancreas. It is also said to help with the regeneration of neurological tissue and in treating leukemia. Metaphysicians say alexandrite crystals may be used to "help one to return to the origin of time and to connect with the beginning of one's development."[1]

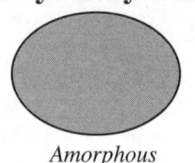

Amber

Crystal System

Amorphous

*Around thee shall glisten the loveliest Amber,
That ever the sorrowing sea-bird hath wept.*
The Fire Worshipers, *Sir Thomas Moore*

Color: Light yellow to brown, red, nearly colorless, milky white, blue, black, greenish
Mohs' Hardness: $2 - 2\frac{1}{2}$
Cleavage: None
Refractive Index: 1.54
Dispersion: None

Streak: White
Specific Gravity: Usually 1.05 -1.09, maximum 1.30
Fracture: Weak; Conchoidal, and brittle
Birefringence: None
Pleochroism: None

Crystal System: Amorphous
Chemical Composition: Approx. $C_{10}H_{16}O$, mixture of resins
Transparency: Transparent to opaque
Fluorescence: Bluish-white to yellow-green

Amber is not a mineral but is of organic origin. It is the fossilized, hardened resin of the pine tree, formed in the Eocene period about fifty million years ago. Various species of tree generated sap, so there are several kinds of amber with varying compositions. Amber is often characterized by its inclusions. These may be fine bubbles or trapped prehistoric insects. Amber exposed to sun and sea water may be opaque and chalky with a frothy appearance. Other opaque amber takes on the look of bone or ivory.

One of the first substances of personal decoration, amber has been used as an amulet from earliest recorded history. Animal carvings found in Denmark which date from the Mesolithic period, about 7000 B.C, are the oldest three-dimensional works of art found in Northern Europe. Excellent examples have also been found in excavations of prehistoric sights at Indersoen, Norway. In Neolithic times this golden gem served as a symbol of the sun. Another sun sign is the crowing rooster. These two symbols were combined in the form of an amber egg engraved with a rooster. This powerful talisman has represented new birth, new dawn and a new life throughout history. Entrapped insects also led to the belief that pieces of amber are the resting places of departed souls.

The early Germanic historian Tactitus wrote of people in the Baltic regions worshipping the "mother of the gods." This deity took the form of a wild boar carved of amber. An amber boar was also a sacred totem of Celtic tribes in ancient Britain. The Celtic sun god, Ambres, derives his name from the fossilized resin. Phoenician sailors recorded their importation of amber on some of their first voyages to the Baltic Sea.

Amber's supposed origin has led to a wealth of myths and lore. A Greek name for the gem was *lyncurius*, as it was thought by some to be the congealed urine on the lynx. The Greeks also held the belief that amber was solidified sunshine which had broken off as it sank into the sea. The Athenian general Nicias said it was the "juice" from the rays of the sun.[1] He theorized that is was a liquid produced when the sun's rays struck the soil with great force. This "unctuous sweat" would flow to the sea and solidify when touched by sea water.[2] The Romans adopted this conviction and called amber *succinum* meaning sap-stone, from the root word succus meaning gum. Succinite is the modern chemical name for the principle resin in amber.

In Greek mythology, the god Eridanus was the keeper of a sacred river. These waters were said to be the origin of amber, as its banks were covered with the gem. The location of this fabled river has never been established. Some say it is the river Po, while others claim it is a tributary of the Atenian river Ilissus. Another popular Roman tale explained the source of amber as melted honey which congealed when it was dripped by bees into the sea.

The Chinese also believed there was a connection between bees and amber. The origin may be from an ancient observer noting a bee trapped in the petrified resin. The following myth ensued: "Somewhere are cliffs, the cliffs of Ning Chou in which dwell thousands of bees. When the Cliffs crumble, the bees come out. People burn them and make them into amber."[3] Ancient European myths state that this golden gem is gathered in the Gardens of the Hesperides, where golden apples bring immortality to those who eat them. The story continues that amber is the dew from these apples, and those who possess it could live forever. "Sophocles believed amber was the tears of the Meleagrides, sisters of the hero Meleager, who, according to legend, were transformed into birds."[4] Edwin Streeter presents the following couplet from Sir Thomas Moore's *The Fire Worshipers* as an example of this Greek tradition.

> Around thee shall glisten the loveliest Amber,
> That ever the sorrowing sea-bird hath wept.[5]

The Greeks also called amber *elektron*, the root for the modern word electricity. The first record of its electrical properties was made as early as 600 B.C. by the Ionic philosopher Thales of Miletus. This observation is considered by many to be the beginning of the study of electricity. Amber can be given a static electric charge by friction. This fact gives rise to the Arabic word for the gem, *kahroba*. *Kah* is the Arabic word for straw and *ruba* their word for robber or thief. Kahroba translates to straw-robber or straw-attractor. The historian Theophrastus speaks of "the Fossil Amber, a stone produced in Liguria, possessing the property of attraction, but rare, and found in only a few countries."[6] The ability of the gem to attract small particles of lint or straw was a test used to verify amber's authenticity.

The Romans were familiar with amber by way of trade with the Baltic region. Originally called *glaesum*, it was highly valued. Pliny the Elder is credited as the first to recognize and record that the gem was in fact petrified tree sap. He wrote that a small carved figure would "fetch more than a healthy slave."[7] The white or waxy amber was considered worthless and was used for incense. He notes that the early Greek writer Callistratus called the golden hued clear material *chryselectrum*.

Amber was considered a stone of benefit to all ages. Roman women adorned their babies with amber beads to protect them from evil spirits. These amulets were worn by children and adults as a charm against insanity and to cure the "ague."[8] Citizens of Rome were known to carry amber beads in their hands to cool them during the summer heat. The warmth of the hand would cause the gem to emit a balsamic odor, acting as a perfume which masked summer body odor.

From the time of Hippocrates (c. 400 B.C.) amber was used as medicine. Pliny states that peasants, in what is modern Northern Italy, wore amber beads in necklaces to protect them from diseases of the throat and tonsils. His *Natural History* includes many medicinal uses for amber. It was said to attract a flame and, therefore, if worn on the neck, it would cure fevers. If powdered and mixed with oil-of-roses or honey, it would cure dimness of vision, and if taken directly or mixed with water or gum mastic, it would remedy diseases of the stomach, ear, and the eye. Oil-of-amber was used as a decongestant salve applied to the chest, and taken internally, it was prescribed for asthma and whooping cough.

The scholar Jean de Mandeville makes reference to a gem called *ligure* in his fourteenth century *Le Lapidaire*.

> Ligure is a stone coming from an animal called a lynx; sometimes it comes from its urine, which hardens to stone. Other times it is the color of saffron or amber attached to black. Other times it is red like a carbuncle, but hardly glows at night. Sometimes it is green and changes color according to the variety of shadow. The water in which this stone is washed helps inflammations of the stomach and abdominal pain, and brings back color to the face. It attracts straw to the wearer, as does Payet and amber. This water, if drunk, will break stone and destroy gravel.[9]

Although compared to amber in the text, the descriptions and attributes seem to refer to amber itself. The colors mentioned, the supposed lynx origin and the attraction of straw, makes one think that Mandeville's ligure is a form of amber. Prior to the advent of chemical and optical analysis of gems, their division into categories and the listing of them as separate stones was common.

The use of amber as a healing agent continued into the Middle Ages. Scholars studied ancient texts and expanded upon the virtues of this petrified resin. It was seen as a cure for the plague, heart disease, impotence, vertigo, goiters and as a relief for stomach pains. Application of the stone to a wound was said to staunch the flow of blood. This may account for its use by Rabbis as the handles of their circum-cision instruments. The gem was also thought to have an affinity for yellow skin and, as such, was a cure for jaundice. The amber would draw the yellow color out of the patient as well as the sickness itself. Camillus Leonardus states in his *Speculum Lapidum* of 1502 that, "If used as a perfume, it is said to provoke the menses in women, to cure epilepsy, to drive away serpents, and to heal their bite if mixed with the marrow of a stag, and fastens loose teeth."[10] The pre-established origins of amber were questioned by

Bartholomæus Glanvilla in his *De Proprietatibus Rerum*. Glanvilla regarded amber as a kind of jet and says it is useful in driving away adders, and is contrary to friends.[11]

Many gems have been thought to foretell coming events or to bring revelations. In medieval times amber was said to disclose the presence of poisons. Goblets made of block amber were used to reveal anything toxic in liquids they contained. A change of color in the stone was thought to predict a loss of affection on the part of the giver. Richard Tofte published the following account in 1615 of an unfaithful love revealed:

> Thy tokens which to me thou sent
> In time may make thee to repent;
> Thy gifts do groan (bestow'd on me)
> For grief that they thee guilty see.
> The amber bracelet thou me gave
> (For fear thou shouldst shortly wave)
> From yellow turned is to pale,
> A sign thou shortly will be stale.[12]

The author failed to note that all amber changes in time with exposure to air, sun, or the human body. Leonardus proposed another use for the gem which he believed made many a husband feel more secure.

> If laid on the left breast of a wife when she is asleep, it makes her confess all her evil deeds. If we would discover whether a woman has been corrupted, let it be laid in water for three days, and then shewn to her, and if she is guilty, it will immediately force her to make water.[13]

The aromatic properties of amber have been the source of interest since the ancient Romans. Burning, heating, mixing it with a salve or dissolving it in a solvent are all methods used to release its stored fragrance. In Germany amber is called *bernstein*, the "stone that burns," because of the ease with which it is ignited. As a fumigant it has been prescribed as a treatment for numerous ailments as well as a cure for tonsillitis, inflammation of mucus membranes and sinus blockage. "The smoak of it drives away devils and dissolves spells and enchantments"[14] states Leonardus. The smell was said to aide women in labor by reducing their discomfort and taking their mind to a calmer place. In the Orient this fumigation is done by throwing powdered amber onto a hot brick.

A mixture known as oil-of-amber is referred to in many pieces of literature. This is most often finely powdered amber dissolved in alcohol or mineral spirits. In 1548, Johann Meckenbach took credit for discovering the process of making this tonic. Although known for centuries, Meckenbach announced it as a new cure and named it *Oleum Succini of the Pharmacopeia*. This concoction has maintained its reputation as a cure for gout, rheumatism, whooping-cough, hysteria, bronchitis, infantile convulsions and as an anti-spasmodic for asthma. An official report of the United States Mining Bureau published in 1935 reported oil-of-amber as a viable ingredient in pharmaceutical products. In China a mixture described as liquid acid-of-amber and opium is used as a sedative, anodyne and antispasmodic. *The Family Dictionary* by Dr. W. Salmon, published in 1696, lists the following use for the gem:

Diderot's Encyclopedia *states this sign meant oil-of-amber to eighteenth century chemists.*

> For falling sickness, take half a drachm of choice amber, powder it very fine, and take it once a day in a quarter of a pint of white wine, for seven or eight days successively. [Continued treatment would be to] take bits of amber, and in a colsestool put them upon a chafing dish of live charcoal, over which let the patient sit, and receive the fumes.[15]

In his *The Natural History of Gems* of 1870, King makes the following claim, "That the wearing of an amber-necklace will keep off the attacks of erysipelas (a strep infection of the skin) in a person subject to them, has been proved by repeated experiment beyond all possibility of doubt."[16]

Modern believers in the medicinal power of gems still prescribe amber for a variety of purposes. It serves as a symbol of renewal in marriage and continued fidelity. The gem enhances one's ability to express feminine energies for both men and women. It stands as a major agent for purification and may be used to cleanse the environment in birthing and "re-birthing" rooms. When worn, carried, or ingested as a elixer, it purifies the mind, body, and spirit. It also enhances the ability to recall specific instances in past lives. Holistic practitioners recommend amber powder or alcohol based oil-of-amber as a treatment for kidney and bladder ailments, goiter and diseases of the throat. "Amber allows the body to heal by absorbing and transmuting negative energy into positive energy."[17]

"The Tree that Exudes Amber" from Ortus Sanitatis, *the first illustrated book on drugs, by Johannis de Cuba published in 1483.*

The first mention of amber in western literature is in Homer's *Odyssey*.

> Eurymachus
> Received a golden necklace, richly wrought,
> And set with amber beads, that glowed as if

38
Lapidarium:

With sunshine. To Eurydamas there came
A pair of ear-rings, each a triple gem,
Daintily fashioned and of exquisite grace.
Two servants bore them.

The Myth of Phaeton and the Heliades: Greek

Phaeton, son of Phoebus Apollo, the sun god, asked his father to allow him to drive his team of wild horses and chariot of the Sun. His journey started well, but Phaeton was careless in controlling the steeds. The horses bolted and pulled the chariot too close to the Earth, setting it ablaze. The entire planet was in flames, the forests burned, the land was parched, and rocks melted. Phaeton's careless behavior led to the origins of volcanoes and vast desserts. The heat was so intense that the residents of Africa were burned black. To prevent the Earth's total destruction, Zeus struck Phaeton dead with a lightning bolt. The Sun retreated from the Earth and Phaeton's body fell into the River Eridanus. The nymphs of the stream pulled his body from the water and buried him on the river bank. In time his three sisters, the Heliades (aka Electrides), came to search for his grave. Upon finding it, they vowed to stay with their dead brother and morn him for eternity. As the Heliades wept, their bodies took root in the river bank and were covered by the bark of neighboring trees. Their arms became branches, their bodies trunks, and the three sisters were gradually transformed into trees. Their tears continued to flow and, as they hardened in the sun, were turned to amber. These amber tears fell into the river and traveled out to sea. When the tides come to the shore, the evidence of their sorrow is spread upon the sand.

The Myth of Juraté and Kastytis: Aistian

The Aistians are ancient ancestors of the Lituanians, *Aestiorum gentes*, meaning the "Honorables."
Along the coast of the Baltic Sea, near the mouth of the Sventoji River, lived and worked a courageous fisherman named Kastytis. The fairest of all the goddesses, a mermaid named Juraté, lived in a palace of amber deep beneath the sea. Kastytis would cast his fishing nets into the sea, a kingdom ruled by fair Juraté. The goddess, angered by this intrusion, sent her mermaids to warn him to cease fishing in her domain. Kastytis refused and continued to cast his nets. Juraté decided to come to the surface and confront Kastytis directly. When she saw how handsome

and courageous he was, she immediately fell deeply in love with him. Juraté decided that to keep her love, she would bring him to her palace to live. Perkunas, god of thunder and father of all gods, was angered to find an immortal in love with a mortal. He knew that Juraté was promised in marriage to the god of water, Patrimpas. To end Juraté's ill-advised love, Perkunas sent a lightning bolt to kill her mortal lover and destroy the amber palace. To further punish her, Juraté was chained to the palace ruins and left to morn the loss of her love for eternity. The tears she wept were of pure amber. When storms stir the Baltic Sea, fragments of her palace are washed onto the shore. Pieces that resemble tears are to be treasured as they are the tears of Juraté. Washed from her eyes, they are as clear and pure as the love she lost.

The Myth of Freya and the Necklace: Norse

Freya is the Norse goddess of love, beauty and fertility. This blond, blue-eyed young woman had a weakness for beautiful jewels. She was the Queen of the Aesir and wife of the sun god Odur, father of her two lovely daughters. They lived in her palace, Folkvanger, in the land of Asgard. One day Freya was out for a walk along the border of her kingdom. Her kingdom shared a border with the kingdom of the Black Dwarfs. As she walked along, she noticed four of the dwarfs were busy making a beautiful necklace. The necklace glistened and shone as golden and bright as the sun. Freya stopped to inquire about this wondrous object. She was told it was the Brisingamen necklace and of great value to the dwarfs. With this Freya knew she must have it. "I will give you a treasure of silver if only I may have it." she exclaimed. "I have never seen anything as beautiful." The dwarfs scoffed at her offer and told her that all the silver in the world could not purchase this prize. Knowing she could not live without it, she asked: "Is there any treasure which you would trade for the necklace?"

"There is one way you may own the Brisingamen," answered the dwarfs. "If you will wed each of us for one day and one night, the necklace shall be yours." Bewitched by this wondrous treasure, Freya was seized with madness. Forsaking her husband, forgetting her children, and failing to remember she was Queen, Freya agreed to the unusual terms. No one in Asgard knew of these weddings. No one but the mischief-maker Loki.

After four days and nights of Freya's unholy betrothals to the four dwarfs, she returned to her palace to live with her shame. She hid the necklace in her bed and vowed to keep her deeds a secret. Always looking for trouble, Loki came to her husband to tell him what had happened in the kingdom of the dwarfs. Odur demanded Loki provide proof of these terrible tales. To provide evidence, Loki set about stealing the necklace. He turned himself into a flea and flew to Freya's bed chamber. Loki bit her on the cheek, causing her to stir, and giving him the chance to remove the necklace from its hiding place. Loki rushed to Odur and presented his evidence of Freya's infidelity. Odur tossed the necklace in a rage, left the kingdom, and set off for far distant lands.

Freya woke the next morning to find her necklace and her husband gone. Stricken with grief, she went to Valhalla to confess her sins to Odin, father of the gods. To reach Valhalla she had to pass through the valley of Glaesisvellir. At the entrance to the holy city was an amber grove called Glaeser, with trees which dripped beads of amber. (The word *glaeser* is derived from the German word *gles* or *glaes* which latter was the derivation of the modern word glass.)

The kindly Odin forgave Freya but demanded a penance for her evil acts. Odin retrieved the Brisingamen from Loki and commanded Freya to wear it for eternity. She was further ordered to spend all of time wandering the world in search of the husband she had shamed. As she wanders the world, she weeps. The tears which fall from her eyes land on the soil and are turned to gold in the rocks. The tears which fall into the sea are turned to amber. The wide variety of locations which contain gold or amber give proof of Freya's world-wide search. The fact that amber continues to wash onto the shores proves that her search continues. In Baltic countries there is a belief that an amber necklace will choke one who lies or is unfaithful. Amber is worn as a reminder not to do evil.

The Myth of Amberella: Lithuanian

The beautiful maiden Amberella lived on the shores of the sea with her fisherman father and his wife. Amberella went into the sea for a swim. She was drawn into a powerful whirlpool and pulled to the bottom of the sea. As she reached its depths, she was captured by the Prince of the Seas and made his bride. She was imprisoned in his underwater palace, a palace of great glowing amber. When Amberella begged to be released and returned to her parents, the prince flew into a terrible rage. He grasped Amberella, mounted his white foaming horse, and rose to the surface in a terrible storm. As the prince and Amberella rose from the sea, her parents saw her struggling to be free. She was dressed in a wonderful gown, draped with amber beads, and adorned with an amber crown. Amberella and the Prince sank back into the sea, convincing her parents she was lost forever. In a final gesture, Amberella tossed a few bits of amber onto the shore to show her mother and father how much she loved and missed them. Now, when the Prince of the Sea becomes angry, the sea churns and storms rage. From the depths, Amberella still tosses amber onto the shore to remind her parents of her love.

Extraordinary amethyst
Violet-colored in beauty,
The Lapidary, *Marbode of Rennes*

Amethyst

Crystal System

Hexagonal

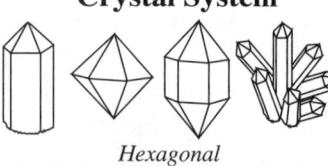

Color: Violet, pale red-violet **Mohs' Hardness:** 7 **Cleavage:** None **Refractive Index:** 1.544 - 1.553 **Dispersion:** 0.013 **Pleochroism:** Very weak; violet, gray-violet	**Streak:** Colorless **Specific Gravity:** 2.65 - 2.66 **Fracture:** Conchoidal, with "herring bone" pattern ridges, very brittle **Birefringence:** 0.009

Crystal System: Hexagonal (trigonal) hexagonal prisms
Chemical Composition: SiO_2 silicon dioxide
Transparency: Transparent
Fluorescence: None

Amethyst is a variety of quartz which may be found as large individual crystals or in clusters of smaller material. It is distinguished by its lavender to deep purple color, the deeper color being the most desirable. Heated to 550°-560° centigrade, they turn dark yellow to reddish brown and are called citrine. These are more expensive and richly colored than natural citrines.

The birthstone for February, amethyst has long had the reputation of resisting drunkenness. This attribute may be traced to the ancient Egyptians. They used the stone as the representative of the zodiac sign

Lapidarium:

of the goat. The goat was considered the enemy of vines and vineyards, and therefore the antidote of wine. They believed the wearer of an amethyst could not become intoxicated. This virtue was continued by the Greeks, who gave the stone its name, *amethustos*. The name translates to 'not drunken.' This quality may have come from the fact that the bowls of drinking vessels were carved from large crystals. If filled with water, the contents would appear to be wine, but intoxication would be impossible. The stone was said to be most effective if held under the tongue. Aristotle said the amethyst was valuable in hindering the "ascension of vapors" by drawing the vapors into itself and then diluting them.[1] The origin of the word amethustos may be the Hebrew word *achlamath* or, as Von Hammer suggests, the Persian *shemest*.[2]

The Egyptians called the stone *hemag* and also considered it the symbol of intellect. Cleopatra's signet ring of amethyst was engraved with a figure of *Mænad*, the Persian god of "The Divine Idea," source of enlightenment and love.[3] She was also said to benefit from its sobering effect, as related by Asclepiades in his *Anthology*.

> A Mænad wild, on amethyst I stand,
> The engraving truly of a skilful hand;
> A subject foreign to the sober stone,
> But Cleopatra claims it for her own;
> And hallow'd by her touch, the nymph so free
> Must quit her drunken mood, and sober be.[4]

The *Book of the Dead* makes reference to a heart carved of amethyst as the symbol of wisdom in life. The Egyptians also believed that if an amethyst were tied around the neck with peacock's 'hairs' and the feather of a swallow, it would protect the wearer from sorcery and also cure the gout. A similar claim is attributed to Pliny the Elder. He writes,

> The lying Magi claim . . . that if the names of the
> Moon or Sun be engraved upon them, and they be
> hung about the neck from the hair of a baboon, or
> the feathers of a swallow, they are a charm against
> witchcraft. They are also serviceable to persons
> having petitions to make to Princes: they keep off
> hailstorms and flights of locusts with the assistance
> of a spell which these doctors teach.[5]

William Jones claims in his nineteenth century work on precious stones that the exact same tradition was followed by the Incas, half way around the world. "Peruvians believed that if names of the sun and moon were engraved upon it, and it was hung round the neck with the hair of a baboon or the feathers of a swallow, it was a charm against witchcraft."[6] It is hard to understand how these Andean natives would have access to baboons. It is also curious that the exact prescription for a charm would be found in two such distant places. As has been found in many resources, information attributed to one author or culture is transposed to another.

Other ancient cultures have added to the lore surrounding this purple quartz. Roman matrons found that wearing an amethyst would help retain the affections of their spouse. The Emperor's soldiers would wear it as an amulet for protection in battle. It was said to be even more powerful if worn in a setting of copper. The ancient Hebrews called the stone *allamah*. They believed it signified the ability to control virtue by way of dreams or visions. The gem is listed as the ninth stone of the twelve gems in Aaron's Breastplate. The amethyst was given to the tribe of Daniel, symbolizing judgment.

Christian lore assigns one of the twelve Foundation Stones, listed in Revelations XXI, to each of the apostles. One of the earliest scholars to make such an association was the tenth century writer Andreas, Bishop of Caesarea. He states that amethyst is the stone of Matthias and gives the following reason:

> By the amethyst, which shows to the onlooker a fiery aspect, is signified Matthias, who in the gift of tongues was so filled with celestial fire and the fervent zeal to serve and please God, who had chosen him, that he was found worthy to take the place of the apostle Judas.[7]

The Roman Catholic Church also sees amethyst as an important gem. The color and purity of hue are associated with the wine transfigured to Christ's blood in the sacrament of the Mass. Goblets fashioned from this beautiful quartz have been used for centuries in the celebration of the Eucharist. The stone has been the signature gem of Bishops and is still worn in many of their ceremonial rings. The color signified royalty and dignity to the priesthood. It is also the stone dedicated to St. Valentine. The color purple was thought to signify true deep love. Legend says the saint wore an amethyst ring engraved with the figure of Cupid.

The attributes of the amethyst have been expanded upon in literature and have grown with time. An eleventh century German writer states "an amethyst owned by a man attracts to him the love and affection of noble women, and protects him from thieves."[8] Camillus Leonardus listed many virtues of the stone in his *Speculum Lapidum*. He writes that it has the power to control evil thoughts, quicken intelligence, and render men shrewd in business. Camillus' prescribed method of maintaining sobriety differs with other authors, ". . . for being bound on the navel, they restrain the vapour of the wine, and so dissolve the ebriety."[9] The fourteenth century *Le Lapidaire* of Chevalier Jean de Mandeville adds to this method of protection, "If it is combined with a sard stone on the navel of an intoxicated man, the amethyst will remove the drunkenness."[10] Leonardus also claims it preserves solders from harm, gives them victory over their enemies and "prepares an easy capture of wild beasts and birds."[11] As with other gems, the amethyst was belived to preserve the wearer from contagion. An old cure for headaches prescribes binding an amethyst crystal to the temple with silk.

The Scottish writer Lucien describes a city of gems in his *Vera Historia* that demonstrates the reverance given the stone.

> The walls were of emerald, the temples for the gods were formed of beryls, and the altars in each were of a single amethyst block of enormous size. The city itself was of gold, a fine setting for these marvelous gems.

The fact that amethyst crystals of great size are sometimes found have led to claims of massive architectural uses. It is thought that in many cases, if these objects were actually observed, these masses of violet may be colored glass.

The amethyst has maintained its popularity in modern times as an powerful amulet. It is described as the "stone of spirituality and contentment."[12] The gem is said to provide a connection between the earth and other worlds. It also brings serenity, composure, calm, and tranquility. Many still wear this crystal for its supposed power to change negative energy into positive energy, and its ability to increase ones power of intuition. The medical uses for amethyst have included treatment for insomnia, pain relief and as an elixir to treat arthritis. A passage in *The Book of Seals* reinforces what has been the primary belief about this popular gem. "A

> *The February born may find*
> *Sincerity and peace of mind,*
> *Freedom from passion and from care*
> *If they the Amethyst will wear.*[14]

bear, if engraved on an amethyst, will put to flight demons, and preserve the wearer from drunkenness."[13]

The Myth of Amethyst: Greek

The ancient myth of Amethyst was related in a French poem written in 1576 and dedicated to King Henry III. It was later retold in the publication *Œuvres Poétiques* by Belleau in 1878 edited by Marty-Laveaux, Paris 1878.

Bacchus, god of wine, was feeling offended by neglect. His bad temper was notorious in the pantheon of the gods. In a foul mood, he vowed to take revenge on the first person he encountered. The god decided he would order his tigers to devour any such unfortunate soul. A mortal, the beautiful and pure Amethyst, passed by Bacchus and his tigers on her way to worship at the shrine of Diana. As the beasts sprang toward Amethyst, she invoked the name of the goddess. Without time to save Amethyst from any harm, Diana did what she could to save the innocent maiden. Amethyst was instantly turned into a pillar of pure white stone and made immune to the teeth of the tigers. Bacchus, recognizing his cruelty, sought to make amends for the demise of the fair maiden. He collected his best vintage and poured the wine over her petrified body. The juice of the grape imparted the lovely hue we associate with the stone that bears her name.

Aquamarine: see Beryl

Color: Light blue, blue, blue-green	**Streak:** White
Mohs' Hardness: $7\,^1/_2 - 8$	**Specific Gravity:** 2.67 - 2.84
Cleavage: None	
Refractive Index: 1.577 - 1.583	**Fracture:** Conchoidal, uneven, brittle
Dispersion: 0.014	
Pleochroism: Definite; nearly colorless-light blue, Blue: sky blue	**Birefringence:** 0.005 - 0.009

Crystal System: Hexagonal (triagonal); long prisms
Chemical Composition: $Al_2Be_3(Si_6O_{18})$ aluminum beryllium silicate
Transparency: Transparent to opaque
Fluorescence: None

Aquamarine is the name reserved for blue beryl. The most desired color in the retail trade is the deep blue. Subtle differences in shade may make the price of stone vary radically.

The variety name aquamarine for blue-green beryl does not appear in ancient texts. The gem is described, but it is included under the name sea-green beryl. "The specific term by which we know it now apparently was first used in an important gemological work by Anselmus de Boodt in his *Gemmarum et Lapidum Historia*," published in 1609.[1] The word probably derived from the common Italian word *acquamarina*, meaning literally "sea-blue."

Crystal System: Beryl

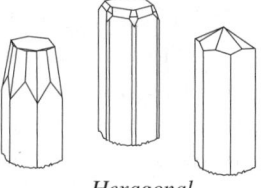

Hexagonal

Beryl

Color: Gold, yellow-green, yellow, pink, colorless
Mohs' Hardness: $7\frac{1}{2}$ - 8
Cleavage: None
Dispersion: 0.014
Streak: White
Specific Gravity: 2.67 - 2.84
Fracture: Conchoidal, brittle
Refractive Index: 1.570 - 1.584
Birefringence: 0.005 - 0.009
Pleochroism: Golden: weak; lemon yellow, yellow
Heliodor: weak; golden yellow, green-yellow
Morganite: definite; pale pink, bluish-pink
Green: definite; yellow-green, blue-green

Crystal System: Hexagonal (triagonal); long prisms
Chemical Composition: $Al_2Be_3(Si_6O_{18})$ aluminum beryllium silicate
Transparency: Transparent to opaque
Fluorescence: Morganite: weak; violet-light red

Beryl comprises a family of gems. Colorless material is called goshenite. Traces of impurities present during crystal growth add color. These color varieties include emerald (green), aquamarine (blue to blue-green), Morganite (pink or pale yellow-orange and named for the banker and gem collector J.P. Morgan), and heliodor or golden beryl (yellow to orange). Emerald has such an extensive lore of its own that it will be treated in a separate entry.

The attributes of beryl, from the Latin *beryllus*, were first recorded in *De Virtutibus Lapidum* by Damigeron in the second century B.C. "This stone is good besides for damage to the eyes, and for all sickness if it is put in water and given as a drink."[1] Pliny the Elder's *Natural History* also lists the stone as an excellent cure for eye diseases. For minor ailments, the eye was to be washed with water in which a beryl has been immersed. To cure serious injuries Pliny prescribed placing the powder of this gem in the eye each morning. Another old Roman legend states that beryl absorbs the atmosphere of young love. "When blessed and worn it joins in love, and does great things."[2] It was considered the most appropriate "morning gift" to be given to a bride by her groom. The custom was for the groom to present a token of love to his betrothed following the consummation of their marriage.

Writers of the Middle Ages claimed beryl was the most popular and effective of the "oracle crystals." Many methods of using the stone as a divining tool were described in ancient literature. One method involved suspending the stone by a thread over a bowl of water, just touching the surface. The inner edge of the bowl dedicated to this process was lettered with the characters of the alphabet. The diviner was to hold the top of the thread and allow the beryl to strike certain letters which would spell out the answers to questions. Another method was to cast a crystal into a bowl of pure water. The resulting disturbances in the surface would reveal messages to a seer concentrating on the liquid.[3] Its powers of revelation were also said to help one in a search for lost or hidden things.

To give a beryl potency as an oracle stone it was recommended that it be charged or consecrated. Sixteenth century scholar Sir Reginald Scot describes such a ceremony in his *Discoverie of Witchcraft*,

> A child, born in wedlock, was to take the crystal in his hands, and the operator, kneeling behind him, was to repeat a prayer to St. Helen, that what he wished would become evident in the stone. The finest stone would manifest an image of the saint in an angelic form, and answer any question asked of her.[4]

The ceremony was to be performed at sunrise on a day with "fine clear weather." Scott adds that many testaments consist of the appearance of visions, but only if these instructions were followed faithfully. The gem also was said to provide insight into the rites of ceremonial magic.

Some authors feel the recommendation of beryl as a scrying material may result from a semantic confusion. The low Latin word for magnifying glass is *beryllus*, from this the German word *brille*, a pair of glasses, is derived. In sixteenth century England window panes were referred to as *berills* and mirrors were called *berral-glas*. From this we may deduce that references to beryl globes as objects of divination may simply mean a glass ball was used.

Beryl is one of the stones connected to the twelve apostles. It serves as a symbol of the apostle Thomas in Christian tradition. This is based on a treatise by the tenth century writer Andreas, Bishop of Caesarea, one of the earliest writers to associate the Foundation Stones of Revelations twenty-one with the apostles.

> The beryl, imitating the colors of the sea and of the air, and not unlike the jacinth, seems to suggest the admirable Thomas, especially as he made a long journey by seas, and even reached the Indies, sent by God to preach salvation to the peoples of the region.[5]

Beryl has been associated with death and the dead for centuries. Albertus Magnus (1193-1280) claims in his writings that they have a "horror of death." He reports they are particularly potent against demons and evil spirits, but the power is lost if the stone touches a corpse. To work most effectively in calling up the spirits of the dead, the stone must be engraved with an eagle or the plant *artemisia dracunculus*.[6]

A pupil of Magnus, Thomas de Cantimpré (1201-1270), expanded upon the virtues of beryl in his treatise *De Rerum Natura*. He states that the gem cures quinsy and swollen glands in the neck if that part is rubbed with the stone. He also recounts Pliny's use of powdered beryl as a cure for eye injuries. He adds that the patient must recline during treatment and stay so for a considerable length of time after each morning application. A cure for hiccoughs is also included in his catalog of mineral cures. If a beryl is steeped in water and the water drunk, the annoying malady will disappear.[7]

Numerous other attributes of beryl have been recorded by various authors. A dream which includes beryl implies a climb to a position of honor. It is the symbol of truth, charity and faith, and it brings a connected virtue of enhancing marital love. It brings insight, cleverness, benevolence and candor to one who owns the gem. It is also said to protect travelers from danger, illness and ambush. The gem may be considered the appropriate stone for one in the legal profession. It has been reported to ". . . help against foes in battle or in litigation; the wearer was rendered unconquerable and at the same time amiable, while his intellect was quickened."[8] It also is said to treat disorders of the heart and spine, concussion, and damage to the brain. Damigeron writes, "When its spirit is broken [may mean powdered], it takes away pain in the liver and in breathing."[9] The fourteenth century lapidaire of Chevalier Jean de Mandeville includes a passage regarding the attributes of the "water sapphire." The fact that he includes a separate entry for the sapphire reinforces the belief that his water sapphire is not a sapphire at all. His description leads to the conclusion that the gem is aquamarine.

> The water sapphire is a stone the color of blue salt and is called in India "Syrices." One finds them in several places and forms around the world. Those from the East are the rarest; some are translucent and some are opaque. They are very potent and against seductions, and if anyone is trapped, and he touches the sapphire on four facets, and he touches the bolts on the door of any given trap, soon he will be free, and the sapphire will restore the sinner to God; it protects against enemies and ill-will and is very valuable to a person who has fevers or other hot spots on his body; it comforts the mind and reduces sweating and swelling; and if one puts it on his stomach, it alleviates indigestion

and cramps; and if one puts it on his heart, it relieves melancholy and vain thoughts; and cured of an illness, one says, "I'm immune (noli me tangere)"; but if one touches the stone too much, it loses its potency; it must be treated carefully and delicately.[10]

The sixteenth century writer Camillus Leonardus lists numerous attributes of the beryl.

It renders the bearer of it cheerful; preserves and increases conjugal love; being hung to the neck, it drives away idle dreams; it cures distempers of the throat and jaws, and all disorders proceeding from the humidity of the head, and is a preservative against them; being taken with equal quantity of silver, it cure the leprosy.[11]

Certain colors of beryl were given additional metaphysical characteristics. The golden beryl generates sympathy and aides in one's ability to maintain a level of sincerity. The aquamarine protects travelers like other beryls, but it is considered particularly effective in the protection of fisherman and mariners. The name of this blue-green gem is derived from the Greek words for sea water. The use of the name aquamarine was first established in de Boodt's *Gemmarum et Lapidum Historia* of 1609. The Romans believed in the sea-green beryl's ability to protect one from the perils of the sea. They also thought it gave the owner a youthful vigor and cured any person of laziness. Pliny's description pays tribute to this gem of vitality. ". . . the lovely aquamarine, which seems to have come from some mermaid's treasure house, in the depths of a summer sea, has charms not to be denied." Pliny's *Natural History* contains an inventory of various colored beryls and the names given by the Romans. A review of this list reveals that many of today's varieties were included, as well as one non-beryl gem. The only modern variety not distinguished was pink, known today as Morganite.

Beryllus of the Romans

Pliny's Name	Certain Identity	Probable Identity
Sea Green	Aquamarine	
Chrysoberyllus	Golden Beryl	
Chrysoprasus		Chrysoprase
Hyacinthosontes	Deep Blue Beryl	
Aeroides	Pale Blue Beryl	
Other Varieties	Common Beryl	

Bloodstone

Crystal System

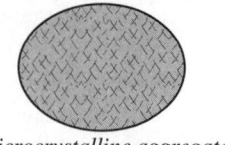

Microcrystalline aggregates

Color: Dark green with red spots Mohs' Hardness: 6 $^1/_2$ - 7 Cleavage: None Refractive Index: 1.535 - 1.539 Dispersion: None	Streak: White Specific Gravity: 2.55 - 2.65 Fracture: Conchoidal Birefringence: up to 0.006 Pleochroism: None

Crystal System: Hexagonal (trigonal) microcrystaline aggregates
Chemical Composition: SiO_2 silicon dioxide
Transparency: Opaque
Fluorescence: None

Bloodstone is a dark-green variety of chalcedony with inclusions of iron which produce red or brownish spots within the stone. White or yellowish spots may also be evident. Stones without reddish "blood" spots are usually referred to as a green chalcedony. The gem is most often cut as a cabachon.

Bloodstone has been known by a variety of names. Modern gemologists also call this stone plasma. It has also been known as jasper, bloody jasper, Babylonian gem, and in ancient times, *heliotrope*. The term heliotrope comes from two Greek words meaning sun-turner. The gem was believed to alter the reflected rays of the sun, giving them a red color. This single characteristic was expanded upon by the ancient historian Damigeron. He states,

> Now, if it is put in a silver basin full of water and placed against the sun, it turns to it and makes it as if bloody and cloudy.... the air becomes cloudy with thunder and lightning and rain and stones, so that even those experienced in the power of the stone are frightened and perturbed, such divine powers does this stone have.[1]

The stone was said to make water boil, stop anger, bring strength and courage, and promote mental health. If thrown in pure water, bubbles of gas were said to rise to the surface and the water would become blood-red. This test for authenticity was important to the ancients. A true sample was valued as a "touching stone." It was believed that if the stone was placed in contact with tainted food or drink, it would detect the poison's presence. It was considered hardly possible for a person to be so poor that they could not keep such a stone to guard their family.[2]

A statement in the *Leyden Papyrus* contradicts the statement that this stone may be possessed by all. The *Papyrus* praises the stone as an amulet and ranks it with great value.

> The world has no greater thing; if any one have this with him, he will be given whatever he asks for; it also assuages the wrath of kings and despots, and whatever the wearer says will be believed. Whoever bears this stone, which is a gem, and pronounces the name engraved upon it will find all doors open, while bonds and stone walls will be rent asunder.[3]

Other beliefs concerning this common stone abounded in the Middle Ages. Rubbed with the juice of the herb heliotrope, the stone would make the wearer invisible.[4] A similar tale prescribes mixing powdered bloodstone with the same herb and applying it to the hands and face to cause the subject to vanish. "To this notion Dante alludes where he sees the damned running about under a hail of fire, 'No hope of hiding-hole or Heliotrope.'"[5] The stone was reputed to stop internal or external hemorrhaging. The methods favored were to cool the stone in water and apply it with pressure to external wounds, or to apply the cooled stone between the shoulder-blades for internal hemorrhages. Other practitioners recommended cooling the stone in water and placing it in the right hand of the patient. This medicinal use was reinforced when early Spanish explorers found native Americans useing it for the same purpose. Franciscan friar Bernardino de Sahagun recorded numerous cures of hemorrhage caused by plague when he was treating the Indians. In his *Booke of Thinges That are Brought from the West Indies* (1574) he writes,

> The stone must be wet in cold water, and the sick man must take him in his right hand, and from time to time wet him in cold water. In this sort the Indians do use them. And as touching the Indians, they have it for certain, that touching the same stone in some part where the blood runneth, that it doth restrain, and in this they have great trust, for that the effect hath been seen.[6]

Spaniards, Mexicans, and Indians in New Spain also cut the stones into the shape of a heart to guard against diseases of the organ. It was also used to heal and cool inflammatory illnesses and fevers. Another testamony as to the stone's efficacy was recorded by Vasari (1514-1578). He wrote of a visit to the artist Luca Signorelli who was working on an altar piece for the church at Orezzo. During the visit, Vasari was seized with giddiness and hemorrhage and fell fainting to the floor. Signorrelli brought a bloodstone amulet and placed it between his shoulder-blades. Vasari states with certainty that this action stayed the flow of blood and saved his life.[7] From that day forward, Vasari carried the healing stone with him, and the malady never reoccured.

A long-standing legend has been passed down in the Christian tradition. The bloodstone was said to be a green jasper which was at the foot of the cross at Christ's crucifixion. One version of the story says that drops from the Savior's wounds were splattered onto the stone. Another says the red spots are from drops which fell from the tip of a Roman soldier's sword. Gem carvers have used the stone as a medium for their art. The face or figure of Christ is carved into the stone to serve as a religious icon. The most extraordinary examples are those which feature Jesus's face with carvings are arranged so the red spots appear to flow from the wounds of the crown of thorns. Lost carvings have been found by the faithful

This is the alchemical sign for bloodstone. It also serves as a symbol for hematite.

According to Diderot's Encyclopedia *this sign was use by eighteenth century chemists in France to mean bloodstone or hematite.*

> *Who in this world of ours their eyes*
> *In March first open, shall be wise;*
> *In days of peril firm and brave*
> *And wear a Bloodstone to their grave.[8]*
>   *or in an earlier version*
> *Who on this world of ours her eyes*
> *In March first opens may be wise,*
> *In days of peril firm and brave*
> *Wears she a Bloodstone to her grave.*

Lapidarium:

and thought to be of natural occurance. Records of the Roman Catholic Church show requests for verification of such wonderous finds and the cannonization of the finder. Charlitans have commissioned such carvings, buried them, and unearthed them as miraculous acts.

Vedic texts of India relate a story of the origin of bloodstone. Vala, a demon god, was slain and dismembered by the demigods. The parts of

A bloodstone is used to stop a nose bleed in this illustration from the Ortus Sanitatis, *a fifteenth century medical text by Johannis de Cuba.*

his body were strewn about the earth and universe to create the various gemstones we know today. Agnideva, demigod of fire, stole the complexion of Vala and transformed it into the seeds of bloodstone. These seeds were dropped primarily into India's Narmada River. Other seeds washed up upon the lands occupied by the lower caste while the rest were spread around India and the world. Wherever they settled, deposits of bloodstone originated.

Contemporary advocates of crystal power give bloodstone many healing powers. It is said to be an intense healing stone and a "stone of courage." The gem is reported to have cleansing energy, especially regarding physical imbalances in blood, bowel, and stomach. It is also said to aid in purification, calming, and soothing the mind and emotions. The belief continues that it is able to neutralize toxins in the body and help in the elimination of the same. Bloodstone is still used to stabilize blood flow and is gaining favor as a marrow builder in the treatment of leukemia.

Carbuncle: see Garnet

Carbuncle is generally accepted as the archaic term for garnet. The term carbuncle has also been used to refer to various other red gems throughout history. Today, gemologists only use the name to refer to red cabochon cut garnets. The origin of the word is Latin, *carbunculus* meaning little spark, and was used by Pliny to refer to all "glowing" red gems. Descriptions and lore may be found under the name Garnet (page 108).

Carnelian

Crystal System

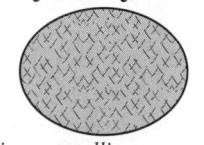

Microcrystalline aggregates

Color: Flesh-red to brown-red (Indian, brown tints reddened with exposure to sun) **Mohs' Hardness:** $6\frac{1}{2}$ - 7 **Cleavage:** None **Refractive Index:** 1.535 - 1.539 **Dispersion:** None	**Streak:** White **Specific Gravity:** 2.55 - 2.65 **Fracture:** Uneven, shell-like **Birefringence:** to 0.006 **Pleochroism:** None

Crystal System: Hexagonal (trigonal) fibrous aggregates
Chemical Composition: SiO_2 silicon dioxide
Transparency: Translucent to opaque
Flourescence: None

The brownish-red to deep red form of chalcedony is known as carnelian. The term is usually limited to the red translucent variety. Opaque brownish material is classified as sard.

    This "stone of August" was one of the first used in ancient Egypt. Carnelian is the modern spelling of the older name cornelian. The archaic form stems from the Latin *cornum*, meaning *cornel berry* or *cornelian cherry*. The change in spelling dates to the fifteenth century. A mistaken belief the name originated in the Latin *carneolus* or *carnem*, meaning flesh, caused the letter 'a' to be substituted.
    The fact that wax does not easily stick to carnelian, and that it is relatively soft, resulted in its use for carved cylinder seals in ancient Egypt. The stone also was used as a burial amulet, readily sculpted and engraved. Papyri have revealed ancient burial practices involving the gem. Chapter 156 of the Egyptian *Book of the Dead* includes the following passage.

    This is the Chapter of the Buckle of Carnelian which is put on the neck of the deceased. The blood of Isis, the virtue of Isis, the magic power of Isis, the magic power of the Eye are protecting this great one; they prevent any wrong being done to him." The chapter continues, "the buckle of carnelian is dipped into the juice of aukhama, then inlaid into the substance of the sycamore wood and put on the neck of the deceased. Whoever has this chapter read to him, the virtue of Isis protects him; Horus, the son of Isis, rejoices in seeing him, and no way is barred to him.[1]

There also exists a prescription that recommends carving the contents of Chapter 29 of the same *Book* onto the stone. The amulet should be carved in the form of a heart. A scarab, symbol of eternal life, was also often carved of carnelian. The belly of the scarab may have the name of the deceased engraved on it. The gem was set in a ring of clay and placed on the heart of the cadaver.

Carnelian has been identified as the first stone of the twelve set in the Breastplate of Aaron. First called *odem*, this stone is identified by Theophrastus in 300 B.C. as a red stone which is either carnelian or sard. The gem was said to be engraved with the name of the tribe Reuben. The children of Israel were said to have brought this stone with them out of Egypt. As they wandered through the desert, they engraved seals in carnelian of sacred symbols and figures.

The traditions of Arabic peoples include the use of carnelian as an amulet. It is thought to overcome envy. The belief exists that if one is envied, they will be drained of the quality which another covets. If one wears a carnelian engraved with the following sacred poem, they will be saved from this draining force.

> In the name of God the Just, the very Just!
> I implore you, O God, King of the World,
> God of the World, deliver us from the devil
> Who tries to do harm and evil to us through
> Bad people, and from the evil of the envious.[2]

Moslems hold the stone dear. It is said the prophet Mohammed wore a silver ring on his finger set with a cylinder seal of carnelian on his little finger. It is thought that any person who owns a cylinder seal of this stone can never be separated from God. One of Mohammed's imâms, Jafar, declared the virtues of the stone and said that all desires of a man who wore carnelian would be met. The name of Ala, another of the prophet's Imâms, and his successors was engraved on amulets in Persia. These were said to preserve the wearer's dignity and serious manner during arguments.

> *Carnelian is a talisman,*
> *It brings good luck to child and man;*
> *If resting on an onyx ground,*
> *A sacred kiss imprint when found.*
> *It drives all evil things;*
> *To thee and thine protection brings.*
> *The name of Allah, king of kings,*
> *If graven on this stone, indeed,*
> *Will move to love and doughty deed.*
> *From such a gem a woman gains*
> *Sweet hope and comfort in her pains.*
>   Wolfgang von Goethe

The range where this stone is found, and the ease with which it is worked, has led to the existence of lore regarding carnelian to be wide spread. In Japan beads of uneven cut and length are commonly buried with the dead. Buddha is said to have offered a vase of carnelian to one of the four Kings of Heaven. Aboriginals of South West Australia wear beads of this stone strung on opossum yarn around their waist to cure all ills. By custom, no woman is allowed to touch these talismans. Zulus claim that no man wearing carnelian will be harmed by falling houses or walls. This same belief is found in native American and central European lore. The nineteenth century author Arakel gave as proof "no man who wore a carnelian was ever found in a collapsed house or beneath a fallen wall."[3] In Arab countries, toothpicks of the gem are used to prevent bleeding gums. The Bghai tribes of Burma use carnelian to form fetishes and believe they must be fed blood. "Spirits good and bad dwelt in stones and if we don't give them blood to eat, they will eat us!" states a Burmese legend. An old tale of Burma says, "A man in one family died, and his widow in wrath commanded their son to throw away the magic stone; the stone returned shortly, bringing two other stones with it, so the widow and her son were resigned."[4]

Carnelians were recommended by the *Lapidario of Alfonso X* for the cure of a weak voice or timid speech. The subject was directed to hold a warm stone which would give the courage lacked so they could speak boldly. This is in line with the common belief that red stones act as a stimulant. This attribute has caused the stone to be adopted as a charm for actors and other public speakers. The thirteenth century *Book of Wings* by Ragiel recommends, "a man richly dressed and with a beautiful object in his hand" should be engraved on carnelian.[5] This amulet could then be used to stop the flow of blood and grant honor to its wearer. The ability of a red stone to control bleeding is based in a stones "sympathy" for a substance based on its like color. Chevalier Jean de Mandeville states in his fourteenth century *Lapidaire*:

> Carnelian is a red-colored stone, light and dark; if worn on one's finger or neck, it will bring peace and concord and give honor and victory. It will restrain the bleeding of a wound or of a nerve; it will also restrain the flow of a woman's bleeding and will appease anger, as well as the enemy of he who wears it."[6]

The sixteenth century scholar Camillus Leonardus echoes Mandeville when he states, "It restrains menstruous fluxes, and stops the hemorrhoids. It cures the bloody flux; and being worn about the neck, or on the finger, it assuages strife and anger."[7] The philosopher De Laet wrote in 1647 that it has the power to stop bleeding from the nose. He states that rings were cut from carnelian and worn for this purpose. King states, "Such are still made and worn in Italy, and with the same idea."[8] A carnelian set in a ring caused one who wore it, "to be of a cheerful heart, free from fear, and nobly audacious, and was a good protection against witchcraft and fascinations."[9]

Modern practitioners use carnelian to relieve lethargy, depression, low self esteem, fear and anxiety. It is also used as an amulet to stabilize emotions in the home and promote love between parents and children. Its medicinal applications include the treatment of neuralgia, gall stones, kidney stones, pollen allergies, and colds. An elixir, made by soaking the stone in pure water, is used to hasten the healing of cuts and abrasions.

Crystal System

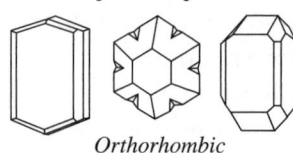

Orthorhombic

Cat's-eye: Chrysoberyl

Color: Golden yellow, green-yellow, brown with cats'-eye phenomenon **Mohs' Hardness:** $8\,^1/_2$ **Cleavage:** Imperfect **Refractive Index:** 1.746 - 1.755 **Dispersion:** 0.009	**Streak:** White **Specific Gravity:** 3.70 - 3.74 **Fracture:** Weak **Birefringence:** 0.009 **Pleochroism:** Strong; reddish-yellow, yellow, light green, green

Crystal System: Orthorhombic; twinned, intergrown triplets
Chemical Composition: $Al_2(BeO_4)$ beryllium aluminum oxide
Transparency: Translucent
Fluorescence: Usually none Green; weak, dark red

Cat's-eye is a generic term for any gem which shows a band of light across a cabochon cut stone. The eye phenomenon is caused by the presence of fiber inclusions within a gem. The term is often misused in the gem trade as an adjective, so stones are referred to as cat's-eye tourmaline, cat's-eye quartz or cat's-eye apatite. The proper use of the name describes cat's-eye chrysoberyl or precious cat's-eye. The color may be yellow, golden or the most valued, yellowish-brown or honey.

Arabs believed that cat's-eye chrysoberyl had a property which caused its wearer to become invisible in battle. As an amulet it protected one from witchcraft and death. They also believed it may be use as a test of fidelity. If a man is about to leave on a trip and has doubts about the faithfulness of his wife, he makes her drink milk in which a cat's-eye has been washed. This elixir is not expected to keep her virtuous, but if she commits adultery in his absence, no children will result to burden him with her sin. The natives of Ceylon use this rare stone as a charm against evil spirits. Throughout history stones which display an eye were used to counteract the evil-eye. They have also been favored in the treatment eye disorders. Modern practitioners claim an increase in night vision, a natural connection with the stone's name-sake.

Vedic texts of India relate a story of the origin of cat's-eye. Vala, a demon god, was slain by the demigods. When arrested and bound by his enemies, Vala gave off a thunderous war cry. This cry was transformed into the seeds of cat's-eye. These seeds fell into the sea and caused huge waves which washed the seeds onto the shore. Wherever they settled, they formed deposits of cat's-eyes; the finest were said to have landed near Sri Lanka's famous Vaidurya Hill. The gems mined in this region are the most revered in India and have come to be known as *Vaidurya* stones. The legend continues to say that the remainder of the seeds were carried to the heavens to impregnate the clouds. Carried to earth by rain and comets, these seeds formed smaller mines in scattered locations.

Chalcedony

Crystal System

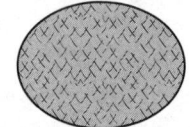

Microcrystalline aggregates

Color: Bluish, white, gray
Mohs' Hardness: $6\frac{1}{2}$ - 7
Cleavage: None
Refractive Index: 1.535 - 1.539
Dispersion: None

Streak: White
Specific Gravity: 2.55 - 2.65
Fracture: Uneven, shell-like
Birefringence: up to 0.006
Pleochroism: None

Crystal System: Hexagonal (trigonal) fibrous aggregates
Chemical Composition: SiO_2 silicon dioxide
Transparency: Dull, translucent
Fluorescence: Blue-white

Chalcedony is the name for a species of minerals characterized by its cryptocrystalline (microscopically crystallized) quartz structure. The term is used to differentiate this species of materials from crystalline quartz. Chalcedony is found throughout the world and is one of the most common and least expensive gems. Varieties within the species have attributes which vary widely. They include agate, bloodstone, chalcedony moonstone, carnelian, carnelian onyx, chrysocolla, chrysoprase, jasper, onyx, petrified wood, plasma, prase, sard and sardonyx.

Chalcedony is referred to in ancient writings as a distinct stone. The name was reserved for milky white to pale blue-gray translucent gems. The origin of the name is uncertain, but the obvious belief is that it originated with the seaport Chalcedon near Byzantium. The Greek name is *calkedon* and the Latin *charcedonius*. Damigeron's entry regarding this gem is the shortest of the fifty stones discussed in his *De Virtutibus Lapidum*. The complete text reads, "The stone chalcedony is bored by and [set in?] iron: he who wears it conquers."[1] An early Christian work, written in Greek in about the fourth century, contains a verse about chalcedony. The presumed author, Physiologus, states, "Chalcedony is a stone which shines with a faint paleness. It comes between the hyacinth and the beryl. Anyone who carries it will, it is said, be successful in lawsuits."[2] Italian peasant women have worn spherical beads of milky-white chalcedony for centuries. They believe these *pietra lattea* have the ability to increase the milk supply of lactating mothers. Similar beads have been found in Northern Italy which date to the iron age and may have been worn for the same purpose.[3] White beads of chalcedony have been used by many cultures. They are, "used as a sacred stone by the Native Americans, promoting stability within the ceremonial activities of the tribes."[4]

The fourteenth century *Lapidaire* of Chevalier Jean de Mandeville contains the most extensive list of attributes to be found in one volume.

> Chalcedony is a white or blue stone, of a pale color of different sorts; there are three different kinds. One serves against bad weapons and deceptions and gives victory in argument and in battle. Another helps with merchandise and other needs; working against illusions and frivolity stemming from melancholy, giving good eloquence in speaking, protecting the bodily virtues, resisting venom, and delivering one from tempests. The last protects one from fire and water and should be mounted in gold.[5]

The virtues listed by Mandeville seem to serve as a foundation for many latter scholars.

The *Speculum Lapidum* of the physician Leonardus contains a prescription for the wearing of chaldedony which reinforces Physiologus' advice to members of the legal profession. "If a person carries about him one of them perforated, with the hairs of an ass run thro', he will be successful in civil causes and contentions."[6] The scholar Josephi Gonelli made an interesting observation regarding this stone in 1702. He explained that it dissipated the, "evil humors of the eye, removing the diseased condition of the organ which caused apparitions to the seen."[7] He mistakenly attributed the affect to a supposed alkalinity in the stone. He did however, seem to recognize that "phantoms and visions" may be the result of eye disease.

As recently as the nineteenth century strings of this common gem were hung about the neck to dispel melancholy. It was widely believed that, "One perforated, with the hair of an ass run through it, would overcome all contentions and preserves from tempests and sinister events."[8] The perforation of the gem and the stringing of it on an ass' hair is a treatment of the gem that survived throughout the centuries.

Modern crystal healers claim it may be used to alleviate hostilities, irritability and to enhance generosity. It is used medicinally to aide in the assimilation of minerals, reduce excess build-up in veins and to cure dementia and senility. A particular variety called rose chalcedony (whetstone) is claimed to aide in the development of feminine energy and to help sexual problems in males and females. Metaphysicians claim it induces healing in the female reproductive organs.

Chiastolite

Crystal System
Orthorhombic

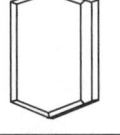

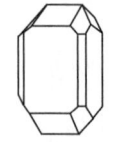

Color: Dark band on a white, gray, reddish, or light brown background	**Streak:** White
	Specific Gravity: 3.13 - 3.20
	Fracture: Uneven
Mohs' Hardness: 7 - 7 $^1/_2$	**Birefringence:** 0.008 - 0.013
Cleavage: Distinct prismatic	
Refractive Index: Varies:1.628 - 1.640 low to 1.641 - 1.647 for high	**Pleochroism:** Strong; brownish to red and green
Dispersion: 0.016	

Crystal System: Orthorhombic, but almost square in cross section with basal faces
Chemical Composition: $Al(AlSiO_5)$ aluminum silicate
Transparency: Opaque in this variety of andalusite
Fluorescence: Green to yellowish-green

Lapidarium:

> Chiastolite is an opaque variety of andalusite that grows in cigar-shaped crystals. A slice across the mid-section of the crystal reveals a light-colored cross against a green background. This design is caused by a regular arrangement of carbonaceous impurities along the axis of the crystal. Although this obscure crystal shows a striking phenomenon, it is rarely made available as a finished gem.

Diagram of a sliced crystal exhibiting chiastolite's distinctive internal markings

The shape displayed in the cross-section diagram of this mineral (andalusite) is what has made this a revered Christian talisman.

Chiastolite is named for its distinctive internal markings which resemble the Greek letter X (chi). It is also called "macle," from the Latin *macula* meaning spot or blemish. The cross design has caused this mineral to be considered mystical and religiously significant. Found near the Shrine of St. James in Santiago de Compostella in Spain, the chiastolite has been revered as a holy relic in petrified form.

The stone is said to staunch the flow of blood from any part of the body if it is worn touching the wound. If suspended from the neck, it is said to cure any kind of fever or inflammation. Worn by a lactating mother the crystal was said to increase the flow and quality of breast milk. Also known as the *lapis crucifer* (cross-stone) by the ancients, it was welcomed in any community. A popular legend states that evil spirits are driven from the entire neighborhood of the wearer.[1] Now used as sign of devotion, it signifies both peace in death and the promise of re-birth to Christians. Metaphysicians recommend the crystal as a medium to induce astral projection and mind travel.

Crystal System

Orthorhombic

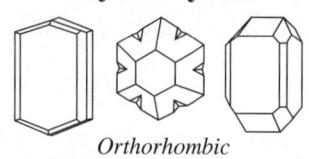

Chrysoberyl

Color: Golden yellow, green-yellow, brown
Mohs' Hardness: $8\frac{1}{2}$
Cleavage: Imperfect
Refractive Index: 1.746 - 1.755
Dispersion: 0.009

Streak: White
Specific Gravity: 3.70 - 3.74
Fracture: Weak
Birefringence: 0.009
Pleochroism: Stong; reddish-yellow, yellow, light green, green

Crystal System: Orthorhombic;twinned, intergrown triplets
Chemical Composition: $Al_2(BeO_4)$ beryllium aluminum oxide
Transparency: Transparent
Fluorescence: Usually none, Green: weak; dark red

The various forms of chrysoberyl are both rare and of high value. The stones may be transparent and used as facet material, or cloudy and cut en-cabochon to exhibit an eye. The transparent yellow or honey-colored variety is called golden beryl; the chatoyant cloudy material is commonly referred to as precious cat's-eye. Another variety, which is purple, is called alexandrite. The attributes and lore of cat's-eye and alexandrite will be treated separately for the purposes of this text.

    The ancients did not recognize chrysoberyl as a separate gem. The name was derived in modern times from the Greek root for golden. Twentieth century crystal healers have established many qualities for this gem. They call chrysoberyl the "stone of immortality." It is said to bring forgiveness among family and friends, bring generosity and charitability and promote understanding. In treatments involving crystals it is used to add effectiveness to other gems.[1]

Chrysolite: see Peridot

Chrysolite is a name which has fallen from fashion. The term is rarely used by modern gemologists, but some reserve it for light yellow-green or greenish-yellow peridot. Through the centuries the name has been applied to peridot, topaz, forsterite, fayalite, chrysoberyl and other yellowish stones. Most of these associations have been made in error. Today peridot is the name preferred by gemologists, believing chrysolite is confusing given the other chryso-prefixed gems. For the purposes of this text, early scholar's references to chrysolite will be included under the gem peridot.

Rich, brilliant, like chrysoprase glowing
Was my beautiful Rosalie Lee.
Rosalie Lee, *Thomas Holley Chivers*[1]

Crystal System

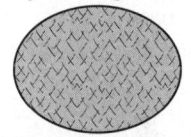

Microcrystalline aggregates

Chrysoprase

Color: Green, apple green
Mohs' Hardness: $6\,^1/_2 - 7$
Cleavage: None
Refractive Index: 1.530 - 1.539
Dispersion: None

Streak: White
Specific Gravity: 2.58 - 2.64
Fracture: Rough, brittle
Birefringence: up to 0.004
Pleochroism: None

Crystal System: Hexagonal (trigonal) microcrystalline aggregates
Chemical Composition: SiO_2 silicon dioxide
Transparency: Translucent to opaque
Fluorescence: None

> Chrysoprase is translucent apple-green chalcedony. Naturally colored material is the most valuable of all the chalcedonies and is often mistaken for jade. Much of the chrysoprase offered for sale is actually dyed pale chalcedony. The ancients did not distinguish chrysoprase from prase, but modern gemologists reserve the term prase for yellowish-green gems. Prase displays a color that shows little customer appeal and is seldom found for sale.

The name chrysoprase is derived from the Greek *khryso'prasos* meaning "golden green." Although known to the ancients, few legends are connected to this variety of chalcedony. The gem is mentioned once in the *Bible* by the Apostle John. The Book of Revelations lists chrysoprase as the tenth foundation stone of the New Jerusalem.

Christian lore states that chrysoprase is the stone of St. Thaddeus. This is based on the writings of the tenth century Bishop of Caesarea, Andreas. One of the earliest writers to associate the Foundation Stones of Revelations with the twelve apostles. Andreas writes:

> The chrysoprase, more brightly tinged with golden hue than gold itself, symbolizes St. Thaddeus; the gold (*chrysos*) symbolizing the kingdom of Christ, and the *prassius*, Christ's death, both of which he preached to Abgar, King of Edessa.[2]

Albertus Magnus relates the story that Alexander the Great wore a prase in his girdle when he led his men into battle. The story states that on one occasion, while bathing in the Euphrates, he laid the girdle aside and a serpent bit off the stone and dropped it in the river. Even Magnus, known for his acceptance of these legends, lists the story as a possible fable.[3] Chevalier Jean de Mandeville gives the following reference in his fourteenth century *Lapidaire*. "Chrysoprase comes from India, it has a green color, mixed like leek juice, and sometimes with golden drops. It is hard to find and gives graces to he who wears it, and it is good for the

> *Midst other treasures to adorn the ring*
> *This gem from Afric's burning sand they bring.*
> *Parent of gems, rich India from her mines*
> *The Chrysoprase, a precious gift, consigns.*
> The Lapidary, *Marbode of Rennes*

eyes." [4] Camillus Leonardus makes the following claim in his treatise on gems: "Its principal virtue is to cherish the sight. It gives assiduity in good works; it banishes covetousness."[5] The gem was also said to make a condemned man invisible, and able to escape execution, if he would only hold one in his mouth.

A gem not readily available in Europe, it was often misidentified. The stone was brought to the continent during the crusades and claimed to be emerald. Cups of green glass and chrysoprase reside in various churches and cathedrals. Each has been claimed at one time or another to be the Holy Grail.

The attributes of the stone have increased in modern lore. It is said to enhance mental abilities and give the wearer insight into self honesty and personal problems. It is used to awaken the imagination and reveal the hidden talents of its owner. Chrysoprase is recommended medically to treat reproductive organs and increase fertility. It is also said to give vitality and strength by increasing the assimilation of Vitamin C.

Citrine

Crystal System

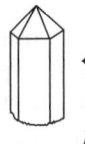

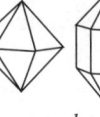

Hexagonal

Color: Light yellow to gold-brown	**Streak:** Colorless
Mohs' Hardness: 7	**Specific Gravity:** 2.65 - 2.66
Cleavage: None	**Fracture:** Conchoidal, very brittle
Refractive Index: 1.544 - 1.553	
Dispersion: 0.013	**Birefringence:** 0.009
	Pleochroism: Natural; weak, yellow- light yellow Heat treated; none

Crystal System: Hexagonal (trigonal) hexagonal prisms with pyramids
Chemical Composition: SiO_2 silicon dioxide
Transparency: Transparent
Fluorescence: None

> Citrine is a yellow variety of quartz which may occur naturally or is created by heat treating amethyst. The yellow-brown variety is known as *cairngorn,* after the city in Scotland where they are found. Dark reddish-brown material is known as *sang de beouf,* French for ox-blood. Also called "topaz quartz," the color has led to its misrepresentation as topaz. Commercial names such as "Saxon topaz," "Spanish topaz," and "citrine topaz" have become common but are considered a misrepre-sentation by gemologists. Other misnomers include "Palmyra topaz" for pale stones and "Madeira topaz" for reddish-brown gems.

The citrine serves as a birthstone for November, sharing this title with the topaz. The name for this yellow gem is derived from the French word *citron,* meaning lemon. Known as the "merchants' stone" or "money stone," it is said to bring prosperity to its owner. Little additional lore has been recorded for this common gem. Ancient lapidaries did not distinguish the stone from yellow corundum or topaz.

The name *citrini* was given to yellow corundum by early writers, and the attributes for this stone have been attached to the citrine. A nineteenth century text gives the following reference: "The Citrini protected the wearer from dangers in traveling, secured him from pestilential vapours, and gave him favour with princes."[1] The same author lists citrine as an alternative spelling for the corundum variety.

Modern crystal power practitioners list many more powers for this stone than are found in early texts. As a beautiful and inexpensive gem, it is commonly used as an amulet. "It is one of two minerals which does not hold or accumulate negative energy but dissipates and transmutes it, working out problems on both the physical and subtle levels."[2] As with other yellow gems, the citrine is used to treat digestive disorders, kidney and bladder diseases, and imbalances in the thyroid gland. Soaking the citrine in pure water renders an elixir which is reported to release toxins from the body.

Coral

Crystal System

Microcrystalline aggregates

Color: Red, pink, white, (rare black and blue)
Mohs' Hardness: 3 - 4
Cleavage: None
Refractive Index: 1.486 - 1.658
Dispersion: None

Streak: White
Specific Gravity: 2.6 - 2.7
Fracture: Irregular, splintery
Birefringence: None
Pleochroism: None

Crystal System: Hexagonal; microcrystalline
Chemical Composition: $CaCO_3$ calcium carbonate (magnesia, organic substance)
Transparency: Opaque
Fluorescence: Weak

Coral is not a mineral but a substance exuded by tiny marine animals called coral polyps. The calcium carbonate material they secrete hardens and forms in branch-like deposits. Coral polyps grow throughout the world's tropical waters which have a temperature above 20° Celsius. Usually found in shallow water, some varieties are found as much as 1,000 feet below the surface. Coral deposits may occur as pink, orangy-pink, orange, blue, red and black. The black variety has become rare and expensive. The poisonous effects of the dust released when black coral is cut has made harvesting and cutting quite risky. The popularity of the various colors is often dictated by local custom and lore. Red "precious coral" has remained the most popular. Deep-red, called "ox-blood," is a highly prized color, but in some countries pink or "angel's skin coral" can be the most expensive. Gem quality coral is uniform in color and able to take a high polish.

The Greeks are the source of the name and original lore concerning coral. The Greek word for pebble, *korallion*, is thought to be the origin of the modern name. The Greek gem poet Orpheus states that coral is "the gift of Minerva." He relates the fable that it "originated when the newly-severed Gorgon's head was laid down by Perseus on the sea-weeds, which the issuing gore turned to stone."[1] He continues by claiming it baffles witchcraft, counteracts poisons, and protects one from tempests and robbers. He also lists this gem of the sea as the farmer's friend. If powdered and mixed with ground seed-corn, it was supposed to protect crops from thunderstorms, blight, caterpillars and locusts. Damigeron wrote in the second century B.C. "It makes him who wears it unconquered, powerful, unable to be touched, free from fear and care, giving orders easily and having easy access to the great."[2]

The Roman historian Pliny the Elder expanded on these beliefs in his *Natural History*. He wrote that coral quieted storms at sea and preserved the wearer from tornadoes and lightning strikes. The Romans also recommended a collar of flint and Maltese coral as a cure for hydrophobia in dogs. This ornament was thought to keep the spell of the evil eye, the supposed cause of the malady, from affecting canines. The hanging of coral branch necklaces around the necks of children was thought also to protect them from the evil eye's curses and to preserve their young teeth. Coral gems were to be cut from freshly gathered material and of richly colored to have any power. It was also believed that if cut coral was broken unintentionally, it would lose all its potency.

Roman physicians prescribed the tonic "tincture of coral" for a number of illnesses. It was used to cause perspiration to reduce fevers, act as a diuretic and drive off "bad humors" from the body. The elixir was prepared by boiling branch coral in melted wax and steeping the resulting product in alcohol.[3] It is thought that the alcohol was probably the active ingredient rather than the coral. A poultice consisting of saffron and coral was wrapped in the skin of a cat and tied around a patient's neck. It was said to have marvelous curative powers. Adding an emerald to the concoction was reported to magnify its ability to drive off the most severe fever.[4] The gem was considered a remedy for a variety of human ills if it were powdered and mixed with water or wine. "Consecrated by God in a holy place this is a great defence for you by day and night; at any hour of the day or night the Coral stone is a great safeguard."[5]

Many writers have added to the lore of coral over the centuries. The thirteenth century theologian Albertus Magnus wrote that he had "proved" the ability of the gem to slow the flow of blood, cure madness and give wisdom to its wearer. He also stated that wearing red or white coral "stilled tempests" and allowed one to cross broad rivers in safety.⁶ Marco Polo wrote in the same century of the use of coral as personal adornment in the Himalayas and as decoration in Tibetan temples. He related that the priests held coral, as well as amber and turquois, in high regard. Coral had religious significance for the lamas; its color represented one of the incantations of Buddha. Camillus Leonardus gives a personal testimony as to coral's protection of children, "I have had it from a creditable person, and have often experienced to myself, that it will prevent infants, just born, from falling into an epilepsy."⁷ An additional caution to women survived from the Middle Ages to the early twentieth century. They were advised to hide their coral from their husbands. The stone was said to change color in sympathy with a woman's monthly cycle; modesty dictated that this change should not be observed by one's mate.⁸ The custom is based in the belief that a gem could reflect the spirit of a living person. The German physician Johann Wittich, writing in the sixteenth century, gives this account of coral's prognosticative powers.

In Diderot's eighteenth century Encyclopedia *the sign below is found as a symbol for coral.*

> Wittich was called to attend to a young man named Bernard Erasmus. As the youth sickened, a red coral which he wore turned
> first whitish, then dirty yellow, and finally became covered with black spots. Wittich directed the young man's sister to remove the coral for death was surely at hand. His prediction was born out, as in a few hours Erasmus was dead.⁹

In another sixteenth century work, *Discoverie of Witchcraft*, Sir Reginald Scot repeats the beliefs of the ancient Romans. "The coral preserveth such as bears it from fascination or bewitching, and in this respect they are hanged about children's necks."¹⁰ The *Speculum Lapidum* gives a more specific reference: "Being carried about one, or wherever it be in a house or ship, it drives away ghosts, hobgoblins, illusions, dreams, lightnings, wind, and tempests."¹¹ A roman Catholic addition to the lore of coral is the use of coral bells. These special bells are said to frighten away evil spirits by their delicate jingle. Eighteenth century French writers recommended the wearing of a *pater de sang* or "blood rosary." These necklaces became popular as a safeguard against uncontrolled hemorrhaging.

Coral has been a valued gem throughout the world. Hindu physicians stated that "true coral" tasted both sweet and sour. They believed that it could have a positive effect on mucous membranes, the composition of bile and "morbid secretions."[12] The king of Benin is the only person allowed to own coral in this African kingdom. The monarch presents bead strings of the gem as a reward for service, but the coral must be returned upon the recipient's death. It is so precious that if lost or stolen, all involved may be put to death. The ancient Persians believed coral did not gain its color or power until after it was removed from the sea. They also thought that genuine coral could be distinguished from an imitation because it had the smell of the sea. Coral is a prized material in the funerary practices of Arab peoples. "They think that to leave their dead without ornaments of coral, is to give them over to the hands of mighty enemies."[13]

Vedic texts of India relate a story of the origin of Coral. Vala, a demon god, was slain and dismembered by the demigods. The parts of his body were strewn about the earth and universe to create the various gemstones we know today. The intestines of Vala were taken by the celestial serpent Vasuki. He deposited them in the oceans around the world, creating the seeds of coral which now grow in the sea.

Native Americans of the Southwest held the color red in high regard. They had collected red fragments from spiny oysters found in the Gulf of Mexico for centuries. Spanish explorers introduced coral in trade with the Pueblo by the mid sixteenth century, and soon it spread to the Hopi and Zuni. All three of these peoples now place coral only second to turquoise as a precious object. Anthropologists generally agree the love of anything associated with the sea, particularly coral and shell, may be attributed to the high value placed on water in this arid region.

Contemporary beliefs about this gem include its ability to promote passion, intuition, imagination, and romantic love. As with many red stones, it is said to strengthen circulation and control bleeding. These

> Whilst rooted 'neath the waves' the Coral grows,
> Like a green bush its waving foliage shews:
> Torn off by nets, or by the iron mown;
> Touched by the air it hardens into stone;
> Now a bright red, before a grassy green,
> And like a little branch its form is seen;
> The Lapidary, *Marbode of Rennes*

skeletal remains of sea creatures are also said to stimulate the regeneration of bone tissue and act as a cure for disorders of the spinal canal.

The diamond is the crystalline Revelator of the achromatic white light of Heaven.
Lily Adair, *Thomas Holly Chivers* [1]

Diamond

Crystal System

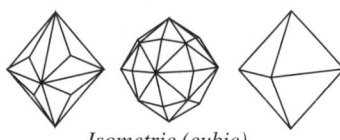

Isometric (cubic)

Color: All **Mohs' Hardness:** 10 **Cleavage:** Perfect **Refractive Index:** 2.417 - 2.419 **Dispersion:** 0.044	**Streak:** None possible **Specific Gravity:** 3.51 - 3.54 **Fracture:** Conchoidal to splintery **Birefringence:** None, often anomalous **Pleochroism:** None

Crystal System: Isometric (cubic) mainly octahedrons, also rhombic dodecahedrons, cubes, twins, plates
Chemical Composition: C, crystallized carbon
Transparency: Transparent
Fluorescence: Variable; Colorless and yellow, mostly blue, Brown and green, often green

Diamond is pure carbon with atoms arranged in a cubic system. The resulting mineral is the hardest substance known to man. Due to trace mineral impurities, diamonds occur in all colors, the most popular is colorless. This is not, however, the highest valued. Red, deep green, and deep blue stones are considered collector stones and are rarely offered to the public for sale. Brownish to very dark gray stones, called 'black diamonds,' have numerous industrial applications, primarily as abrasives. Their extreme hardness allows them to 'cut' any other material.

Diamond shares a chemical kinship with man, as both are carbon based. This modern "King of Gems" has not always been the most valued or most popular. The stone has a shorter history of use by man than many other gems. In some parts of the world it took many centuries for the stone to be distinguished from colorless quartz and corundum. It was not until the fourth century B.C. that the diamond was distinguished from other gems by stone engravers in India. References to the diamond seem to pre-date this discovery, but the word which is the root of our modern name had many applications. The Greek word *adamas* translates as "invincible" or "of extreme hardness." The first known use of the word in Greek literature was by Hesoid in the eighth century B.C. Scholars agree that Hesoid's reference was to iron and not to a precious stone. Adamas was used for centuries by Hellenic authors to refer to hard metals and, later, an imaginary substance harder than iron. The poet Theocritus (228 B.C.) calls Pluto "the adamas of Hades."[2] It was in the second century B.C. when the Roman magician Damigeron used the Latin form, *adamus* or *adamantinus*, to name this colorless stone of unparalleled hardness.

The Hebrew word *yahalom* (*shamir* in later rabbinical writings) served a similar purpose in Old Testament scriptures. The twenty-eighth chapter of Exodus describes the breastpiece of the high priest Aaron as containing such a material. The third stone in the second row is described as yahalom. It is also recorded that these stones were engraved with the names of the tribes of Israel. It is highly unlikely that the Jews were able to engrave in diamond. Other references to diamond are found in Jeremiah, Ezekiel, and Zachariah. In each of the instances, the material is used as a graver. Jeremiah 17:1 states, "The sin of Judah is written with a pen of iron; with a point of diamond it is engraved on the tablet of their heart, and on the horns of their altars." Yahalom has also been translated as emery and in some instances corundum. It is known that both of these substance were in use at the time as engraving points. Another term which has been confused by translators is *halom*, used in a metaphorical or poetic sense for hammering. To "drive in" or "make an impression" may at times be considered to mark or engrave.

From ancient times, until eighteenth century discoveries in Brazil, India was the only significant source of diamonds. A twentieth century translation of the *Artha Sastra* of Kautilya, *The Lesson of Profit*, revealed the economic and legal history of India in the fourth century B.C. This remarkable record confirms that diamonds served as an important com-

> *Those who in April date their years,*
> *Diamonds shall wear, lest bitter tears*
> *For vain repentance flow. This stone*
> *Emblem of innocence is known.*[3]

modity in the court of King Chandragupta Maurya (320-298 B.C.) and that the stones were the subject of taxes and customs duties. This Indian work refers to another text called the *Ratnapariska*, translated as *The Estimation of Value of Precious Stones*, which shows a great knowledge of diamonds existed. Through nearly one thousand years this information evolved into the *Ratna-Sastra*, a technical manual used by noblemen, officials, merchants, and even poets. Sixth century manuscripts called the *Ratnapaiska* of Buddhabhatta and the *Brihatsamhita* of Varahamihira were known to medieval European scholars.

These Hindu treatises list the characteristics of the diamond and also a system of classification by color. A typical stone is described as,

> A six-pointed diamond, pure, without stain, with pronounced and sharp edges, of a beautiful shade, light, with well-formed facets, without defects, illuminating space with its fire and with the reflection of the rainbow, a diamond of this kind is not easy to find in the earth.[4]

The text adds that the form of the gem includes eight flat and similar sides with twelve sharp straight edges. This is undoubtably an early description of the characteristic octahedral crystal. This form of the stone is rare in nature and was the most prized and ideal diamond. Since they were not cut, the stones were used in the form in which they were found. Not all shapes were favored. Triangular stones were said to caused quarrels, a square diamond caused the wearer to have hallucinations and a five-cornered stone should never be worn as it would bring death.

The color of diamonds also held great significance to the ancients in India. They related caste and position to each of the gems' hues. *Brahmins*, the priest caste, could own colorless diamonds. Yellow stones the shade of the *kadali* flower could be owned by the *Vaisyas*, landowners and suppliers of food. The *Kshatriyas* caste, knights and warriors, could own red diamonds as a sign of their station in life. Warriors were also sometimes paid in red diamonds for services rendered to their lord. *Parsees* of Persian origin worshiped the color red and prized such gems. The treasuries of many kings and princes were said to hold great quantities of these stones. The Kshatriya could also possess the stones which were the brown color of the hare. Laborers and artisans, the *Sudras*, were assigned ownership of dark gray gems, diamonds the "dark tone of a sword."[5] The price of diamonds was fixed at two hundred *rupakas* for a one-carat perfect crystal. Gray stones were assigned a value of one-quarter that of colorless diamonds. It is obvious that few Sudras would have owned diamonds. The wage of a Sudra was one and a quarter rupakas a month. The color of a diamond would also determine what attribute the stone could convey. The Brahmin stones gave power, friends, riches and good luck; the Kshatriya prevented the approach of old age; the Vaisya stone brought success; and the Sudra gave good fortune. "A diamond, a part of which is the color of blood or spotted with red, would quickly bring death to the wearer, even if he were the Master of Death."[6]

Other properties are praised in these Indian writings. The optical properties of clarity, transparency, fire and iridescence were additional points to be judged. The authors knew a perfect octahedron displays the best optical qualities. Some "would illuminate the sky with all the fire of the rainbow." Specific gravity is a property which seems to have been known by these early authors. The fifth and sixth texts of the *Ratnapaiska* and the *Brihatsamhita* refer to *laghu,* or specific lightness. The diamonds of greatest value were given a mythical weight. "If a diamond possessing all these (optical) qualities floats on water, that is the stone to be desired above all other jewels."[7] The concept of weight evaluation is explained in an anonymously authored sixth century text titled *Agastimata*. The ultimate weight to size ratio is expressed as a density of three *pinda*. This equates to a volume of three *yava* to a weight of three *tandula*. A stone of three yava and two tandula would have lower density, inversely a stone of three yava and four tandula would have a greater density. Stones which exceeded the three pinda ideal by one quarter were devalued by one half. A stone which exceeded the ideal by one half would have only one quarter the value. Variations in specific gravity would, therefore, result is stones of different value. All colorless stones with an octahedral shape where called diamonds, but a colorless quartz octahedron would be a much less valuable diamond. This system allowed the gem merchants of India to rate true diamonds against their imposters.

The use of specific gravity as a criteria may have led to the inclusion of other materials in the class diamond. Magnetite is octahedral and may be the "gray diamonds" listed is the caste system. This natural magnetic ore has a specific gravity of 5.18, one and one-half that of the diamond. The ore is an abundant mineral in parts of India. A "black diamond" described by Pliny the Elder as heavy, soft, and able to be drilled may actually have been siderite iron ore. Either of these minerals may account for the myths that diamonds had magnetic powers. Pliny wrote that diamond was like an anti-magnet able to remove the power of a magnet. The red spinel has a specific gravity of 3.6, a density very close to diamonds 3.52. Such a gem would have the ideal pinda required of a true diamond by the Hindus. This may account for the reported abundance of red diamonds in royal treasuries, a color of diamond known today to be extremely rare. It is a near certainty these stones, reserved for the Kshatriyas and revered by the Parsees, were in fact octahedral crystals of red spinel.

It may been seen that it was not the hardness of diamond that brought it status in India, but its form and optical properties. The status of the octahedron form is demonstrated in the Hindu hierarchy of gems and their association with the gods. The white octahedron was consecrated to Indra, god of storms, thunder, and lightning. The name for lightning in India is *vajra*, the same word is used to denote diamond. The black manifestation of these geometric forms was dedicated to Yama, god of death. The twin crystals of this black diamond were thought to resemble a coiled serpent, Yama's physical form on earth. Vishnu, god of the heavens, controlled all crystals. He was particularly associated with yellow, or Kadali colored gems.

Vedic texts of India relate a story regarding the origin of diamonds. Vala, a demon god, was slain and dismembered by the demigods. The parts of his body were strewn about the earth and universe to create the various gemstones we know today. Vala's bones were stripped of flesh and crushed to fragments. Wherever these fragments fell, they grew into diamond crystals of various types and colors. In the Vedic texts different demigods preside over exceptionally fine stones of different colors. White diamonds are ruled by Varuna, lord of the oceans. Yellow diamonds are ruled by Indra, King of the Heavens; copper-hued stones are the protectorates of the Murats, the wind gods; and greenish diamonds are under the control of Surya, sun-god. The fire-god, Agni, rules brown diamonds, and blue tinged stones are under the lord of ancestors, Aryama. Vedic traditions place some guidelines on who should wear which stones and for what reasons. Only the king is privileged to wear fancy colored stones of yellow, chartreuse or pink. Anyone else wearing these colors

> *Then first were diamonds from the night*
> *of earth's dark centre brought to light*
> *And made to grace the conquering way*
> *Of proud young beauty with their ray.*[8]
> Loves of the Angels: Second Angel's Story
> Sir Thomas Moore

would come to harm. Women should not wear diamonds according to these religious writings, because they have a mystic potency that causes women to be unhappy. A well formed rough diamond brings the following virtues: prosperity, longevity, marital happiness, children and livestock, and good harvests. Diamonds of good quality are also thought to protect from poisonous serpents, tigers, thieves and general poisons. Finally, these Sanskrit writings state that the wearer of diamonds is protected from floods and fire, as well as being given the gift of a clear and lustrous complexion.

Hindu physicians claimed diamonds could be found in six flavors: sweet, sour, salty, pungent, bitter and acrid. They claimed a potion could be made from the gem which would stimulate and strengthen all bodily functions. It was reasoned that since the stone possesses all the flavors known, it must possess the ability to cure all diseases known. The author Buddhabhatta's writing states, "he who wears a diamond will see dangers recede from him whether he be threatened by serpents, fire, poison, sickness, thieves, flood, or evil spirits."[9] A gem of such universal virtue and relative rarity was immediately elevated to great stature when introduced to Rome. Damigeron states, "When you have obtained it, put it in a silver shetstone, and when you have consecrated it, it will make you unconquerable by enemies, opponents, and evil doers, and overbearing men, for you, will be shone terrible in every way[10]. Pliny the Elder wrote of all things on earth, "and not only of precious stones, it is to the diamond that we attribute the highest value." The Roman poet Manilus, a contemporary of Pliny's, writes of the diamond in his *Astronomica*, "point of a stone more precious than gold."[11] It is clear that the Romans honored the diamond for the mythical qualities Indian traders related to their customers.

Its mystical properties and great hardness made diamond the ideal example of the quality adamas. It was considered to have all power over disease and to be an invincible stone. Damigeron adds, "they are useful for all kinds of magical operations . . . You should, therefore, make a bracelet of gold, silver, iron, and bronze, twist it, and wear it around your left arm. This stone brings great help from God."[12]

An early Christian work, written in Greek in about the fourth century, gives attributes for the adamas and adds extensive moral lessons and metaphors of Christian teaching. The presumed author, Physiologus, states,

> This is a stone called adamas found on a certain mountain in the east. Such is its nature, that you should

search for it by night, not day, since it shines at night where it lies, but it does not shine by day, since the sun dulls its light. Against this stone, neither iron, fire or other stones can prevail.[13]

Thought to have little aesthetic value, its mystical attributes were not enough to maintain the diamond's status. The spread of Christianity saw the diamond downgraded to a stone of value only in pagan superstition. Medieval lapidaries were aware of the virtues listed by Pliny, but they ranked diamond as seventeenth in value among precious stones. Psellus, writing in Constantinople in the eleventh century, stated that the diamond was hard, difficult to pierce, and only of value in reducing fever. All clear crystals were considered to be permanent ice. Diamond, due to its hardness, was thought to be only superior rock-crystal. Hindu authors wrote that rock-crystal was *kacha*, "unripe" stone and that diamond was *pakka*, "ripe."

Astrologers of the Middle Ages began to revive the diamond's ancient attributes. Since very few people owned one, or had even seen one, it was easy to give the stone many magical powers. Marbod, who called the stone *adamans*, wrote that it is a magic stone of great power and that it drives away nocturnal specters and is a cure for insanity; it should be set in gold and worn on the left arm. St. Hildegarde (1098-1179), the Abbess of Bingen, wrote that holding a diamond in one hand while making the sign of the cross with the other would stimulate a diamond's curative powers. She also said the diamond was the enemy of the devil because it resisted his power day and night. The stone was said to be capable of curing madness, protecting crops from natural disasters, and guarding houses against fire and lightning.

Marco Polo reported in *The Book of Marvels* that diamond's extreme hardness and optical qualities made it an ideal talisman. The stone was said to be able to keep away all dangers and bad luck. Such statements caused the gem to be more and more popular in the West. The diamond was regarded as a valued stone, but it still did not enjoy the stature of some other gems. The Spanish naturalist Garcias ab Horta wrote upon his return from India that, "Here (in India) diamond is regarded as king of precious stones. Yet if we apply the criteria of value and beauty it is

> *The diamond has become notoriously common since every tradesman has taken to wearing it on his little finger,*
> Against the Grain, *Joris Karl Huysmans (1884)[1]*

certain that, for us, it is the emerald that holds first rank, followed by the ruby."[14] The great Renaissance jeweler Benvenuto Cellini gave diamond the same third place in his ordering of gem values. The hardness of the stone was still its blessing and its curse. Without the knowledge and ability to fully fashion the gem, it would be hundreds of years before the diamond could gain its place as king of gems.

In the late Middle Ages the diamond became a symbol of invincibility, courage, and virility. These qualities made it the exclusive privilege of men. Saint Louis (1214-1270) declared the stone to be forbidden to women. He decreed the only woman worthy of wearing such a grand gem was the Virgin Mary. Nearly two hundred years later the gender ban was broken. Agn'es Sorel, Mistress of Charles VI, was the first woman recorded to wear diamonds. By the late fifteenth century both men and women in the House of Burgundy were wearing diamond jewelry. From this time diamonds have been among the most valued stones in royal treasuries.

Legends concerning diamonds increased as more stones came into Europe. During the Middle Ages doctors prescribed powdered diamonds for various ailments. When many patients died, the practice diminished. It later became known as a powerful poison. Catherine dé Medici (1519-89) was known to use diamond dust to eliminate rivals. It was known as the "powder of succession" by her numerous enemies. Benvenuto Cellini reported that Pierluigi Fanese, son of Pope Paul III, tried to poison him using diamond powder. He was imprisoned in Rome in 1538 and wrote in his journal of the supposed attempt on his life.

Eating a noon meal while in prison, Cellini felt a grating sensation on his teeth. Fearing an assassination, he carefully examined his food. To his dismay he found fine particles of what he suspected were diamond shards. To test his theory he tried to crush some of the bits between the point of his knife and the stone sill of the cell's window. When the bright particles were able to be pulverized, he was relieved that what he had consumed was not deadly diamond. After his release he learned that an enemy had indeed given a diamond to a local gem cutter so that it could be powdered and placed in Cellini's food. Confronting the cutter, Cellini learned the man had been in need of money and had substituted a pale citrine for the valuable stone. Cellini believed the dishonest lapidary had inadvertently saved his life.

The Portuguese writer Zacutus tells another story of the deadly consequences of ingesting diamonds. A servant was said to have swallowed three rough diamonds in an attempt to steal them from his master. The unscrupulous servant soon became violently ill and died of massive internal hemorrhages. It was believed the gem would fracture teeth if it were placed in the mouth or burst internal organs in swallowed. One tale gives an interesting reason for diamonds' toxic effect. It was said the place where diamonds were created is guarded by venomous snakes; as they slither across the stones, their skin is pierced by the sharp points of the crystals. By this contact with the serpent's blood, the venom is transferred to the gems.

To refute the claims of diamonds' deadly effects, the Spanish naturalist Horta gave the following evidence. He wrote in 1563 that diamonds' safety was shown by the many slaves who swallowed diamonds to smuggle them from the mines with no ill effect. Other writers claimed that dia-monds were an antidote to poisons, particularly the "poison" of ground glass. It was also said that diamonds absorbed the ills of plague and pestilence. As evidence it was noted that many of the poor died of plague, but few people who owned diamonds met the same fate.

The gem was also used as a discovery agent for poisonous liquids. The stone was said to darken in their presence. How this was accomplished was explained by the Italian physician Gonelli in 1702. He stated that diamond was too dense to absorb the minute particles which poisons emanate. The emanations would congeal on the surface of the stone, causing it to lose its luster and darken. Even a poisoned patient gave off the same particles which resulted in the same effect. Since diamond was a cold substance, the warm toxic moisture given off by the body would condense on the stone. Chevalier Jean de Mandeville states in his fourteenth century *Lapidaire*, "It makes venom disappear, because if venom is taken, it is eliminated in the form of sweat."[15]

Diamond was used as a juror for those who chose to wear one. The stone was said to grow dim if worn by one who was guilty and gain brilliance if worn by the innocent. Mandeville wrote of this phenomenon, "It happens often that the good diamond loses its virtue by sin and for incontinence of him who bears it."[16] The old English tale of Hind Horn and Maid Rimnald contains a similar story. Hind Horn loved the king's daughter and wished to marry her. The king had forbidden the marriage and Hind Horn went to sea to escape his wrath. Upon his departure the princess gave him a ring of seven diamonds as a pledge of her love. The story continues, "One day he looked his ring upon, He saw the diamond pale and wan."[17] Knowing that paleness foretold of unfaithfulness, Hind

> *Leave these gems of poorer shine,*
> *Leave them all, and look on mine!*
> *While their glories I expand,*
> *Shade thine eyebrows with thy hand.*
> *Mid-day sun and diamond's blaze*
> *Blind the rash beholder's gaze.*—[18]
> The Chorus Song of the Fourth Maiden:
> Bridal of Triermain, *Sir Walter Scott*

Horn made a hasty return to his home. He arrived in time to prevent the princess Rimnald from marrying another and make her his wife. As often happens, they lived happily ever after.

The fourteenth century alchemist Pierre de Boniface asserted that diamond made the wearer invisible. A mystic living at the same time, Rabbi Benoni, believed the gem attracted planetary influences, made the wearer invincible, and caused somnambulism. An anonymous fourteenth century Italian text restates an ancient belief that diamonds were created by lightning strikes, but the same manuscript states they may be consumed or melted by thunder. Mandeville listed numerous attributes of the gem.

> All diamonds give victory and make he who wears them strong and powerful against his enemies and protect the limbs and other bones. It works against quarrels, dissension, fancies, and vanity of the mind. They work against enchantment and sorcery, they heal lunatics and those whom the devils have worked upon...
>
> Diamonds also give graces to adults and children and if it is given from one friend to another, it has much greater strength and virtues than if it had been bought.[19]

He also wrote extensively regarding the nature, quality and sources of diamonds.

> The diamond is from the foam of water, and grows in the northern parts of India, it is the color of polished iron, the biggest is hardly bigger than a bean, Diamonds also grow in dew from the heavens, in various mountains and in several parts of the earth, such as Arabia, Cyprus, Macedonia and in several other countries. It also grows in various crystal mountains,

and those are of a white color like crystal, but they are murky. Others grow also in the rock of Marcadoit and are of an iron color. Others grow in the high and marvelous mountains in which there are gold mines and are of a yellow color, some are the color of water, others are of a violet color, others are pale, others are murky and white as in India, and these are the hardest and the murkiest and those which grow in other places are the softest and the lightest. There are many diamonds which have three sides, other have four, others five, others six. There are fewer with four sides than the others, they are more beautiful and more expensive, not because they have more properties than the others but because the wise Indian say that the virtue lies not in the shape but in its being, its nature and its essence.[20]

Not all who had studied the diamond were as complimentary to the stone. The sixteenth century Italian Lapidary of Girolamo Cardano cautions against placing too much faith in the powers and protection of a diamond.

It is believed to make the wearer unhappy; its effects, therefore, are the same upon the mind as that of the sun upon the eye, for the latter rather dims than strengthens the sight. It indeed renders fearless, but there is nothing that contributes more to our safety than prudence and fear; therefore it is better to fear.[21]

Despite its detractors, the diamond has become one of the most valued of all gems. Its supposed medicinal value has been maintained throughout history. The story is told of an Austrian nobleman who was plagued with terrible nightmares. He was cured by wearing a diamond set in gold on his arm so that the diamond came in contact with his skin. This is reported to show that the cure was not magical since the stone had to touch the skin, proving that the cure came from a natural force emanating from the stone. As recently as 1880 Jacobs and Chatrian wrote in their book about diamonds that these beliefs were still alive in 1830. People would borrow or rent diamonds from rich families to, "apply them to afflicted parts of the body."[22] A presciption for the use of the gem in healing was recorded in the 1879 edition of the *Mani Málá* by Rajah

Sir Surinda Mohun Tagore. On a day considered auspicious for the occasion, the diamond was to be dipped in the juice of the *kantakára* plant and subjected to a whole night of heat from a fire of cow or buffalo dung. In the morning, it was immersed in cow urine and again heated by the fire. This was to be repeated for seven days and the diamond would be purified. The stone was then buried in a paste of legumes mixed with asafetida and rock salt. It was then heated twenty-one times until reduced to ashes. These ashes would be dissolved in a liquid to make a potion which would, "conduce to longevity, general development of the body, strength, energy, beauty of complexion, and happiness,"[23]

Tales of the diamond's ill effects have also persisted. As recently as 1995, a diamond was blamed for a major outbreak of disease. When the ebola virus struck Kitwit, Zaire, a story regarding the source of the disease spread as quickly as the virus. It was rumored that the doctors in a local hospital were performing surgery on a woman, and during the surgery a diamond was found in the woman's abdomen. It was said the dishonest surgeons took the diamond and sold it for their own gain. This deed was the reported cause of the disease, the death of surgeons and nursing staff, and the concentration of ebola in the hospital.

Throughout the centuries gems have been given gender, but few have been said to reproduce. Stories of how diamonds "grow" have come from various sources. Mandeville writes,

> They grow together, male and female, and are nourished by the dew of heaven; and they engender commonly, and bring forth small children that multiply and grow all year. I have oftentimes tried to experiment that if a man keep them with a little rock, and water them with May dew often, they shall grow every year and the small will grow great, for right as the fine pearl congeals and grows great by the dew of heaven, right so doth the true diamond; and right as the pearl of its own nature takes roundness, so the diamond, by virtue of God, takes squareness.[24]

A Chinese legend from the third century B.C. states that the gem grows under the sea, and may be found inside of certain fish. Girolamo Cardano wrote in the sixteenth century, "Precious stones are engendered by juices that distill from precious minerals in the cavities of rocks. The diamond, the emerald, the opal from gold."[25]

Tradition states that whether a diamond acts positively or negatively may depend on how it was obtained. Its talismanic powers were said to be lost if it came to be owned through purchase. Mandeville wrote, "The diamond should be given freely, without coveting and without buying, and then it is of a greater virtue and it makes a man stronger and firmer against his enemies and heals him that is a lunatic, and those whom the fiend pursues or torments."[26] To obtain or grant all these favors, one should consider a diamond only as a gift.

Solomon and the Shamir: Arab Legend

When King Solomon conceived the temple of Jerusalem, he commanded Satan to fashion the masonry. The work commenced with great energy, but it was accompanied with a great deal of noise. The king's people complained bitterly that they could not bear the din. To find a remedy, Solomon sought the council of his scribes. They recommended the services of a wizard named Sahr. This wise man said he knew of a substance which would do the stone cutting efficiently and quietly, but he did not know its source. He did know it as a special abrasive stone which could cut any other stone.

Sahr devised a plan to find this special stone. He first ordered an eagle's nest filled with eggs and a bottle shaped vase of strong glass be brought to him. He placed the eggs in the bottle and had it sealed. He then directed his servants to place the bottle in the nest and the nest returned to its original location. When the eagle returned it could not break the bottle to retrieve its eggs. The eagle left in frustration and returned the next day with a stone in its beak. With a single strike of the stone against the glass, the bottle was split in two. This was accomplished without a sound and without even harming the eggs.

Sahr summoned Solomon and brought him to the eagle. Solomon, who knew the languages of all beasts and birds, asked the eagle for the source of the cutting stone. The eagle replied, "O Prophet of God, the stones may be found in a mountain to the West. The place known as Samur Mountain. Solomon went to this mountain and soon had a plentiful supply of the stones of Samur, a stone called *shamir*. This allowed the splitting, shaping, and polishing of the temple blocks to continue without a sound. The Arabs say this is the story of the discovery of diamonds.

Rabbinical writers agreed that shamir was the stone which allowed Solomon to construct the temple walls. Shamir is further described as being no larger that a grain of barley corn, but able to split any substance it

The Story of Hope

The 44.5 carat, dark-blue diamond known as the Hope Diamond is one of the world's most celebrated diamonds. While far from the largest, it is finest example of a blue diamond. Similarities in color with the 13.75 carat Brunswick Blue Diamond provide almost conclusive proof that the two stones are the result of the recutting of the French Blue Diamond, stolen in 1792 and never recovered.

The Hope was purchased for $90,000 in 1830 by Henry Phillip Hope when in first appeared on the London diamond market. After his death it became the property of his nephew, Henry Thomas Hope. Now carrying the family name, the stone was displayed at the 1851 Crystal Palace Exposition.

When Henry Thomas' wife died in 1887, she willed the diamond to her grandson, the Duke of Newcastle. A condition of his acceptance of the gem was a name change. He agreed to adopt the name Hope: Henry Francis Hope Pelham-Clinton Hope. The great value of the gem convinced him he could live with the awkward name. Lord Hope, as he was called, married the American actress Mary Yohe in 1894. By 1906 Lord Hope was nearly bankrupt and was forced to sell the diamond to cover his debts.

The stone was purchased by Abdul Hamid II, Sultan of Turkey, in 1908. Threatened by revolution, he sold it in 1911 to Pierre Cartier who subsequently sold it to Edward B. McLean, owner of the *Washington Post,* for $154,000. The diamond was a gift to his wife, a woman made fabulously wealthy by the her Camp Bird Mine in Ouray, Colorado.

Following Mrs. McLean's death in 1947, the New York City gem merchant Harry Winston purchased the Hope for $179,920 and presented it to the Smithsonian Institution in Washington D.C. in 1962.

The legend of the stone includes a dozen violent deaths and the ruin of two royal families. Despite its checkered past, Mrs. McLean never considered the diamond to be bad luck. Neither Harry Winston or the Smithsonian seem to have suffered any ill effects from possession of this fabulous gem.

touched or was near. It is also said to be the seventh of the ten miraculous things created at the end of the sixth day of creation.

The Legend of the Valley of the Diamonds
A Legend of Wide Distribution

One version of the legend is told by Epiphanius (ca. 315-403) Bishop of Salamis. This story is said to have originated in the Hellenistic East in the first century B.C. at the time diamond trade with India began. It is also given as the legend of the source of the jacinth.

> In the Scythian desert there is a deep valley surrounded by high and rocky mountains. From the summit one cannot see the bottom of the valley, which is lost in the fog as though in impenetrable depths. The kings of the surrounding countries send their people into the mountains bordering the said valley to extract the treasures of precious stones heaped in the farthest depths. But to accomplish this task they must resort to trickery. They kill and flay sheep, then cast quarters of raw flesh into the depths where the incalculable treasures lie. Soon eagles appear from their aeries; they swoop down through the fog, seize upon the flesh, and carry it back to their nest. The precious stones adhere to this flesh and the king's people have only to rob the eagle's nests to gather them.[27]

A variation from the oldest Arab treatise on mineralogy (ca. 350 B.C.) is ascribed to an unnamed author who claimed he accompanied Alexander the Great on his travels through India.

> Other than my pupil Alexander (Alexander the Great), no one has ever reached the valley where the diamonds are found. It lies in the East, along the great border of Khurasan, and it is so deep that a human eye cannot see to the bottom. When Alexander reach the bottom of the valley, a multitude of serpents prevented him going farther, for their glance proved mortal to men. So he resorted to the use of mirrors; the serpents were caught by the reflection of their own eyes and so perished. Alexander then adopted

another ruse. Sheep were slaughtered, then flayed, and their flesh cast into the depths. Birds of prey from the neighboring mountains swooped down and carried off in their claws the flesh, to which countless diamonds adhered. Alexander's warriors hunted the birds, which dropped their booty, and the men merely had to gather it where it fell.[28]

This same tale also appears in the memoirs of a Chinese princes of the Liang Dynasty around 500 A.D. Pliny describes the origin of the diamond to a valley in Ethiopia. The location of the legendary valley was transferred to Ceylon, now Shri Lanka, in writings of the fifth to sixth century. Marco Polo's *The Book of Marvels* (1298) contains another version of this same legend, but the location is set in the mountains of India. Variations of the tale spread throughout the medieval world. The third voyage of Sinbad the Sailor in the *Thousand and One Nights* contains another version of this story.

Sinbad is cast into an inaccessible valley, a valley which is so deep no one is able to climb in or out. Gem merchants of the region cast large slabs of meat into the depths to attract vultures. Sinbad ties a piece of the meat to his belt and is plucked from the valley floor by a large bird of prey. The bird carries Sinbad to his nest and ultimate safety. In payment for his rescue, Sinbad gives the merchant a large diamond he brought from the valley.

The Diamond and the Blood of the He-goat: Roman

Book twenty of the *Natural History* discusses war and piece, and the sympathy and antipathy of inanimate objects. After a discussion of fire and water, magnet and iron, the diamond and the blood of the he-goat are given as examples. Pliny states, "The diamond, this rare joy of opulence, invincible and resistant to any other form of violence, can be broken by the action of the he-goat's blood." In book thirty-seven Pliny recalls the fact that *adamas* is synonymous with "invincible force," he then sights the diamond as the best example of harmony and discord that governs the universe.

This invincible force, which despises the two most powerful natural elements, iron and fire, is broken by the he-goat's blood, but only when the diamond has been dipped in the fresh, warm blood of the animal and struck with many blows; for even then it breaks everything except the most solid anvils and iron hammers.[29]

This legend is written as a statement of the supposed physical law of sympathy and antipathy. The earliest written reference occurs in Pliny's *Natural History*, but it is believed to be the record of a pre-existing tale. It was continued in medieval theological literature as a parable of purity versus lust. The purest heart, represented by the diamond, could be broken or conquered by the lust for flesh, symbolized by the goat's blood. Another Christian interpretation reverses the roles of the elements involved. Christ was challenged by the pain of crucifixion, but the pure blood of the Savior overcomes the hardened or adamas sinner. Eighteenth century naturalists include this tale as a method of "cutting" a diamond.

The Raven, The Snakes, and The Diamond: A Rabbinical story in the Talmud

> R. Jehudah of Mesopotamia use to tell: Once while on board of a ship, I saw a diamond that was encircled by a snake, and a diver went to catch it. The snake opened its mouth, threatening to swallow the ship. Then a raven came, bit off its head, and all water around turned into blood. Then another snake came, took the diamond, put it in the carcass, and it became alive; and again it opened its mouth, in order to swallow the ship. Another bird then came, bit off its head, took the diamond and threw it on the ship. We had with us salted birds, and we wanted to try whether the diamond would bring them to life, so we placed the gem on them, and they became animated and flew away with the gem.[30]

The Doors of the Temple of Bacchus
From the novel *Gargantua and Pantagruel* by Francois Rabelais

Pantagruel and his companion came to the kingdom of the Dive Bouteille when on their epic voyage. In this far land, they enter the temple of Bacchus by means of an automatic door. These seemingly magical doors contain a polished steel plates, magnets, and gigantic Indian diamonds. Rabelais does not explain the exact mechanism, but states that the action works due to diamonds ability to remove, or counteract, the attraction of magnets to steel. In the story, a diamond is removed by those wishing entry and the doors spring open.[31]

The fourteenth century author Rabelais was believed to know of such natural phenomenon. He was not only a fiction writer, but a scientist and respected physician.

The deep-green emerald, in whose fresh regard
Weak sights their sickly radiance do amend.
Lover's Complaint, *William Shakespeare*

Crystal System

Hexagonal

Emerald

Color: Emerald green, yellow-green, dark green Mohs' Hardness: $7\,^1/_2 - 8$ Cleavage: None Refractive Index: 1.576 - 1.584 Dispersion: 0.014	Streak: White Specific Gravity: 2.67 - 2.75 Fracture: Small conchoidal, uneven, brittle Birefringence: 0.005 -0.009 Pleochroism: Definite; green, blue-green to yellow-green

Crystal System: Hexagonal (triagonal); hexagonal prisms, columnar
Chemical Composition: $Al_2Be_3(Si_6O_{18})$ aluminum beryllium silicate
Transparency: Transparent to opaque
Fluorescence: Usually none

> Emerald is the bright green variety of beryl. It is found in metamorphic rock, unlike other beryls which are found in pegmatites. This limits the size of most emerald crystals while other beryl's crystal may be quite large. Emeralds are brittle and easily fractured during handling and setting. Most have inclusions which are called the jardin of the stone. Jardin is the French word for garden. This is in reference to the moss or branch like appearance of many of these flaws. Flawless emeralds are rare and very valuable.

The stone whose color is said to have "entranced humankind" is the emerald. Its distinct green color, symbol of new life and the promise

ribing the emerald in its matrix, another *garalari*, or "enemy of poison." In Mexico the Aztec called it *quetzalitzli*, "stone of quetzal" after the brilliant green plumes of the bird *quetzal*, a symbol of royalty. The modern name is derived from an ancient Persian word, translated to Latin as *smaragdus*, and corrupted to *esmeraude*, *emeraude*, and *emeralde*. The current English spelling dates to sixteenth century manuscripts. Inventory records show the stone was known and offered for sale in the markets of Babylon as early as 4000 B.C.

The earliest references to emerald in western literature are assigned to Aristotle. He wrote that emerald increases the owner's importance in presence and speech during business, gives victory in trials and helps settle litigation, and comforts and soothes eyesight. The Roman magician Damigeron stated in the second century B.C. that, "It influences every kind of business, and if you remain chaste while you wear it, it adds substance to both the body and the speech."[1] Aristotle also stated, "An emerald hung from the neck or worn in a ring will prevent the falling sickness (epilepsy). We, therefore, commend noblemen that it be hanged about the necks of their children that they fall not into this complaint."[2]

The Greek historian Theophrastus, a student of Aristotle, states in his third century B.C. text that the gem emits light when submerged in water. He states the Emerald of Cyprus is, "very rare and of small size; it has the same peculiar property for it renders water of the same color with itself. It soothes the eyes, and people wear seals of this stone in order that they may look at them."[3] By its supposed location and description, it is thought by many scholars the stone Theoprastus was refering to was the copper based mineral chrysocolla. The Roman historian Pliny the Elder expanded on the virtues of emerald by stating "when viewed from a distance .. (emeralds) ... appear all the larger to the sight."[4] He continues, "people carry seals made from it, so as to see better." He gives an extended account of this virtue.

> Indeed, no stone has a color that is more delightful to the eye, for whereas the sight fixes itself with avidity upon the green grass and the foliage of the trees, we have all the more pleasure in looking upon the emerald, there being no green in existence more intense than this. And then, besides, of all the precious stones, this is the only one that feeds the sight without satiating it. Neither sunshine, shade nor artificial light effects any change in its appearance; it has always a softened and graduating brilliancy.

He continues by giving an example of a practical application of emerald's effect on the eye.

> If the sight has been wearied or dimmed by intensively looking on any other subject, it is refreshed and restored by gazing upon this stone. And lapidaries who cut and engrave fine gems know this well, for they have no better method of restoring their eyes than by looking at the emerald, its soft, green color comforting and removing their weariness and lassitude.[5]

This stone which soothes eye fatigue was also used as an eye wash. Powdered and mixed with water, it was said to be effective in reducing irritation and stopping the flow of fluid from the eye. Hindu physicians used it as a cure for eye troubles adding that the best are cold and sweet to the taste.

The stone was also said to have softened the possible negative effects of what was being seen. The Emperor Nero was said to have viewed the gladiatorial contests "upon a smaragdus." This tale was expanded upon by later authors to say that he viewed the games "through" an emerald. Such a large and clear gem would have been a rare find, even for a Roman Emperor. The first and only source of emeralds for the ancients was Egypt. These stones were known to be small and heavily flawed. It is more likely that Nero's "emerald" was actually a polished slab of jasper. The Romans were known to use green jasper as a mirror. To state the contests were viewed "upon" a stone suggests Nero watched the bloody contests indirectly on such a green mirror. The red produced in the arena would certainly have been tempered if seen reflected upon a green filtered surface.

The ancient historian does provide evidence that the Romans recognized emeralds as a form of beryl. Pliny states, "Beryls are the same nature as emeralds, or at least very similar." However, this astute observation as to emerald's nature is clouded by the fact that the Romans classified nearly all green gems as smaragdus. In all they listed twelve varieties. The "emeralds" found in Pliny's *Natural History, Book 37* and the stones they may represent are listed in the following table.

Smaragdus of the Romans

Pliny's Name	Definite Identity	Probable Identity	Possible Identity
Scythian		green sapphire	
Bactrian		emerald	
Egyptian	emerald		
Cyprian		chrysocolla or malachite	copper-stained quartz
Aethiopian	emerald		
Hermonian			green turquois
Persian			green turquois
Attican		smithsonite	
Median		malachite with azurite	turquois
Chalcedonian		bornite (peacock ore)	
Sarcicon		bornite (peacock ore)	
Cloras	green alabaster		

Other green "beryls" listed by Pliny

Pliny's Name	Definite Identity	Probable Identity	Possible Identity
tanos		green turquois	
Chalcosmaragdus		malachite with sulfides	
pseudo-smaragdus			jasper with malachite
Limoniatis		emerald	

In describing a stone which was said to be half jasper and half emerald, Pliny said it was "unripe," or in an incomplete stage of metamorphosis. This may have referred to a stone which was half green and half another color. It is obvious the ancients were able to discern differences in stones. They certainly would have been aware of the opacity and transparency of various gems.

Pliny's statement also demonstrates the common belief that stones grew and ripened. This opinion was reinforced by the location in which many stones were found. This myth included the view that stones actually lived and that tropical heat was essential for their full maturity and development of beauty. Later writings continued to reinforce this notion. The seventeenth century English lapidary Thomas Nicols quoted the ancient gemologists, "Every gem, saith he, hath a matrix formed out of some stone or other, in which matrix, by the distilling of certain nutritive juices, it is nourished." The same story was repeated in South America with the discovery of emeralds in this new tropical location. Seventeenth century Spaniard Garcilaso de la Vega wrote, "Emeralds take their tincture from the nature of the soil from whence they are produced, ripening there with time like fruit in their proper seasons."[6]

Pliny also included a tale of the brilliance of emeralds in his historical accounts.

> On the Island of Cyprus, near the tomb of Hermias, there was in ancient times a marble statue of a lion. In its head were set two emeralds as eyes, looking out to sea. They glittered so brightly, piercing deeply into the water, that all the tunafish of the coast were

> frightened and fled from the nets and the other paraphernalia of the fishermen. For a time the fishermen were nonplused, but, finally realizing what had happened, they put other eyes into the lion's head, removing the emeralds.

The use of gems in the eyes of statuary was common with the Greeks and Romans. The stones were said to bring greater life to the images.

Damigeron gave an account of how the gem should be prepared for optimum power.

> Whoever obtains and consecrates it shall never lose his liberty. This is how the stone should be treated:– Having obtained it, order it to be carved in the shape of a scarab beetle, with Isis on its belly, standing; then perforate it longitudinally. Then wear it, having consecrated it and mounted it on a brooch; and prepare a suitable place for it, and it will adorn you and your belongings, and you will see the power of the stone given to it by God.[7]

It is suggested by this tale that Egyptian customs and lore had been passed to the Romans by their suppliers of emeralds. A lengthy poem by Marbod, Bishop of Rennes, on gemstones contains a summary of emerald lore known in twelfth century Europe. This medieval master work became the source of information for nearly all succeeding writers. Marbod, mirroring Aristotle's work, states emerald makes one persuasive in pleading a cause and "comfortable with words." He states the attribute is limited, however, to those born between June 22 and July 23, and those with the moon in good aspect. Expanding on this trait, Marbod wrote that the gem promotes candor, sharpens wits, confers dignity, imparts discretion and renders the wearer agreeable and amiable. For this reason the gem has been favored by businessmen and barristers. Emerald was also said to be the gem which is symbolic of divine glory, goodness and courage. It imparts benevolence of the heart, confers grace, promotes constancy of the mind and brings or imparts "holy joy." Following Marbod's advice, the emerald became a popular stone with church officials. Its supposed ability to bring fame, dignity, and honor seem to contradict the use of the stone by fifteenth century Catholic clergy as a symbol of humility.

The fourteenth century scholar Chevalier Jean de Mandeville reports an additional power to the green color of the stone. "Emeralds are a very green stone, because to speak of its greenness, it makes the air green."[8]

Mandeville's *Le Lapidaire* also included a listing of sources and the powers granted by the gem.

> There are many kinds of emeralds, and they are found in many places and in many different parts of the world, especially in Litharia (Cyprus) and in Egypt; they are also found in places where there are minerals such as tin and copper, but then they are stained because of the metals. The stones from Litharia are worth more than the others; they do not change their color nor clarity with nightfall, nor overcast weather and they can reflect a man's face like clear water. Emeralds give riches and dignities, and make men honest and wise with speech; they also cure fevers, vision problems, skin diseases and spots. Also, not man or woman who live carnally should wear them, because they will deteriorate and become murky. Emeralds increase wax-making honeybees, turn tempests away, and calm malevolence.[9]

Mandeville's *Lapidaire* includes a second entry for this gem, possibly denoting the importance he placed on the stone.

> Emeralds are green and come from the River of Paradise. Those from Egypt are fewer and are of a nicer color. One can see from one side to the other. There is a kind of people having the name Armespilles, who go to war and remove the emeralds from the griffons. The flat emerald is very good for seeing and protecting oneself. Heron had one in which he looked at himself and knew by the strength of the stone that which he wanted to conquer. Emeralds bring riches to those who wear them around their neck. It protects against bad women, who cannot be killed by any man, and who want to find a man in order to bring him to his death. The emerald heals the eyes of someone who looks at it; it takes away tempests and lechery; and it gives the virtue of thinking of God to he who wears it, and it should be encased in gold.[10]

Numerous other attributes were reported about the deep green beryl. It was said to strengthen intellect, impart wisdom and improve memory. The most effective memory increase was to occur if the stone was worn on the index finger. A flat emerald was considered very good for self-admiration and it was said to purify the thoughts of those who gazed into it. The gem was reported to be the enemy of slovenliness, a cure for idleness and an enemy to sin. Thoughts were said to be purified by gazing into an emerald. It would then bring purity and prevent evil from entering the owner's mind. The stone was favored in the Church for these same reasons and was said to help the owner to resist temptation and realize the hope of immortality.

Christian tradition associates the emerald with St. John. One of the earliest writers to make such an association was the tenth century writer Andreas, Bishop of Caesarea. He based this on the Foundation Stones listed in Revelations 21 and their supposed connection with each of the apostles. His treatise makes the following proposition:

> The emerald, which is of a green color, is nourished with oil, that its transparency and beauty may not change; we conceive this stone to signify John the Evangelist. He, indeed, soothed the souls dejected by sin with a divine oil, and by the grace of his excellent doctrine lends constant strength to our faith.[11]

The color green, and, therefore, the emerald, has long served as a symbol of obedience, spring and rebirth. The ancient Egyptians called the emerald "the lovers' stone," as it was said to increase love. This gem, which also brings fertility, was considered an appropriate gift for young lovers. The stone revived conjugal love and ensured the truth of a lover's pledge. It was also said to be capable of revealing adultery. Worn by a faithful wife, it would help to preserve her chastity in her husband's absence.

> Who first beholds the light of day
> In Spring's sweet flowering month of May,
> And wears an Emerald all her life
> Shall be a loved and happy wife.[12]

It would also report any of her indiscretions, if she were unfaithful the stone would shatter. Albertus Magnus relates a story about Bela, King of Hungary, and his flirtatious queen. Upon returning from a journey, the king embraced his wife and his exceptional emerald broke into three pieces. The two breaks were thought to have revealed the lady's sins during her husband's travels. Others thought that a true and pure emerald would fracture if worn during intercourse.

The color of the gem was said to act as a barometer of love. "Assigned to Venus, it stood for the highest form of love and testified to the fidelity of lovers, the color waxing deeper or waning paler as love came and went."[13] A poem based in this tradition is credited to "the late Mrs. McLean" by William Jones.

> It is a gem which hath the power to show
> If plighted lovers keep their troth or no.
> If faithful, it is like the leaves in spring:
> If faithless, like those leaves when withering.
> Take back again your emerald gem,
>   There is no colour in the stone;
> It might have graced a diadem,
>   But now its hue and light are gone.
> Take back your gift, and give me mine—
>   The kiss that seal'd our last love vow—
> Ah, other lips have been on thine—
>   My kiss is lost and sullied now!
> The gem is pale, the kiss forgot,
>   And more than either you are chang'd;
> But my true love has altered not—
>   My heart is broken, not estrang'd.[14]

Those who suffered from love lost, and, therefore, dulled emeralds, may have found Jean de Mandeville's prescription for restoring the stone helpful. He states, "If this stone loses its clarity, dunk it in wine and then rub it well with a sheet and olive oil, and it will soon regain its clarity."[15]

Literature is filled with references to emerald's power in love. The gem has served as a symbol of virility and of wanton thoughts. Although it is a powerful talisman of love and desire, it was also said to subdue sexual passion. Opinions of who should wear the gem have been established according to gender. The Parisian oracle Baron d'Orchamps states, "Emeralds should not be worn by women before they are fifty, but men may wear them without danger at any age."[16]

The stone has also served as a talisman of travelers. It averts ambush, protects one from natural disasters, and gives victory in combat. It was said to be particularly effective in the protection of mariners, fisherman and anyone who ventures onto the seas. This was accomplished primarily through its connection to the goddess Isis, the moon, and the tides. To be of greatest effect the stone should be engraved with the symbol of Gemini and worn on the left arm attached by a green cord. An added benefit to the wearing of an emerald in this manner was its ability to dispel any evil spirits which may attack during a trip. Emerald was also thought to drive away water demons and mermaids.

The emerald was said by some to be a favorite of wizards and magicians. It was used as a protective agent for the wearer. As an "enemy of enchantments"[17], it acted to deflect the power of conjurations away from the magician and toward his subject. Other authors state that emerald is the enemy to conjurers and tricksters, magicians "could not weave a spell" if one were near.[18] The emerald was also believed to give insight into future events and that to dream of the gem meant there was much to look forward to. Some said the full power of prognostication was released if the oracle held an emerald under his tongue. Magnus reported that the best stones for this purpose are the light-colored variety which were found in the nests of griffons. He further stated that such stones were especially valued in India. Another method of divination was prescribed in the seventeenth century. A ring set with an emerald was to be suspended over a bowl of water with letters of the alphabet engraved around the rim. The ring would "answer questions" by swaying against the letters in the fashion of a Ouija board.[19]

The emerald has long been considered a protection against venom. It would cure bites of poisonous insects and protect the wearer from serpents. Mandeville reports a protective power of the gem. "True emeralds shown to a snake will make the snake go blind."[20] Sir Thomas Moore, in *Lalla Rookh,* included this same attribute in two poetic lines.

> Blinded like serpents, when they gaze
> Upon the emerald's virgin blaze.[21]

The Arabian gem dealer Ahmed Teifahi wrote in 1242 of an experiment he performed to test the stone's power over serpents.

> After having read in learned books of this peculiarity of the emerald, I tested it by my own experiment and found the statements exact. I chanced that I had in my possession a fine emerald of the zabâbi variety, and with this I decided to make the experiment on the eyes of a viper. Therefore, having made a bargain with a snake-charmer to procure me some vipers, as soon as I received them, I took a stick of wood, attached to the end a piece of wax, and embedded

my emerald in this. I then brought the emerald near to the viper's eyes. The reptile was strong and vigorous, and even raised its head out of the vessel, but as soon as I approached the emerald to its eyes, I heard a slight crepitation and saw that the eyes were protruding and dissolving into a humor. After this the viper was dazed and confused: I had expected that it would spring from the vessel, but it moved uneasily hither and thither, without knowing which way to turn; all its agility was lost, and its restless movements soon ceased."[22]

Other sources of poison were also thought to be defeated by the emerald. Hindu literature lists emerald as a protection against all form of animal and vegetable poisons. The sixteenth century writer Franciscus Rueus prescribed the weight of eighty barley-corns of emerald should be powdered and served in a beverage as an antidote. Arab physicians recommended placing one emerald in the mouth and one on the stomach as a cure. An old Rosicrucian belief states that if an emerald is mounted in gold and worn on the solar finger of the left hand, food poisons can be detected. This method of discovery was believed most effective when the sun was in the constellation Taurus. The other eleven months of the year one must have been expected to live in jeopardy.

The emerald was credited with increasing mental abilities. It reveals things that were previously unknown to its owner. Holding the stone in the mouth made it useful in calling upon the Devil. It would compel the dark angel to reveal the answers to any question asked. To dream of the green stone would foretell fame and give the ability to recognize the truth. Cardano stated that the gem sharpens the wit and quickens intelligence. He further reasoned that this also made people more honest. He declared that "dishonesty is nothing but ignorance, stupidity, and ill-nature."[23] He added that the stone makes men economical, and, therefore, rich. Cardano later wrote that he remained skeptical of this power. He observed that he, nor any other owner of an emerald he knew, had become wealthy. Other writers proposed that the gem brought wealth by bringing wisdom which would then bring frugality and honor.

The curative properties of the emerald have been documented in numerous texts. Teifashi wrote that the stone cured haemoptysis and dysentery if worn over the liver and gastric problems if placed over the stomach. The Arabian physician Abenzoar (a.k.a. Abû Meruân) recommended six grains of 'tincture of emerald' as an internal treatment for dysentery, and also claimed a cure of mild cases if the stone was worn.[24] Hindu authors included similar remedies and more. They considered the emerald to be a laxative which also diminished the secretion of bile and stimulated the appetite. It was supposed to promote general good health and destroy the demonic influences which bring on disease. The eleventh century author Psellos states, "that it has power, when mixed with water, to heal leprosy and other diseases."[25] The Spanish physician Michaele Paschali claimed to have documented such cures in his sixteenth century experiments. The stone was also said to stop hemorrhages. The physician would also know immediately if the disease being treated was too strong. Contact with an incurable disease was said to cause emeralds to fracture. The gem is also said to hasten labor and protect against miscarriage if tied to a woman's thigh. It is listed as a cure for skin diseases, gall bladder ailments, belching, colic, discomforts of gas, hemorrhoids, kidney stones, and miscellaneous digestive ills. The ills of malaria, cholera, and the plague could also be eliminated. In Moslem countries, all the medicinal effects are magnified if a verse from the *Koran* was engraved on the stone. The Mogul of Delhi once owned a 78 carat talismanic emerald inscribed with Persian characters which read, "He who possesses this charm shall enjoy the special protection of God."[26]

The Origin of Emeralds: India

Forbes' Oriental Memoirs gives the following story as the fabled origin of emeralds in India.

A person was watching a swarm of fire-flies in an Indian grove, one moonlight night. After hovering for a time in the moonbeams, one particular fire-fly, more brilliant than the rest, alighted on the grass and there remained. The spectator, struck by its fixity, and approaching to ascertain the cause, found, not an insect, but an emerald, which he appropriated, and afterwards wore in a ring.[27]

The Origin of Emeralds, Vasuki and the Bile of Vala: India

Vedic texts of India relate another story of the origin of emeralds. Vala, a demon god, was slain and dismembered by the demigods. The parts of his body were scattered about the globe to create the gemstones we know today. The following story explains both the existence of emeralds and their distribution.

The bile of the great demon Vala was taken from his torn body by the serpent king, Vasuki. In his exuberance he turned abruptly and split the heavens in two with one swipe of his mighty tail. His long body, head crowned with a hood of jewels, appeared like a bridge across the sky. Garuda, the eagle-king and carrier of Supreme Lord Vishnu, came flying through the heavens encountering the serpent-god Vasuki. Sworn enemy of serpents, Garuda blocked Vasuki's path with his mighty claws. Frightened by this attack, Vasuki dropped Vala's bile to earth near Mount Manikya. This mountain was known for its beautiful forests and fragrant trees. (It is said that this great mountain split, and as the continents moved apart, it formed the mountain ranges of South America and South Africa, an explanation for emeralds being found in these two locations). The balance of Vala's bile fell to the landlocked areas beyond the Himalayas. (Probably modern day Afghanistan and Pakistan).

The green bile was transformed into the seeds of emeralds, and wherever they fell, the great emerald mines formed. Garuda picked up some of the scattered seeds, but their power was so great that even he was overcome with fainting spells. As he swooned he scattered small quantities of the seeds in many regions of the earth. Most of the areas where Vala's bile fell are lush and beautiful, but rugged and difficult to access.

The Emerald Tablet: Egypt and Greece

The *Tabula Smaragdina* is the name of a Greek literary work which is reported to be a translation of ancient Egyptian writings. The original inscriptions were said to have been discovered by Alexander the Great on a tablet in the sepulchre of Hermes. Legend says this original tablet was formed of a single massive slab of emerald. Early alchemical literature frequently mentions such a tablet and claims it was written by an Egyptian king named Hermes Trismegetus or Hermon. The name Hermes Trismegetus, meaning "Hermes three times great," is the Greek name for the Egyptian moon god, Thoth. He is also the god of time, wisdom and credited by the Greeks as the author of all Egyptian literature.

Hermon was said to be the "founder of chemistry," and the Hermetic arts. He supposedly authored at least forty-two books of magic and mystery. One of these, the emerald tablet, was said to have contained all the important knowledge of alchemy, including the formula to "make" gold. The alchemists of the Middle Ages spent much of their time seeking Hermon's original texts. It is likely that the original emerald tablet, if it truly existed, was a slab of green igneous rock or a cast piece of green glass.

The Table of Solomon

A massive emerald table has been the subject of stories throughout history. Originally called the Table of Solomon, it is described as a slab of flawless emerald, bordered with rows of pearls and supported by gem studded gold feet. Containing supernatural powers, it was presented to the King of Israel by genies to win his favor. Little is recorded, besides its existence, in regards to its specific purpose. The history of the table is in its travels as a spoil of war. It was supposed to have been removed from Jerusalem by the armies of the Emperor Vespasian (9-79 A.D.). Legend says it was transported to Rome and placed in the Temple of Concord to serve as an altar. Its next home was said to have been in Spain, part of a treasure taken in the sack of Rome by the Goth armies of Alaric. Its last documented location was in Toledo in 712 A.D. It was in this year the Saracen warriors of Musca captured the table and reportedly took it to Damascus. Following this move no record of Solomon's Table exists. Tales of such a large emerald may be attributed to the use of the term smaragdus for any deep green stone. Theophrastus described a royal gift given to the Pharaohs as containing a smaragdus six feet in length and four feet in width. He also lists four emeralds placed in the obelisk of Zeus. They were said to measure six feet in length, six feet in width at one end, and three feet at the other. From such descriptions, one may be certain these 'emeralds' must have been some other more common stone. Theophrastus included a disclaimer by reminding the reader that his accounts are translations based on lost writings of the ancient Egyptians.

The Ring of Polycrates: Greek

The ruler of the Aegean island Samos in the sixth century B.C. was Polycrates. He listed as one of his most prized possessions an engraved emerald ring of gold made by the famed jeweler Theodorus. Knowing of the emerald's ability to protect travelers at sea, Polycrates tossed the ring into the sea as he set off on a naval campaign. He reasoned that he would sacrifice part of his fortune to bring victory at sea. As the ring drifted toward the bottom, it was swallowed by a passing fish. A few days latter the fish was caught by a fisherman from Polycrates' home port. This catch was a fine specimen and worthy of the monarch. Upon the king's return, it was prepared and served to him at a banquet. When the fish was cut open, the emerald ring was found and returned to its owner. Polycrates' attempt to cheat fate was foiled. On a later voyage his rule ended. Polycrates was lured to the mainland by Oroetes, the Persian governor of Lydia, and crucified in 522 B.C. Similar stories of a objects lost at sea and returned to the owner in a fish have appeared throughout history. As recently as 1995, a similar story was reported by international news agencies concerning a Danish fisherman and his false-teeth he had lost overboard. The tale was later confessed as a hoax.

The Emerald Buddha: Thailand

One of the greatest green gems in Asia is a famous statue of Buddha in Bangkok, Thailand. The eighteen inch carving rests at the pinnacle of an elaborate ornament in the Chapel of The Emerald Buddha in the Grand Palace of Bangkok. The statue is said to have been the inspiration of Nagasena, a student of Maha Dhamma Rakitta, who wished to create a tribute to the deity in precious stone. He was told the source of such a grand stone would be the mountain Vipulla, fabled for its large jade. The genies which protect the mountain would not allow such a massive jade to be taken for the sculpture. They did suggest an alternative gem, the Keo Amarakata, or crystal smaragdus. Nagasena collected such a crystal and presented it to Visukamma, a heavenly sculptor. Visukamma spent seven days a seven nights creating the lovely statue of Buddha.

The official literature of the temple gives the origin of the statue as Chiengrai in 1464. It also lists its composition as a "single crystal jade." The name associated to this work is from the European tradition of listing brilliant green gems with the adjective emerald. No gemological inspection has been made of this sacred relic, but it is thought by most scholars to be a green jasper.

"Child of the Sun": Pijaos Indians of Columbia

The following story was related by Fray Simon, an early missionary to the region. After an earlier missionary had worked to educate the natives in Christian tradition, Simon tells of a story he believed had been implanted in the people by a demon trying to contradict Christian teachings. It is believed the story predated the coming of missionaries and was related as follows to indict the existing religions.

The demon there began to give contrary doctrines; and among other matters he sought to discredit the teaching of the incarnation, telling them that such a thing had not yet taken place. Nevertheless, it should happen that the Sun, assuming human flesh in the body of a virgin of the pueblo of Guacheta, should cause her to bring forth that which she should conceive from the rays of the sun, although remaining virgin. This was bruited throughout the provinces, and the cacique of the pueblo named, wishing to prove the miracle, took two virgins, and leading them forth from his house every dawn, caused them to dispose themselves upon a neighboring hill, where the first rays of the sun would shine upon them. Continuing this for some days, it was granted to the demon by Divine permission (whose judgments are incomprehensible) that the event should issue according to his desire: in such manner that in a few days one of the damsels became pregnant, as she said, by the Sun." At the end of nine months the girl brought forth a *hacuata*, a large and beautiful emerald, which was treated as an infant, and after being carried for several days, became a living creature– "all by the order of the demon." The child was called Goranchacha, and when he was grown he became cacique, with the title of "Child of the Sun."[28]

> *Of all green things which bounteous earth supplies*
> *Nothing in greenness with the Emerald vies*
> *Twelve kinds it gives, sent from the Cythia clime,*
> *The Bactrian mountain, and old Nilus' slime.*
> C.W. King, 1866[29]

Garnet Group

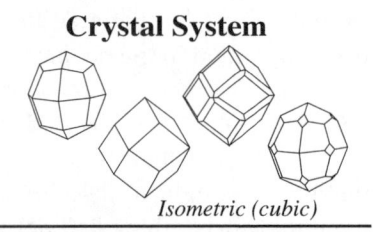

Crystal System

Isometric (cubic)

Garnet is not a single mineral, but a group of minerals which share a nearly identical atomic structure. The stones in the group are chemically different complex silicates; each chemical variation results in a distinctly different mineral. They vary in hardness, color, and transparency. The archaic term for red garnets is carbuncle, from the Latin word for little spark, *carbunculus*.

Almandine or Almandite Species

Color: Red with violet tint **Mohs' Hardness:** $7\,^1/_2$ **Cleavage:** Imperfect **Refractive Index:** 1.77-1.810 **Dispersion:** 0.024 **Pleochroism:** None	**Streak:** White **Specific Gravity:** 3.93 - 4.17 **Fracture:** Conchoidal, splinter, brittle **Birefringence:** None

Crystal System: Isometric; rhombic, dodecahedron, icositetrahedron
Chemical Composition: $Fe_3Al_2(SiO_4)_3$ iron aluminum silicate
Transparency: Transparent, translucent
Fluorescence: None

Lapidarium:

Andradite or Demantoid Species

Color: Green, emerald green Melanite: opaque black Topazolite: lemon yellow **Mohs' Hardness:** $6\frac{1}{2}$ - 7 **Cleavage:** Imperfect **Refractive Index:** 1.856 - 1.895 **Dispersion:** 0.057	**Streak:** White **Specific Gravity:** 3.81 - 3.87 **Fracture:** Conchoidal, splinter, brittle **Birefringence:** None **Pleochroism:** None

Crystal System: Isometric; rhombic, dodecahedron, icositetrahedron
Chemical Composition: $Ca_3Fe_2(SiO_4)_3$ calcium iron silicate
Transparency: Transparent
Fluorescence: None

Pyrope Species

Color: Red with brown tint **Mohs' Hardness:** 7 - $7\frac{1}{2}$ **Cleavage:** Imperfect **Refractive Index:** 1.720 - 1.750 **Dispersion:** 0.027	**Streak:** White **Specific Gravity:** 3.62 - 3.87 **Fracture:** Conchoidal, splinter, brittle **Birefringence:** None **Pleochroism:** None

Crystal System: Isometric; rhombic, dodecahedron, icositetrahedron
Chemical Composition: $Mg_3Al_2(SiO_4)_3$ magnesium aluminum silicate
Transparency: Transparent, translucent
Fluorescence: None

Grossularite Species

Color: Green, yellow, copper-brown
Tsavorite: Rich green transparent
Hessonite: aka "cinnamon stone" brown-orange,
Hydrogrossular: aka "Transvaal jade" opaque green

Mohs' Hardness: 7-7 $\frac{1}{2}$
Cleavage: Imperfect
Refractive Index: 1.725- 1.745
Fracture: Conchoidal, splintery, brittle
Dispersion: 0.027
Birefringence: None

Streak: White
Specific Gravity: 3.50 - 3.73 Tsavorite; 3.59-3.63 Hessonite 3.57-3.73 Hydrogrossular; 3.45-3.50
Pleochroism: None

Crystal System: Isometric; rhombic, dodecahedron, icositetrahedron
Chemical Composition: $Ca_3Al_2(SiO_4)_3$ calcium aluminum silicate
Transparency: Transparent, translucent, Hydrogrossular; opaque
Fluorescence: None

Rhodolite Species

Color: Rose-red or pale violet
Mohs' Hardness: 7 - 7 $\frac{1}{2}$
Cleavage: Imperfect
Refractive Index: 1.740 - 1.770
Dispersion: 0.026

Streak: White
Specific Gravity: 3.74 - 3.94
Fracture: Conchoidal, splinter, brittle
Birefringence: None
Pleochroism: None

Crystal System: Isometric; rhombic, dodecahedron, icositetrahedron
Chemical Composition: $Fe_3Mg_3Al_2(SiO_4)_3$ a mix of iron aluminum silicate and magnesium aluminum silicate
Transparency: Transparent
Fluorescence: None

Spessartite Species

Color: Orange to red-brown **Mohs' Hardness:** 7 - 7 $1/2$ **Cleavage:** Imperfect **Refractive Index:** 1.795 - 1.815 **Dispersion:** 0.027	**Streak:** White **Specific Gravity:** 4.12 - 4.18 **Fracture:** Conchoidal, splinter, brittle **Birefringence:** None **Pleochroism:** None

Crystal System: Isometric; rhombic, dodecahedron, icositetrahedron
Chemical Composition: $Mn_3Al_2(SiO_4)_3$ manganese aluminum silicate
Transparency: Transparent, translucent
Fluorescence: None

Uvarovite Species

Color: Emerald green **Mohs' Hardness:** 7 - 7 $1/2$ **Cleavage:** Imperfect **Refractive Index:** about 1.870 **Dispersion:** 0.027	**Streak:** White **Specific Gravity:** 3.77 **Fracture:** Conchoidal, splinter, brittle **Birefringence:** None **Pleochroism:** None

Crystal System: Isometric; rhombic, dodecahedron, icositetrahedron
Chemical Composition: $Ca_3Cr_2(SiO_4)_3$ calcium chromium silicate
Transparency: Transparent, translucent
Fluorescence: None

This legendary "glowing" gem gained its named from the Latin words *granum* or *granitus* meaning grain or "like seeds," which originally came from the Phoenician word for pomegranate, *punica geranatum*.

> For January stone of Aquarius
> No gem save garnets should be worn
> By her who in this month is born.
> They will insure her constancy
> True friendship and fidelity.[1]
> A later gender neutral version of the same poem
> By those who in this month are born,
> No gem save Garnets should be worn;
> They will ensure your constancy,
> True friendship and fidelity.[2]

Clusters of garnets in their matrix often look similar to the seeds in a pomegranate.

Known for centuries before the birth of Christ, the stone's intense red color led to the belief that it glowed with its own light. The Koran says the light of the fourth Heaven is given by the all pervading glow of its garnet structure. This supposed internal energy was said to give it great potency. The stone was considered powerful enough to make the wearer angry, passionate and even prone to apoplexy. The Romans likened the stone to a glowing coal, symbolizing fire and stimulus. It was also the stone they chose to represent Mars, god of war and bloodshed. "Pliny divides his Carbunculi into male and female, the former of a brilliant, the latter of a duller lustre."[3] It is speculated that his "male" was ruby and the "female" a garnet. In Christian tradition the blood red color suggested its use as a symbol of Christ's suffering on the cross. The stones connection with warmth and blood have caused it to be prescribed for heart stimulation, but the patients were warned against displays of passion for the fear of heart failure.

Carbuncle has been connected to the apostle Andrew. One of the earliest writers to make such an association was the tenth century writer Andreas, Bishop of Caesarea. Using the stones listed as Foundation Stones of the New Jerusalem in Revelations 21, Andreas gives the following rational for his conclusion.

> The chalcedony was not inserted in the high-priest's breastplate, but instead the carbuncle, of which no mention is made here. It may well be, however, that the author designated the carbuncle by the name chalcedony. Andrew, then, can be likened to the carbuncle, since he was splendidly illuminated by the fire of the Spirit.[4]

Testing the genuineness of garnets set in rings. If the gem is genuine flies and wasps will not light on the subject, even though he is smeared with oil and honey. This illustration is from the Ortus Sanitatis by Johannis de Cuba, published in 1483.

It is interesting to note how early writers were able to justify any of the designations they wished to give a particular stone.

Other attributes of garnet have been featured by many writers. Camillus Leonardus writes, "It is not hurt by fire, nor does it take the colour of another gem that is put to it, tho' other gems receive from it."[5] *The Book of Seals* states, "a garnet with the image of a lion engraved thereon (is used)

> *Upon his bloody finger he doth wear,*
> *A precious ring that lightens all the hole,*
> *Which, like a taper in some monument,*
> *Doth shine upon the dead man's earthly cheeks,*
> *And shews the rugged entrails of the pit.*
> Shakespeare in Titus Andronicus

to protect and preserve the honors and health of the wearer."[6] Burton tells of a granutus, called an "imperfect ruby," in his *Anatomy of Melancholy*, "If hung about the neck, or taken in drink, much resisteth sorrow and recreates the heart."[7] An anonymous Bohemian manuscript of 1391 lists garnet and other red stones as a cure for inflammatory diseases and as a calming influence on quarrels, discords and hot temper. The author continued to explain that the color plays an important part in the cure. The red was reported to absorb the heat of fever and passion.[8]

Garnet was considered to possess other curative powers. Powdered and taken internally, or as a poultice, it was a popular antidote for poisons. The stone was reported to prevent over-indulgence and help in resisting the temptation to live luxuriously. If worn it would dissipate sadness, control incontinence, avert evil thoughts and dreams, exhilarate the soul and foretell misfortunes.[9] The old *Book of Dreams* says to dream of garnets meant "the accumulation of riches." "It furthermore brings sleep to the sleepless, drives away the plague and all evil spirits, and attracts to its possessor riches, glory, honor, and great wisdom, and, to crown all, constancy with fidelity to all obligations."[10] Leonardus adds, "The virtue of carbuncle is to drive away poisonous and infectious air, to repress luxury, to give and preserve the health of the body."[11] Another benefit of possessing the gem was its ability to keep a house safe from lightning strikes. Garnet is still used as an amulet to aid the heart, circulation and diseases of the blood. It is praised as the gem of peace, love, passion and sensuality.

> *Garnet is said to be the gem*
> *of Tuesday or "Tuisco"*
>
> *If Tuisco assists*
> *And at birth keeps apace*
> *The bairn will be born*
> *With a soul full of grace.*

The many colors and sources of garnet have led to numerous pseudonyms for the gem. The deep maroon pyrope is called "Bohemian garnet." Its deep red color has caused it to be labeled "ruby." Stones from Arizona are known as "Arizona rubies," those from South Africa are called "Cape rubies," and garnets mined in Montana are sold as "American ruby." The lovely green demantoid variety has carried the name "Uralian Emerald" for many years. These trade names may have charm and add interest to a relatively common gem, but the use of the terms ruby or emerald when trading in garnets is considered an unfair trade practice by modern gemologists.

The various forms, colors, and names for garnet have led to confusion in literature regarding gems. Examples of this may be found in Chevalier Jean de Mandeville's fourteenth century *Lapidaire*. He describes three forms of carbuncle and a separate gem called garnet. One of the carbuncles, is described as green mixed with red, a stone which may in fact have been a tourmaline. The text also reinforced the common belief that garnet was a form of ruby.

> Carbuncles are red, similar to the color of hot coals, and generally have all the properties of precious stones. There are three kinds of carbuncles; the first is called anthrax in Greek, and its color is green intermixed with red, as in fire. This stone is found in parts of India and Libya. The second is called ruby, is less expensive, and doesn't glow much.[12]
>
> The baleze has a paler color than a fine ruby and doesn't sparkle in the same way. It is found in several places: in India, in Ethiopia, in Corinthe, and in Libya. It is the third kind of ruby and can be either red or violet. Both kinds eliminate vain thoughts and frivolities, lessen sadness, make those who wear it happy, cools lechery, procures peace and agreement and gives safe passage amongst one's enemies and through perilous places; snakes and other venomous beasts will not approach.[13]
>
> Garnets are in the ruby family. They are red like pomegranates, they are violet and very precious. When they are placed on a black object, they glisten even more. They give courage to the heart, they chase away sadness, and bring joy. Some people say that

they have the virtues of hyacinths, but that is not true because some hyacinths are called garnets, and are found in India, Ethiopia and other places, in the rivers of Paradise, near Alexandria and Thir [probably the Greek island Thera]. This kind of hyacinth has all the same properties as garnets.[14]

One variety of garnet which is flame-red or golden-orange, a mixture of pyrope and spessartite molecules, is often called a "malaya garnet." There is no connection to the Malay states, but the term is derived from a native word. Malia means "out of the family" and refers to this stone being separate from other garnets. Another translation of the word is considered vulgar and means "prostitute," not a flattering name for this beautiful variation.

The reported glow of the carbuncle has been the source of most of the stories about this birthstone of January. Chaucer, one the greatest poets of all time, describes the jeweled clothing of Richesse in his *Romount of the Rose*.

> Richesse a girdel hadde upon
> The bokel of it was of a stoon
> Or vertue greet, and mochel of might.
> For who-so bar the stoon so bright,
> Of venim [thurtle] him no-thing doute,
> While he the stoon hadde him aboute.
> That stoon was greetly for to love,
> And til a riche mannes bihove
> Worth al the gold in Rome and Fryse.
> The mourdaunt, [a metal tag on the end of a belt, not the buckle] wrought in noble wyse,
> Was of a stoon ful precious,
> That was so fyn and vertuous,
> That hool a man it coude make
> Of palasye and of tooth-ake.
> And yit the stoon hadde suche a grace,
> That he was siker in every place,
> Al thilke [that] day, not blind to been
> that fasting mighte that stoon seen.[15]
> But al bifore ful sotilly [subtilty]
> A fyn carboucle set saugh I.

116
Lapidarium:

> The stoon so cleer was and so bright,
> That, al-so sone as it was night,
> Men mighte seen to go, for nede,
> A myle or two, in lengthe and brede,
> Swich [such] light [tho] sprange out of that stoon,
> That Richesse wonder brighte shoon,
> Bothe hir hedde, and al hir face,
> And eke aboute hir al the place.[16]

The Renaissance goldsmith Benvenuto Cellini told the following tale of a luminous garnet.

> Jacopo Cola, a vine grower, going into his vineyard at dusk one night, noticed a gleam of light in a far-off corner among the vines. Hastening toward the glowing light, he could not locate it, for the radiance had vanished into darkness. Then returning to his home, he was about to give up the search, when again it flashed, and again he carefully approached the gleam and found it at last in the heart of a rough stone at the foot of a vine. Joyfully he took the stone back and showed it to his friends, and one of them, being acquainted with precious jewels through his travels in Rome, bought it for ten scudi, and later sold it to a Venetian envoy, who carried to Constantinople, there selling it to the Sultan for 100,000 scudi.[17]

A fantasy concerning the ruler of the Island of Amboin is repeated by the author Rumphius in 1687. It is the tale of a carbuncle brought by a serpent. The ruler, when a child, had been placed by his mother in a hammock attached to two branches of a tree. While there, a serpent crept up to him and dropped a stone onto his body. In gratitude the parents of the child fed and cared for the serpent. The stone had a warm yellow hue, verging on red. It shone so brightly at night that it would light a room. Rumphius concludes the story by stating this fabulous stone was passed on to the King of Siam.

The treatise *Travels* by Mandeville tells of many fabulous lands. In one, he claims the Emporer had a ruby and a carbuncle, each a foot long. He reports the stones shone so bright, they were the source of all night illumination in his chambers. "In the adventures of the *Golden Fleece* the hall of King Priam is described as illuminated at night by a prodigious carbuncle."[18]

The origin of red garnets and hessonite (a cinnamon to yellow form of grossular garnet, also called "cinnamon stone") are given in the Vedic texts of India as part of the story of the death of Vala. A demonic god, Vala was slain and dismembered by the demigods. The parts of his body were strewn about the earth and universe to create various gemstones, including these two types of garnet. The seeds of red garnets were created from Vala's toenails and spread about the earth by the snake-god. This demigod carried them in its mouth and dropped them in the lands surrounding the Himalayas, where garnet mines are said to have originated. Red garnets are associated with the red lotus and are considered to possess the mystic virtue of increasing wealth and progeny. The hessonite form spread from the seeds created from Vala's fingernails. Winds blew these seeds into the lotus ponds of Sri Lanka, India and Burma. This orange hued stone is considered to contain the colors of honey, blood and moonlight. A flawless example, set in gold, in considered the ultimate talisman. The wearer is assured a long life, many children, and a life of happiness. The cinnamon stone also removes evil thoughts and motives from its owner and grants fame and wealth.

Another use of the stone in Asiatic countries is as a projectile in guns. Indian warriors believed the red gem wound inflict a more deadly wound when it struck its target.

Practitioners of crystal healing in the twentieth century call the garnet "stone of health." The gem is used to removed negative energies from a subject and transfer these energies to a beneficial form. Garnet is also called upon to strengthen love and bring devotion and commitment to couples. The various stones of the garnet group are also given individual attributes and powers.

Almandine garnet— used to stimulate the physical body and the intellect. It also makes the wearer more contemplative, peaceful and charitable. It is effective in the treatment of heart disease, eye disorders and maladies of the liver and pancreas.

Andradite garnet— gives strength, stamina, courage and stability to males. It also acts to attract another who is essential to one's life. This type of garnet activates mysticism and brings heightened experiences to the mind, body and spirit.

Grossular garnet— enhances fertility. It gives strength in the face of legal challenges by providing ingenuity and an ability to compromise.

Pyrope and rhodolite garnet— serves as a stabilizer to maintain a proper balance of bodily rhythms. It can be helpful in the treatment of digestive tract disorders, including heartburn and throat problems.

Spessartite garnet— helps to stimulate the analytical parts of the mind and to activate ingenuity. As a healing stone, it is used to treat lactose intolerance and other problems in assimilating calcium.

Uvarovite garnet— grants peace, quiet and solitude while promoting clarity of thought. It is claimed to be excellent in the treatment of leukemia, kidney and bladder infections, heart and lung disorders, acidosis, and as a purgative of bodily toxins.

Hematite

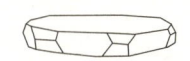

Crystal System

Hexagonal plates

Color: Black, black-gray, brown-red
Mohs' Hardness: $5\frac{1}{2} - 6\frac{1}{2}$
Cleavage: None
Refractive Index: 2.94 - 3.22
Dispersion: None

Streak: Blood-red
Specific Gravity: 5.12 - 5.28
Fracture: Conchoidal, uneven, fibrous
Birefringence: 0.28
Pleochroism: None

Crystal System: Hexagonal (trigonal); platy crystals
Chemical Composition: Fe_2O_3 iron oxide
Transparency: Opaque
Fluorescence: None

The stone which is a dark-gray to black form of iron oxide is known as hematite. It is one of the most common minerals and usually is formed in association with iron ore deposits. Its metallic luster led to its proper association with iron from ancient times. This mineral is also characterized by its brownish-red streak. Facetted black hematite has been sold as "black diamond" and in the form of a bead it is offered as "black pearl." Both misnomers are considered misleading in the jewelry trade.

Hematite's reddish streak when rubbed against a touch stone led to its name. Derived from the Greek *haima*, meaning blood or blood-like, it is also known as the "stone which bleeds." These characteristics and the fact that it is an iron ore caused the Greeks to associate it with Mars, god of war. Moistened and rubbed on the body, its bloody residue was considered a valuable safeguard to warriors on the battle field. The epic Greek myth *Lithica* states the stone originated as the petrified spilled blood of Uranus when wounded by Kronos.

The Babylonian author Azchalias wrote of its virtues in a first century treatise composed for Mithridate the Great, King of Pontus, a renouned lover of precious stones. Azchalias postulated that human destinies were greatly influenced by gems. He stated that one who wears a hematite will be successful in petitions to the king and victorious in law suits and legal judgments. This is the foundation of the use of hematite signet rings as a favored jewelry of the legal profession.

The fact that this gem was held in high regard by ancient peoples is evidenced in its use in Pueblo inlays. The stone is difficult to fashion with primitive tools and splinters or fractures easily. Its appearance in delicate inlays, interspersed with turquois, shows that these artifacts must have had great value to their makers.

The curative powers of the stone were recognized and recorded by numerous ancient cultures. Many of the claims made for this gem have a basis in fact. This dense iron oxide has strong astringent and styptic properties. Hematite powder will act to quickly coagulate blood and remove other fluids from a wound. The ancient Egyptians prescribed it as an ingredient in dressings for wounds. The *Magical Papyrus of Leyden* contains a spell for the treatment of cuts or stings. The spell is to be recited over a mixture of "faeces of cats, hematite, fat of goats and honey."[1] These ingredients were to be ground and placed in a bandage over the wound. The same physicians also used the gem as a treatment for diseases of the eye lids, styes and eye tumors. Hematite powder was dissolved in egg white to treat small growths. If the tumor was large, the same ingredients were to be boiled in water which con-

This is the alchemical sign for bloodstone. It also serves as a symbol for hematite.

According to Diderot's Encyclopedia *this sign was use by the eighteenth century chemists in France to mean bloodstone or hematite*

tained the herb fenugreek. The treatment of mild inflammations involved making a thin solution of hematite powder and water which was applied with a dropper. If continued treatments were necessary, the solution was made successively thicker and applied with a dobber. Taken internally it was claimed as a cure for the "spitting of blood and all ulcerations."[2]

The Romans saw hematite as a class of five stones and not as a single gem. These five stones were given particular medical attributes. Pliny documents each of these prescriptions by quoting the earlier writer Sotacus. The best type was called *Ethiopic* and was used for diseases of the eye and for burns. The second, *androdamus*, a very black and dense material, was said to come from Africa. It was also known as the "conqueror of man." This stone was ground, mixed with water, and administered as a cure for bilious disorders. A third sort, which produced a yellowish streak, was thought to come from Arabia. It was also administered for biliousness and was used to ease the pain of burns. The fourth variety was called e*latite* in its natural state and *melitite* when burned. The final form was a mix of hematite and schist. The last two stones shared a common use. Mixed in a quantity of three grains of stone dissolved in oil, it was administered orally as a tonic for blood disorders.[3] Without knowing the biochemistry involved, the Romans may have found any early treatment for anemia. The iron present in the gem is able to be absorbed into the body and acts to fortify the blood.

Modern practitioners call hematite the "stone of the mind."[4] It is said to help with mental acuity, memory enhancement, original thinking and technical knowledge. The gem is still used in the treatment of anemia and has also been helpful in the reduction of fever. It is either applied to the forehead or taken as an elixir. Leg cramps, nervous disorders, and insomnia are also said to be eased.

Jacinth: see Zircon

Jacinth and hyacinth are names which have fallen from fashion. They are rarely used by modern gemologists, but some reserve them for reddish-brown natural zircon and hessonite garnet. Jargoon (also jargon) is another name used through the centuries for these stones. This name is applied in the modern gem trade to zircons from Ceylon ranging in hues from pale-yellow to colorless. Today zircon is the name preferred by gemologists to refer to these zirconium based gems.

> Jacinth *is dark blue*
> *With moderate verdure [brightness in oldest M.S.]*
> *Its beautiful appearance*
> *Is changed as the weather.*
> *It signifies the angelic life*
> *Endowed with capacity for discernment.*
> The Lapidary, *Marbode of Rennes*

Jade Group

Crystal System

Monoclinic

The word jade refers to two distinct and different minerals; jadeite and nephrite. In many ways jadeite and nephrite resemble one another. Both materials are very tough crystalline aggregates, a toughness unmatched by any other gem. Each may also occur in masses which vary from opaque to translucent material: gray-green, white, grayish white, brown to reddish brown, and black in color. In appearance the two minerals may seem identical, but they differ in chemical make-up and crystal structure. Nephrite consists of minute interlocking crystals of calcium magnesium silicate. Jadeite's aluminum sodium silicate structure more closely resembles interlocking grains. As a result, jadeite is more granular and nephrite more fibrous. Jadeite is the harder of the two and occurs in a wider range of colors, including an intense green that is unknown in nephrite. The ancients did not distinguish between the two minerals and treated both as one gem — jade.

Jadeite Variety

Color: Brown, blackish, violet, green, also white, reddish, yellow
Mohs' Hardness: $6\,^1/_2 - 7$
Cleavage: Imperfect
Refractive Index: 1.660 - 1.680
Dispersion: None

Streak: White
Specific Gravity: 3.30 - 3.38
Fracture: Splintery, brittle
Birefringence: 0.020, often none
Pleochroism: None

Crystal System: Monoclinic; intergrown, grainy and fine fibrous aggregate
Chemical Composition: $NaAl(Si_2O_6)$ sodium aluminum silicate
Transparency: Opaque, translucent
Fluorescence: Green; very weak, gray-blue

Nephrite Variety

Color: Green, also white, gray, yellowish, reddish, brown, often spotted
Mohs' Hardness: $6 - 6\,^1/_2$
Cleavage: Perfect lengthwise
Refractive Index: 1.606 - 1.632
Dispersion: None

Streak: White
Specific Gravity: 2.90 - 3.00
Fracture: Splintery, sharp edged, brittle
Birefringence: 0.026, or none
Pleochroism: Weak; yellow to brown, green

Crystal System: Monoclinic; intergrown, fine fibrous aggregate
Chemical Composition: $Ca_2(Mg,Fe)_5(Si_4O_{11})_2(OH)_2$ calcium magnesium iron silicate
Transparency: Opaque
Fluorescence: None

The Chinese word *yu* and Japanese words *gyoku* and *tama* mean both jade and precious stone. The Chinese character *pao*, for the word precious, is the outline of a house with the symbols for jade beads, shell, and an earthen jar. When these early symbols were established in ancient China, jade was already a valued object. The oldest form of the character for king is a string of jade beads. A jade string is also a modern insignia for rank or power in China. These examples demonstrate the significance of the stone to the people of Asia. Another ancient location which held jade in high regard was Central America. The Aztec word *chalchihuitl* means jade, green stone, and precious object.

Early Spanish explorers of Central and South America found numerous objects made of jade and brought examples back to Europe with them. These returning sailors probably devised the phrase *lapis nephriticus*. It comes from the Greek words *lapis* for stone and *nephros* for kidney. They had observed the gem being used by the natives of the New World as a cure for kidney ailments.

The ancient Chinese symbol for king is a string of jade beads.

The first reference to the stone in western literature is found in the writings of Dr. Monardes of Seville. In 1569 he refered to the *piedra de hijada* meaning "stone of the flank." His account of the stone and its use in Central America became the foundation of European beliefs regarding the gem. Monardes states,

> The so-called nephrite stone is a species of stone, the finest of which resemble the emerald crystal, and are green with a milky hue. It is worn in various forms, made in ancient times, such as the Indians had; some like fish, some like the heads of birds, others like the beaks of parrots and others again round as balls; all, however, are perforated, since the Indians used to wear them attached for nephritic or gastric pains, for they had marvelous efficacy in both these infirmities. Their principal virtue regards the nephritic pain, and the passing of gravel and stone, in such sort that a gentleman who owns one, the best I have ever seen, wearing it bound on his arm, passed so much gravel that he often takes it off, thinking that it may be injurious for him to pass such a quantity; and, indeed, when he removes the stone he passes

much less. . . . This stone has an occult property, by means of which it exercises a wonderful prophylactic effect, preventing the occurrence of nephritic pain, and should it nevertheless ensue, removing or alleviating it. The duchess my lady, having suffered three attacks of this malady during a short period, had one of these stones set in a bracelet and wore it on her arm, and from the time she put it on, she has never felt any pain, although ten years have past. In the same way it has served many, who have realized the same benefit. Therefore, it is highly prized and it cannot now be worn so easily as in former times, as only caciques and noblemen own it, and rightly, since it has such wonderful effects.[1]

Both phrases, *piedra de hijada* and *lapis nephriticus*, are more medicinal than mineralogical. They began to be used to reference any green stone which appeared to relieve pain in the groin. The evolution of the modern word jade may be traced through its first appearance in English in 1598 as *ijada*. The mid-seventeenth century French poet and writer of prose Vincent Voiture used the name *l'éjade*, which demonstrates another link between the Spanish term and the modern name. By 1811 the writer John Pinkerton, the man who coined the word gemology, made reference to the stone as a *jad*. Professor A. Damour was the first to document the difference between the two minerals called jade and coined the word *jadeite* in 1863 to make the distinction.

Jade is found in many different localities. This is evidenced by its prehistoric use as axe-heads and other cutting tools in many areas of the world. Its toughness and ability to hold an edge made it a favored stone for these tools. The earliest carved jade implements are dated at about eight to nine thousand B.C. Early man seems to have made the distinction between jade and other green stones, an ability lost in later generations. Medieval writers had come in contact with jade, but did not recognize it as a separate gem. Marco Polo recorded a visit to nephrite mines in Knotan, Turkestan in 1295. In his account he refers to the stones as *jaspre*. Other writers describe "jasper" material which could be jade, but it is not named separately.

The fourteenth century writer Chevalier Jean de Mandeville included a passage on a stone he called *silente*. The gem he describes is most likely jade. In his *Le Lapidaire* Mandeville states,

> Silente is a dark stone and tends sometimes to be black, sometimes to be green, like jasper. It waxes and wanes like the moon. It protects pregnant women and helps them deliver on time. It brings peace and agreement and helps to reconcile lovers. It raises the morale and lessens inflammations. It is found in India and in Persia.[2]

Sir Walter Raleigh's account of a trip to Guiana reinforced the status of jade and its medicinal virtues. His treatise published in 1596 on the treatment of the peoples of the "Empire of Guiana" contains this reference to the gem:

> These Amazones have likewise great store of these plates of golde, which they recover by exchange, chiefly for a kinde of greene stone, which the Spaniards call Piedras Hijadas, and we use for spleene stones and for the disease of the stone we also esteeme them: of these I saw divers in Guiana, and commonly every King or *Casique* hath one, which theire wives for the most part weare, and they esteeme them as great jewels.[3]

By the mid-seventeenth century the value of the stone was well established as a treatment for maladies of the feet and various ailments of the mid-section. The following reference by Voiture is taken from a thank you letter written to Mademoiselle Paulet. She had given him a jade bracelet to relieve his pain from kidney stones. "If the stones you have given me do not break mine, they will at least make me bear my sufferings patiently; and it seems to me that I ought not to complain of my colic, since it has procured me this happiness."[4] Johannes de Laet wrote in 1647 that he possessed a jade which held all the virtues observed by Monardes. He described his gem as oblong, smooth, moderately thick, with a honey color, and an oily surface. He states that the stone was able to relieve the pain of his wife's kidney stones when it was kept bound to her wrist. Although many believed that jade should be worn constantly, others cautioned of over use. It was said that, "unremitting use weakened the effect, so that when the wearer was suddenly attacked by some disorder for which jade was a cure, his system would have become so habituated to its action that it would no longer work its remedy."[5]

The greatest volume of history and lore surrounding jade is to be found in China, a country which has known the stone for over thirty-five hundred years. Folklore holds that jade is the petrified tears of a dragon which were shed over the conquest of China by the Tartars. The gem is used for a variety of amulets and talismans. Each has a different power and significance depending on the shape the carving takes. Many are said to have greater power if received as a gift. A carving of mountains symbolizes longevity and a wish for a long life. Jade carvings of five bats symbolize the five happinesses: wealth, old age, health, natural death, and love of virtue. A carving of two men is called Two Brothers of Heavenly Love, also known as the "jade twins," and is given to friends.

The Chinese say jade is the concentrated essence of love. A gift for newly weds is a jade of a man riding a unicorn and holding castanets. It was said to assure the early and successful birth of an heir. An amulet in the form of a jade padlock is tied around a child's neck to bind them to life and protect them from infantile diseases. Grain grown near jade is said to be protected from heavy rains or drought. A common parable in China also states, "As pearls can keep off the disaster of fire, jade is able to preserve the fine grain."[6] A phoenix of jade is a favorite ornament for young girls, given when they come of age. Businessmen are known to carry jade amulets to counsel them during negotiations. A piece of the stone carried and handled allows the secret virtues of jade to be absorbed. The stone was considered a bringer of rain and was also carried as a protector as it would ward off attacks by wild beasts. It is also said that the sound made when a jade is struck is akin to the sound of a loved one or a dear friend.

A cup carved in the form of a cock is used for a bride and groom's first toast. The tradition is rooted in a legend of lost love. The story is of a white cock which witnessed the suicide of his mistress. The lady threw herself into a well upon hearing of the death of her lover. The cock, in sympathy for his lady, found death in the same way. The toast is therefore a form of the "till death do us part" pledge of matrimony. A butterfly in jade is a symbol of successful love and is a favored gift of a groom to his bride. This symbolism comes from the story of a young man who, while chasing a butterfly, entered a private garden without permission. To his surprise, he met the girl that he would eventually wed instead of meeting the severe punishment he would have expected.

A butterfly jade is sometimes buried with the dead to ensure the immortality of the soul. This is one form of tomb jade amulets placed with the dead. These and other burial jades are also called *han-yü*, or 'mouth jades', because they were placed in the mouth of the corpse to give protection. The Chinese also believed that powdered jade drunk before death stopped decomposition. A more elaborate use of tomb jades was reserved for royalty. The use of six jade objects of specific color and shape is described for use by a Master of Religious Ceremonies in the *Chow Li*, ancient writings which set the requirements of government. These six jades represent the heaven, earth and four cardinal points of the compass. A round tablet of green jade, the *pi*, represents heaven. A square tube of yellow, the *ts'ung*, is the earth. North is represented by a semi-circular shaped black jade called the *huang*, while a red tablet, the *chang*, represents south. West is shown by a white tablet, the *hu*, and a green tablet called the *kuei* stands for the east. These objects were placed in the coffins of members of the Imperial house. The *kuei* was placed to the left, the *chang* at the head, the *hu* to the right, and the *huang* at the feet. The green *pi* was placed under the back and the yellow *ts'ung* on the abdomen. The placement of these jades represented the brilliant cube, or *fjang-ming*, which serves as an emblem of balance and harmony. The separation and placement of the yellow and green jades represented the soul in balance between heaven and earth.[7]

The mystical Taoist religion accounts for a wealth of lore concerning jade. A description of some of these beliefs is outlined in the 1937 text *Jade Lore*.

> Yü Huang, or the Jade Emperor, is the supreme deity, the personal god to whom millions of Chinese address their earthly pleas. He dwells in the Jade Palace on the Mountain of Jade, described as being 3,000 miles in circumference and 3,000 high. At its foot is found the Jade Lake on whose shore blooms the Jade Tree, the fruit of which confers immortality upon those who partake. This tree is 300 arms across. The Clear Jade is the name ascribed to another heavenly peak where dwell the five Taoist Immortals.[8]

Chinese medicine prescribed the ingestion of jade for various maladies. The Taoist author T'ao Hung Ching directed in his sixth century writings that powdered jade used medicinally must not be ground from carved jade or unwrought material used as tomb jades. The custom was to crush jade into fine pebbles the size of rice grains. In this form the stone could pass through the system unchanged, but the innate virtue of the stone would be absorbed by the patient. Ching claimed this treatment would relieve heart-burn, render the hair glossy, and ease asthma. Jade grains are also used to strengthen the lungs, heart, and vocal chords. The gem is also the major ingredient in a tonic. This "divine liquor of jade" contains equal parts of jade grains, rice and dew-water. The mixture is boiled in a copper pot and carefully strained. It is said to strengthen muscles, make brittle bones supple, calm the mind, enrich the flesh, and purify the blood. If this liquor is administered over a long period of time, the patient will no longer be vulnerable to heat, cold, hunger or thirst. Jade grains, the size of rice, mixed with silver and gold powder are also thought to prolong one's life.

Jade served similar purposes for the people of Central and South America. A powder of the gem was used as a tonic. As mentioned it served as a treatment for the kidneys, liver and spleen. Pre-Columbians placed a piece of jade, or other green stone, in the mouth of nobles upon their death. Called a *chalchihuitl*, it was intended to serve as a surrogate heart in the afterlife. Inferior stones, called *texaxactli*, served the same purpose for those of lower class. Revered by Mayans as 'Sovereign of Harmony,' jade was used to facilitate peace and bring man into accord with his environment. Used by ancient and come current tribes as a sacred stone, it is thought to allow access to the spirit world, bringing confidence, self assurance, self reliance and self-sufficiency. The natives of Brazil consider the Amazon-stone as sacred. These jade ornaments are prized above all other gems. The natives call them *ita ybymbae*, meaning green stones. They are not valued for their beauty, but because they have been handed down from generation to generation over the centuries. These amulets are drilled and polished geometric forms which are worn around the neck. Also called *ita poçanga*, meaning medicine stone, they are worn as protection against many diseases and venomous bites. The same stone forms with nearly the same attributes are also worn by natives of some Caribbean islands. They are called "the smooth stones from the far-off continent."[9]

The Story of Balarama and the Origin of Jade:India

Vedic texts of India give an origin of jade. Vala, a demon god, was slain by the demigods. The parts of his body were strewn about the earth to create the various gemstones we know today. One of the parts of Vala distributed among the demigods was his fat. The Lord Balarama scattered the fat from Vala's body over the areas of China, Nepal, and the lands of the Yavanas and Mongols. Everywhere particles of fat fell to the earth, they were transformed into the seeds of jade. As these seeds matured, the worlds deposits of jade grew and developed the various colors we are familiar with. The sacrifice of Vala gives mankind a valuable charm, fine quality jade is believed to have the greatest ability of all stones to remove negative karmic reactions.

The Story of the Naming of the Yü-t'ien District:: Chinese

In the second century a young man named Yang Yung-po became known for the kind deeds he performed. One charitable task he worked at was to provide water to travelers as they passed his home near a mountain pass. One grateful traveler rewarded Yang with a large vessel filled with cabbage seeds. He was instructed to plant the seeds with care, and he would harvest jade and a good wife. Soon after planting the seeds Yang met a beautiful young woman. After an appropriate time of courtship, he and the maiden were married. As they began their life together, Yang remembered the rest of the stranger's promise. He returned to his cabbage field and recovered five fine pieces of jade. Each piece brought one of the five cardinal virtues: clarity, modesty, courage, justice, and wisdom to Yang and his wife. They both live long and happy lives blessed with prosperity. From this story of good fortune the region where Yang lived took its name. It is the Yü-t'ien District of China. Yü-t'ien means a "field of jade."[10]

The Story of the Dishonored Statesman: Chinese

Nearly two thousand years ago a statesman offended the imperial custom by using the personal name of his sovereign. For this indiscretion he was informed that he would most certainly be struck down by the powers of Heaven. Determined to preserve his own corpse from incineration by lightning, the condemned man obtained a jade talisman. He knew the power of the stone to save the body from decomposition and hoped it would serve as a protective talisman. Upon his death his body was prepared for burial, complete with the tomb jade. At his funeral a storm did rise and the coffin and shroud where destroyed by lightning. The only remnant left of the blasphemous statesman was the small area of his body covered by the jade. From that time forward it has been known that jade also serves as a protection from lightning.[11]

The Legend of the Hei-tiki: The Maori People of New Zealand

The rarest form of jade was known to the Maori as *pounamu*, meaning "green stone." The stone was only found with the aid of a tribal wizard. The wizard would lead a party of jade seekers into the region where jade was traditionally found. Working himself into a trance, the wizard would be led to the location of this precious gem by the spirit of a member of the tribe. The spirit may have been that of a person living or dead. With the permission of the spirit, the wizard would then lead the rest of the party to the location of the pounamu. This special piece of jade would be carved into an amulet in the form of the spirit which led to its discovery. These amulets were known as *hei-tiki* or "carved image for the neck."

These grotesque carvings of the faces of departed ancestors were passed from generation to generation as a reminder of the family lineage. The hei-tiki were passed from one male member of the clan to the next elder in succession. If no male heir existed in a family, the hei-tiki would be buried with the last male member. Tradition held that this amulet may be buried with its owner and, after a respectful period of mourning, only retrieved by his male next of kin. This guaranteed the proper line of ownership and that the hei-tiki would not fall into the hands of a stranger. History records tribal and intertribal feuds were fought over the proper lineage of a particular amulet. The tradition of the hei-tiki seems to have died out about three hundred years ago with the introduction of European influences. Other tiki figures are still carved in jade and native woods.

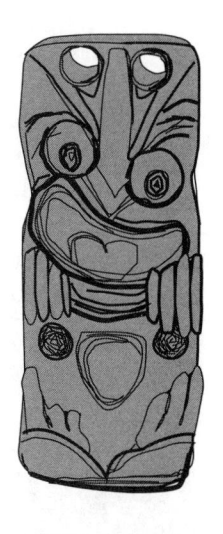

Sketches of the tiki carvings produced by the Maori people of New Zealand. The grotesque face of the top figure is representative of the sacred hei-tiki.

The Jade Commandments[12]: Chinese

> The jade uncut will not make a vessel for use, and if men do not learn, they cannot know the way to go.
> The man of virtue is the jade of the state;
> The superior man guards his body as if carrying jade.
> If jade is not cut and polished, it cannot be made into anything; neither is a man perfect without trials.
> The superior man competes in virtue with jade.
> A real scholar embraces the truth as he would cherish jade.

Who wears a Jasper, be life short or long,
Will meet all dangers brave, wise, and strong.[1]

Crystal System

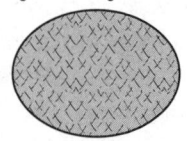

Microcrystalline aggregate

Jasper

Color: All colors, striped or spotted
Mohs' Hardness: $6\frac{1}{2} - 7$
Cleavage: None
Refractive Index: About 1.54
Dispersion: None

Streak: White, yellow, brown, reds; determined by foreign material contained
Specific Gravity: 2.58 - 2.91
Fracture: Splintery
Birefringence: None
Pleochroism: None

Crystal System: Hexagonal (trigonal) microcrystaline grainy aggregates
Chemical Composition: SiO_2 silicon dioxide
Transparency: Opaque
Fluorescence: None

Jasper is the term applied to most semitransluscent to opaque chalcedonies. They are usually deep in color and seldom display a pattern or banding. Jasper is generally dense, quite hard and accepts a good polish. The gem is found in various forms throughout the world and may occur in red, yellow, brown, green, grayish blue or any combination thereof.

Lapidarium:

Used as a generic term for opaque stones in many cultures, the jasper or *jaspis* has been referred to in the most ancient of writings. It is considered a sacred stone by Native Americans and is worn by shamans as a protective gem. The Egyptians used red jaspers to represent the blood of their goddess Isis. Amulets of the gem were said to have the same virtues as the goddess' blood and when worn helped prepare one for the judgment of Osiris upon death. The earliest references in Greek literature are from the third century B.C. writer Theophrastus. He does not list any specific virtues, but simply states "that the stones out of which signets are made for the sake of their beauty alone, are the Sard, the Jaspis, and the Lapis-lazuli."[2] Claudius Galen (130-200 B.C.) a physician to the school of gladiators states, "that the green jasper benefits the chest, if tied upon it."[3] Galen also recommended setting jasper in a ring and engraving it with the figure of a man wearing herbs around his neck as a protection against evil and illness.[4] Roman athletes also used the stone as a talisman to protect them from injury. The historian Pliny the Elder states in his *Natural History* that in the first century A.D. the stone remained popular, "though now surpassed in value by many other gems."[5] The fourth century Cypriot Bishop St. Epiphanius said of jasper that it is valuable in keeping away evil spirits and in driving away poisonous snakes. The Author of *Lithica*, also writing in the fourth century, speaks poetically of the jasper's ability of bring rain.

> The gods propitious hearken to his prayers,
> Whoe'er the polished grass-green jasper wears;
> His parched globe they'll satiate with rain,
> And send showers to soak the thirsty plain.[6]

The medical value of jasper has usually been associated with its color, but the sixth century Greek physician Alexander Trallianos tells of a virtue added by engraving. He wrote that if the figure of Hercules strangling the Nemaean Lion is carved into a jasper and it is set in a gold ring, it will cure the wearer of colic. An example of such a stone is said to reside in the collection of the Biblioteca of Ravenne. The letters KKK are carved into the reverse side of the stone, possibly standing for Kwlikh, the Greek word for colic.

Chevalier Jean de Mandeville describes two major forms of the gem in his fourteenth century *Lapidaire*. He also lists numerous attributes for the stone.

> There are two kinds of jasper (about which I shall speak here): the green, which is the best when there

> are red or golden drops, and the speckled, which is very precious and cures fevers and dropsy and is very helpful to women in labor, because it throws out the baby, dead or alive. It also gives sweetness and increases honor and valor; it removes frivolity and cures bites of venomous animals. Powder from jasper serves to curb inflammations of the blood. If venom is brought before this stone it will soon diminish it sweat. It is not for a woman to wear it, because it will prevent her from conceiving. Jasper should be mounted in silver and worn on the right side of the body.[7]

In a separate entry Mandeville speaks of other forms of the gem.

> There are nine kinds of jasper, they are of many colors and are found in many parts of India. Those which are green like emeralds against the light are the rarest. If there are vermilion drops, they have the most special properties. (Mandeville may be referring to bloodstone in this entry.) Jasper protects men who wear it chastely from fevers and dropsy. It will help a woman giving birth to a child. It makes a man friendly and powerful; it pushes away frivolity and reheats the blood.[8]

The color of a stone was usually more important than its chemical make-up in talismanic use. Red jasper has long been recommended for its ability to stop hemorrhages and reduce pulse rate. A list of valuables bequeathed to George, Earl of Marischal, in 1622 includes this reference to jasper's virtue, "ane jaspe stane for the steming of bluid."[9] Anselmus de Boodt relates the following story of the red jasper's ability in his *Gemmarum et Lapidum Historia* of 1636. A young woman from Prague was referred to de Boodt for hemorrhages she had suffered for six years. As soon as a red jasper was placed on her, the bleeding stopped. After wearing the stone for a while, she thought the cure was complete and removed it. Her symptoms immediately returned and the use of the gem was again prescribed. The hemorrhaging again ceased and never returned. De Boodt states that she eventually was cured to the point that she no longer required the talisman.

A popular shape for deep red jasper is the pear-shaped cut. The similarity to a drop of blood is said to increase the stone's ability to stop bleeding and prevent infection. The power of such an amulet is again magnified if it is hung from a red ribbon. The stone blood drop is a popular amulet in modern Italy and is often brought to the church to be blessed by the local priest. Such an act is a modern example of a pagan object being transformed into a Christian symbol. The Church sanctifies the jasper drops as a symbol of the sacred blood of Christ.

Another Christian symbolism is the assignment of the Foundation Stones, listed in Revelations 21, to each of the twelve apostles. One of the earliest writers to make such an association was the tenth century writer Andreas, Bishop of Caesarea. He states that jasper is the stone of St. Peter and gives the following reasoning:

> The jasper, which like the emerald is of a greenish hue, probably signifies St. Peter, chief of the apostles, as one who so bore Christ's death in his inmost nature that his love for Him was always vigorous and fresh. By his fervent faith he has become our shepherd and leader.[10]

De Boodt also listed an attribute for green jasper. He prescribes the stone as a cure for stomach ills if worn around the neck so it touches the effected area. He also relates the story that jasper is most potent in cures if it is engraved with the figure of a scorpion while the sun is entering Scorpio. It is interesting to note that de Boodt dismissed such a tale as pure superstition, but repeats Mandeville's advice that jasper is most effective if mounted in silver.

Camillus Leonardus in the *Speculum Lapidum* of 1502 writes, "Being carry'd about one, it drives away the fever and dropsy, clears the sight, expels noxious phantasms, restrains luxury, and prevents conception." He also advises, "it ought to be set in gold, because that increases its virtues."[11] It may be noted this is in direct opposition to de Boodt's prescription. The sixteenth century author Cardano states that the stone tends to make the wearer more timid and cautious. He also claims it will make the owner victorious in battle. He explains this seemed contradiction by stating, "The timid usually conquer, since they avoid a doubtful contest if possible."[12]

Bartolomæus Anglicus gives a unique source for the stone in his poetic reference of 1495, "in the head of the adder that hyght Aspis is founde a lytyl stone that is called Jaspis." He continues by relating that "it hath as many vertues as dyvers coloures and veines." Modern literature on the subject of gem healing expands on this five hundred-year-old statement by listing numerous types of the stone, giving separate attributes for each.

Blue jasper— used to help make connection with spirit forms in the after-life.

Breciated jasper— aides with digestion and colon or intestinal disorders. It also reduces stress and nervousness.

Fancy or variegated jasper— aides in granting a positive outlook, happier feelings, and enthusiasm.

Green jasper—opens the heart, expands awareness and strengthens self esteem.

Imperial jasper— enhances healing abilities, clairvoyance, and telepathy. It also is used to help relieve muscular pain, tumors, circulatory problems and eliminate toxins from the body.

Leopard jasper—used as a grounding stone to balance physical, mental, and emotional bodies. It is used to attune one to the balance required in nature and also helps personally to increase organizational skills.

Picture jasper— strengthens the immune system, decreases skin problems, and slows the aging process. It is also said to increase one's intuition and ability to recall past lives.

Rainbow jasper— Produces a calming effect, especially in decision making.

Red jasper— Alleviates fear, anxiety, and stress.

Yellow jasper— Protects one on a journey.

Jet

Crystal System

Amorphous

Color: Black, dark brown
Mohs' Hardness: 3- 4
Cleavage: None
Refractive Index: 1.64 - 1.68
Dispersion: None

Streak: Black-brown
Specific Gravity: 1.10 - 1.40
Fracture: Conchoidal
Birefringence: None
Pleochroism: None

Lapidarium:

Crystal System: Amorphous
Chemical Composition: Compressed lignite (coal), mostly carbon
Transparency: Opaque
Fluorescence: None

Jet is not a crystal mineral, as are most gems, but a very hard variety of coal. It forms by the compression of lignite, a brown coal from driftwood and other sea-water soaked organic material.good polish. The gem is found in various forms throughout the world and may occur in red, yellow, brown, green, grayish blue or any combination thereof.

Jet may rank with amber as the world's first gem material used as talismans. Jet ornaments have been found in paleolithic caves in Switzerland and France and in ancient ruins of the Pueblo Indians. Early fragments are round and drilled through for stringing. This soft gem was able to be shaped and carved with flint tools. Evidence of mining activity has been found in England which dates to 1500 B.C.

The ancient name for the gem, *gagates*, originated from the Greek word for its supposed origin. The Greeks claimed the stone was first found in the fabled river Gagates in Lycia. The name evolved through the Latin for *gagete* into the old French *jaiet*.

The ancient physician Galen searched for the mythical river Gagates in Asia Minor. He did not find the river, but he did report small deposits of jet on the eastern shore of the Dead Sea. Galen recorded that he used these fragments as a cure for swelling of the knees. The fumes from this slow-burning gem are mentioned in the one of the earliest Greek medical texts. This treatise, written in the second century B.C. by Nicander, prescribes the fumigation of bedrooms with the smoke of jet to remove the plague. This cure for the black death should naturally be affected by a black stone.

The Roman historian Pliny the Elder listed numerous medicinal uses for jet. The fumes of burning jet cured hysteria, revealed latent epilepsy, and in some unexplained way acted as a test for virginity. The gem was used in a powdered form as a dentifrice, and mixed with wine, it was supposed to relieve toothaches.

Jet found favor in the early Catholic church as a material for crosses and rosaries. Prayers recited on black rosaries were thought to be particularly effective in driving away bad dreams and hallucinations. It was thought that these maladies were brought about by the devil and his black angels. A black stone, due to the powers of sympathy and antipathy of color, was believed to be most effective in counteracting these dark spirits. Saint Hilda, Abbess of the north eastern English monastery of Streoneshalh from 654-680, was known to use such a rosary. The beads were mined from jet found in the region. This same monastery latter became known as Whitby Abbey, and the jet found nearby is known as Whitby Jet. In the eighteenth century this same material returned to popularity for use in religious articles.

The era of greatest popularity for jet was during the late nineteenth century, the reign of Queen Victoria of England, and the early twentieth century. A widow for the last forty years of her life, Victoria wore the gem as a sign of her long and continual mourning. As a result the stone was popularized for mourning jewelry and favored as the "jewel of widows."

The stone has been used since the Middle Ages to treat delusions and hallucinations and is still used to diminish depression and dispel fearful thoughts. Powdered and mixed with wax it has been prescribed for centuries as a salve in the treatment of scrofula (tuberculosis of the lymph system and joints). Modern practitioners also claim jet increases fertility and relieves problems with female reproductive organs

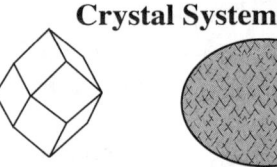

Lapis-lazuli

Crystal System

Isometric   *Grainy aggregate*

Color: Blue to blue-violet	**Streak:** White to light blue
Mohs' Hardness: 5 - 6	**Specific Gravity:** 2.50 - 3.00
Cleavage: None	**Fracture:** Conchoidal, grainy
Refractive Index: About 1.50	**Birefringence:** None
Dispersion: None	**Pleochroism:** None

Crystal System: Isometric; crystals rare, mostly dense, grainy aggregate
Chemical Composition: $Na_8(Al_6Si_6O_{24})S_2$ sulphur containing sodium aluminum silicate
Transparency: Opaque
Fluorescence: Strong; white

Lapis-lazuli is not a single mineral, but a conglomerate rock composed of lazurite, pyrite, and calcite. The lazurite lends the deep blue color and the pyrite adds the distinctive gold flecks found in genuine material. It is generally cut in cabochons, slabs, or carved. Although lapis is often thought of as being deep blue, the finest material has a slight violet cast.

One of the most precious commodities of the ancient world, lapis-lazuli was valued on a par with gold. In some ancient lists of loot taken from conquered foes lapis is listed ahead of gold or other precious stones. Known in ancient Babylon and Egypt, the source for the stone may be some of the same mines used today. The gem has been mined continuously for more than six thousand years at the Kakcha River in the Badakshan region of Afghanistan.

The modern name finds its origins in the two Latin words *lapis*, meaning stone, and *lazuli* meaning blue. The root of the word lazuli is an older form *allazward*, or the Arabic word *L'azulaus*, both meaning sky, heaven or blue. It is thought that the French word *azur* and the English form *azure* came from the same roots.

Some of the oldest writings concerning stones contain references to lapis-lazuli. *Puabi*, a Sumerian queen of the third millennium B.C. was, according to ancient records, the first queen to be dressed in gems and precious metals. Her royal robes were said to be of gold and silver studded with lapis-lazuli. Assyrian tablets of the second millennium B.C. mention the gem as an article of tribute. Descriptions of the garments worn by a Babylonian goddess as she prepared to descend into the underworld include the following; "She adorned herself with ornaments of lapis-lazuli and put on a necklace of great lapis-lazuli stones."[1] The goddess' Sumerian name was *Innini*. She was the daugher of the water god and may have worn the blue gem as a representation of water. Her Accadian name was *Ishtar*, goddess of love and sexuality. A poetic reference to the moon-god *Sin* also demonstrates the high status the stone *uknu*, its Assyrian name, was given by the ancients. Sin is described as, "the strong bull, great of horns, perfect in form, with long flowing beard, bright as lapis-lazuli."[2] The Sumerians believed that anyone who carried the blue gem with them as an amulet carried a god with them. The stone appears in a list of tributes paid by the Babylonians to the Egyptian pharaohs. The gem is also said to have been used by the Hebrews to adorn their ceremonial robes. Legend says a ring of lapis-lazuli, given by an angel of the Lord, was worn by King Solomon. The ring was supposed to have allowed him to control legions of demons which he used to build the temple.

The following symbol was used as the sign for lapis-lazuli in medieval alchemy texts.

The Egyptians used the gem extensively to make amulets, scarabs, and cylinder seals. The funerary masks of the pharaohs included inlaid lapis and the high priests wore lapis images of *Mat*, Goddess of Truth, around their necks. The importation of lapis-lazuli was an arduous task for the ancient Egyptians. Trips to retrieve the stone would last for years and include whole regiments of armed soldiers to escort the traders.

The Berlin Museum holds a papyrus from the about the fifteenth century B.C. which contains directions for the curative powers of three precious stones. Lapis-lazuli, malachite, and what is thought to be red jasper were worked as beads and strung to form a necklace for a child's wear. The cure was affected by the recitation of a specific formula over the child. This incantation called upon the disease to be drawn from the patient, through the beads, and into the air and water, or literally the creatures of the air and water. The text of the papyrus is included here as translated by Dr. Adolph Erman:

> O, ye beads! fall upon the haunches (of the . . .) in the flood; on the scales [?] of the fish in the stream; on the feathers of the birds in the heavens. Hasten forth! *nsw*, fall upon the earth. Let this text be recited over the beads [?], one of lapis-lazuli, the other of jasper [?], the other malachite, which are drawn on a string of . . . and hung upon the neck of a child.
> ([?] *denotes Erman's opinion that the preceeding word is of questionable trans-lation.*)

140
Lapidarium:

The word nsw is not translated by Erman but is thought to refer to a nasal discharge. The formula may possibly be a cure for the croup or other childhood respiratory disease.[3]

The first century Romans spoke of a gem called *sapphiros*. Pliny the Elder's description of the stone states, "Sapphiros contains spots like gold. It is sometimes blue, although sometimes, and indeed rarely, blue tinged with purple. It is never transparent." It is hard to argue that the stone he is describing is not our modern sapphire, but the stone we know as lapis-lazuli. This opaque blue gem was powdered and prescribed by the Greeks and Romans as a general tonic and an effective purgative.

In the Middle Ages lapis-lazuli served as a medicinal stone and as a source of pigment for illuminated manuscripts. Powdered and mixed with oils, the gem was the base of the intense blue paint known as ultramarine. The thirteenth century writer Albertus Magnus refers to lapis-lazuli as, "a blue stone with little golden spots,"[4] which served as a cure for depression and for the *quartern* fever, an intermittent fever which returned each third day. Chevalier Jean de Mandeville gives additional attributes in his fourteenth century *Le Lapidaire*. Mandeville states,

> Lapis-lazuli, the magnet stone, is found in parts of Armenia; it is a blue stone, slightly shiny and opaque, and it protects against illness that causes fainting as a result of a weak heart; it prevents conception if a man or woman carries it on themselves.[5]

Powdered lapis was prescribed as an astringent into the late nineteenth century. In King's *Natural History of Gems* the following description of its use is included. "It is likewise medicinal, for being powdered it heals the sore following pustules and boil is smeared over them, being applied mixed with milk to the ulcerations." The Chinese have used the stone for centuries to make a cosmetic used to paint eyebrows. Twentieth century authors have continued to praise the stone. Listed among its virtues are its abilities to shelter the wearer from physical danger and

Diderot's eighteenth century Encyclopedia *lists two symbols that were used in chemistry to represent lapis-lazuli.*

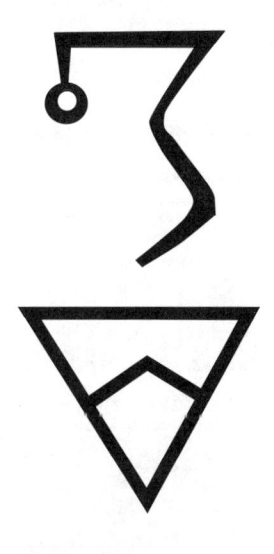

psychic attacks. It is still used to treat disorders of the throat, bone marrow, thymus, and immune system, and it is claimed that it prevents and rectifies RNA/DNA damage.[6]

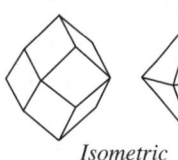

Lodestone

Crystal System

Isometric

Color: Black to brown **Mohs' Hardness:** 5.5 - 6.5 **Cleavage:** None **Refractive Index:** About 1.50 **Dispersion:** None	**Streak:** Black **Specific Gravity:** 5.20 **Fracture:** Octahedral parting **Birefringence:** None **Pleochroism:** None

Crystal System: Isometric, perfect octahedrons or dodecahedrons, or granular masses
Chemical Composition: $Fe^{+2}Fe^{+3}_2O_4$ iron oxide
Transparency: Opaque
Fluorescence: None

Lodestone (or loadstone) is magnetite, or magnetic iron ore, with the additional property of polarity. The stone is not considered a gem in modern times, but was classed with the other precious stones in antiquity.

    The Greek philosopher Plato tells us that the term *magnetis* was first connected to lodestone by the poet Euripides (480-405 B.C.). Prior to this time the common name had been *heraclean* stone. Both of these names had been applied through the stones association with two locations in Lydia, Magnesia and Herakleia, where the mineral was found. The Roman historian Pliny the Elder relates the legend of a young shepherd named Magnes who is credited in Roman lore with the discovery of the mineral. It is said that Magnes first noticed the stone on Mount Ida because the nails of his shoes clung to a particular piece of rock. He brought this magical stone to the city, where it was declared an important find and named for the young man.

Pliny also described a "black adamas" or black diamond; he said it was heavy, soft and able to be drilled. He also attributed to this stone the ability to attract iron. By this description we may conclude that this black adamas may have been a form of lodestone. The diamond and lodestone have long been associated with each other and have some common attributes. The mineral magnetite is known to crystallize in an octahedral form and may be the gray diamonds listed in the Indian caste system of diamond types. These "diamonds" are said to be one and one half times the weight of an ideal diamond. Lodestone has a specific gravity of 5.18, one-and-a-half that of the diamond.

Alexander the Great was said to have issued a piece of lodestone to each of his men as a guard against the spells of *jinns*, or evil spirits. Just as the stone guarded against evil, it was also supposed to attract good and soothe pain. In far eastern countries a piece of lodestone is placed in the throne of the sovereign. The magnetic influence would draw power, favor, and gifts. The mineral has long been connected with stability, good luck, health, strength, love, and sexual attraction. In several languages its name indicates its association with love. The French name for the stone is *aimant*, a participle of the verb *aimer*, to love. The Sanskrit name is *chumbaka*, the kisser, and in Chinese *t'su shi*, the loving stone.

The fifth century Roman poet Claudian tells the story of two statues which had an attractive power. One was of the god Mars which had been fashioned of iron, the other was a lodestone figure of Venus. In his text Claudian describes how at a special festival the two statues would be literally drawn to each other.

> The priests prepare a marriage feast
> Behold a marvel! Instant to her arms
> Her eager husband Cythereia charms;
> And ever mindful of her ancient fires,
> With amorous breath his martial breast inspires;
> Lifts the loved weight, close round his helmet twines
> Her loving arms, and close embraces joins,
> Drawn by the mystic influence from afar.
> Flies to the wedded gem, the God of War.
> The Magnet weds the Steel: the sacred rites
> Nature attends, and th' heavenly pair unites.[1]

> Orpheus, on the use of the stone as a test of faithfulness.
> *If e'er thou wish thy spouse's truth to prove,*
> *If pure she's kept her from adulterous love,*
> *Within thy bed unseen this stone bestow,*
> *Muttering a soothing spell in whispers low;*
> *Though wrapp'd in slumbers sound, if pure and chaste,*
> *She'll seek to strain thee to her loving breast:*
> *But if polluted by adultery found,*
> *Hurl'd from the couch she tumbles on the ground.*[2]

A fourteenth century English treatise by John of Trevisa described the use of lodestone as a "touchstone" for the virtue of one's wife, a custom which had come from as early as the fourth century. "Also magnes is in lyke wyse as adamas; yf it be sett under the heed of a chaste wyfe, it makyth her sodenly to beclyppe [embrace] her husbonde; & yf she be a spowse breker, she shall meve her oute of the bed sodenly by drede of fantasy."[3]

The Italian physician Camillus Leonardus gave an extensive account of the attributes of lodestone in his *Speculum Lapidum* of 1502. He writes;

> For being carried about one, it cures the cramp and gout. In the hour of travail, if held in the hand, it facilitates the birth. If bruised and taken with honey, by purging, it cures the dropsy. And being applied in the same manner, it affords relief to wounds from poisoned iron. The head being anointed with it, it cures baldness.[4]

Leonardus also describes two methods of coun-teracting the power of the magnetic stone. "It seems to contend with the diamond, for when diamond is put to it, it does not attract iron. Garlick likewise binds up its virtue."[5]

> Marbodus, on loadstone's use to evacuate a building for the profit of a thief.
> *If a sly thief slip through the palace door,*
> *And strew unseen hot embers on the floor,*
> *Then powder'd loadstone on these embers spread,*
> *The inmates flee, possess'd with sudden dread.*
> *Distraught with horrid fear of death they fly,*
> *Whilst from the square the vapour mounts on high,*
> *They fly: within the house no soul remains,*
> *And copious spoils repay the robber's pains.*[6]

Malachite

Crystal System

Monoclinic

Color: Light green alternating withgreen and black-green **Mohs' Hardness:** $3\frac{1}{2}$ - 4 **Cleavage:** Perfect **Refractive Index:** 1.660 - 1.910 **Dispersion:** None	**Streak:** Light green emerald **Specific Gravity:** 3.25 - 4.10 **Fracture:** Splintery, scaly **Birefringence:** 0.254 **Pleochroism:** Very strong; colorless, green

Crystal System: Monoclinic; small long prismatic crystals, aggregates with fine needles
Chemical Composition: $Cu_2((OH)_2CO_3)$ basic copper carbonate
Transparency: Opaque
Fluorescence: None

Malachite is a copper ore which serves as an attractive gem. The soft stone takes a high polish, but dulls quickly with wear. It is worked into beads, cabochons, and carved shapes. Large deposits in Russia have caused the stone to be used as an architectural treatment. The walls of rooms in the Imperial palaces are lined with malachite slabs, and some Russian Orthodox churches feature massive columns and alters fashioned of the gem.

Malachite was well known to the ancient Egyptians. They mined the stone in an area between Mt. Sinai and the Suez as early as 4000 B.C. The gem was carved into amulets and scarabs and was thought to be a protective stone for children. If tied to a cradle, all evil was kept away, and the child slept soundly. The image of the sun was engraved into the stone to protect the wearer from the spells of evil spirits. The sun serves as the natural enemy of demons and the spirits of darkness. It was also considered a defense against venomous creatures. Eyptian hieroglyphs describe malachite as an intregal part of their dieties existence. The stone is thought to have been the, "Green bed of the sun, 'Horus on his green' and 'the malachite lake(s)' in which the gods are sometimes said to dwell."[1] The modern name was granted by the Greeks and comes from their word *mallow*. They determined that the stone's deep green colors resembled the leaves of the mallow plant.

The symbol below appears in medieval alchemy texts to represent malachite.

The Middle Ages saw an expansion of the ancients' beliefs in the stone's power. As an amulet it would turn aside fainting spells, prevent hernia and prevent injury from falls. Powdered and mixed with milk it served as a tonic for colic and cardiac pain. Applied directly to the body, it relieved cramps and muscle aches. Mixed with honey and applied to a wound, it would stop the flow of blood.

The *Speculum Lapidum* of 1502 by the Italian physician Camillus Leonardus continues to advise of its value to new mothers. "The virtue of this stone is to defend infants from adverse casualties, and preserve the cradle from hurtful fancies."[2] Modern midwives in Bavaria still prescribe pieces of malachite strung on necklaces to relieve the pain of teething. Cut in a triangle, a Christian form, so that the banding in the stone forms an eye, it is still used in Italy as a talisman against the evil eye.

Modern gem therapists suggest the stone as a protection from newer hazards. It is worn as a guard against the radiation generated by florescent lights, computers, x-ray machines, and microwaves. It is also said to be helpful in the treatment of asthma, arthritis, swollen joints, tumors, and growths.

Moonstone

Crystal System

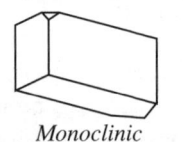

Monoclinic

Color: Colorless, yellow, pale sheen
Mohs' Hardness: 6 - 6 $\frac{1}{2}$
Cleavage: Perfect
Refractive Index: 1.518 - 1.530
Dispersion: 0.012

Streak: White
Specific Gravity: 2.56 - 2.57
Fracture: Uneven, conchoidal
Birefringence: 0.008-0.010
Pleochroism: None

Lapidarium:

Crystal System: Monoclinic; prismatic
Chemical Composition: K(AlSi$_3$O$_8$) potassium aluminum silicate
Transparency: Turbid, transparent
Fluorescence: Weak, olive green

The most prized and popular of the feldspars is the moonstone. The moving "cloud of light" effect of the stone, called *adularescence*, makes this stone stand out from other members of the feldspar family. This *schiller*, as it is called by mineralogists, is caused by tiny crystals of albite arranged in layers within the stone. The stones are usually cut as high-doomed cabochons to concentrate the sheen as a bright spot at the top. This phenomenon may be a band of light forming a distinct eye. These gems are called 'cat's-eye moonstones.' Other stones may exhibit a second band of light at right angles to the first, forming a cross.

The lore of the moonstone is limited. Besides its use as a talismanic cure for insomnia, the literature regarding the stone has generally been a retelling of a single attribute recorded by the Roman historian, Pliny the Elder. He states the stone was known for "shining with a yellow lustre from a colourless ground, and containing an image of the moon, 'which, if the story be true, daily waxes and wanes according to the state of the luminary'."[1] The writer Marbodus repeats the tale of the same phenomenon and adds that the ability of the stone to mimic the phases of the moon is a sacred sign. The sixteenth century writer, Antoine Mizauld, describes experiments he performed with a moonstone owned by Pope Leo X (1475-1521). He meticulously recorded the changing size and position of a point of light in the stone and concluded a direct correspondence existed to the lunar cycle.

Moonstone is a sacred stone in India and is thought to bring good fortune. The stone is only displayed in the gem markets against a yellow cloth, as yellow is considered a particularly sacred color to the Hindu. The moonstone has remained a prized gift for lovers. It is said to arouse passions and give lovers the power to see their future together.

The association of the stone with lunar cycles is said to make it a powerful talisman for women and their sexual organs. It is used to ease childbirth, enhance fertility, promote pregnancy, relieve menstrual discomfort and increase lactation. The stone serves modern gem therapy practitioners as a tool to enhance feminine energy and help to heal emotional distress. Along with its abilities to bring calmness, it is used to treat degenerative conditions of the skin, hair, eyes and fleshy organs.

Obsidian

Crystal System

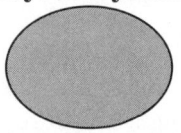

Amorphous

Color: Black, gray, brown, green **Mohs' Hardness:** 5 - 5 $^1/_2$ **Cleavage:** None **Refractive Index:** 1.48 - 1.52 **Dispersion:** None	**Streak:** White **Specific Gravity:** 2.35 - 2.46 **Fracture:** Large conchoidal, sharp edged **Birefringence:** None **Pleochroism:** None

Crystal System: Amorphous
Chemical Composition: Volcanic, amorphous, siliceous glass
Transparency: Opaque, translucent.
Fluorescence: None

A number of natural glasses have been use as gems; the most common is obsidian. Material occurs in a range of colors but is most commonly found in semi-transparent black to an opaque brown. Formed by the rapid cooling of volcanic magmas, it is found in many countries and ranges in size from small pea-sized pebbles to thick slabs which measure hundreds of square feet. Collectors distinguish many varieties: banded obsidian (similar in pattern to agate), onyx obsidian (containing parallel bands), snowflake or flowering obsidian (included with white patches of crystallized silica) and sheen obsidian (displays an iridescence due to minute inclusions). A brown to black obsidian found in Hungary is known to the local people as *Tokay Lux-sapphires*.

The Romans called this natural glass *obsianus*. The name was derived from Obsius, a Roman citizen who discovered a similar stone in Ethiopia. Pliny the Elder referred to it as *obsidianus*, a name which was eventually shortened to its modern form.

Carved and polished slabs and spheres have served as scrying stones throughout history. Obsidian mirrors have been found in archaeological digs in Central and South America. The black mirrors seemed to smoke and then clear when looked into. The surface was said to reflect one's flaws and reveal a clear image of needed changes. The gem served as the symbol of the black *Tezcatlipoca* in the Aztec religion.[1]

The Cakchiquel Indians in Central America call the stone *chay abah*. The writer Fuentes y Guzman gives an account of the place where the tribe's judges heard pleas and pronounced sentences. "On the summit of a small hill overlooking the town—is a circular wall, not unlike the curb of a well, about a fathom in height. The floor within is paved with cement, as the city streets. In the center is placed a socle or pedestal of a glittering substance like glass."[2] This account describes a communal fetish or talisman of obsidian.

Native Americans in California were known to use obsidian knives to ritually scar young boys and girls as a form of initiation. Priests of the same tribes also used obsidian "claws" to slash people as a representation of the bear spirit.

Modern practitioners of gem therapy list numerous attributes for obsidian. Most of the qualities deal with focusing energy and making clear a path of action. They are said to be helpful in healing by clarifying the ailment and cure. Obsidian is thought to be particularly helpful in dealing with digestive and intestinal disorders.

Other qualities are connected to particular forms of obsidian. Those most often mentioned in contemporary literature are listed.

Black transparent obsidian (Apache tears)— the owner will never cry in sorrow and have comfort in times of grief. Used to calm and balance emotions.

Fire obsidian— opens intuition and higher psychic awareness with firm grounding and balance.

Mahogany obsidian— good for stomach and intestinal disorders; an excellent healer on all levels as it tends to absorb negative energy; the most effective obsidian for treating children.

Rainbow obsidian— containing all colors of the rainbow, it is said to balance all energy centers of the body on a subtle level.

Sheen obsidian— acts as a good meditation tool which helps to focus and concentrate abilities.

Snowflake obsidian— helps to bring in more spirit energy; acts as a "grounding" stone and as an excellent healing agent.

The Legend of the Apache Tear: Southwestern United States

United States Army Cavalry troops were in pursuit of a band of Apache warriors. Expecting the Apache to lead them to their village, the troops followed but did not overtake them. Aware of the soldiers, the Apache led the troops away from their camp. When the cavalry realized they were being led aimlessly, they attempted to capture the men of the tribe and force them to reveal the camp's location. Rather than surrender and betray their women and children, the Apache warriors chose to take their own lives by plunging off a cliff. By diverting the troops, the women and children of the tribe were spared. When the grieving women gathered the broken bodies of their warriors, a voice in the breeze whispered, "Their bitter tears shall be turned to beautiful stones, for I should not have made these cliffs so high." The tears shed in mourning over the loss of their brave husbands fell to the ground and turned to stone. These "stone tears" may still be found throughout the American southwest, serving as a reminder of the warriors' selfless sacrifice.

> *On the use of obsidian for scrying by the famous nineteenth century magician Kelly.*
>
> *Kelly did all his feats upon*
> *The Devil's looking-glass, a stone;*
> *Where, playing with him at bopeep,*
> *He solved all problems ne'er so deep.*[3]

Onyx

Crystal System

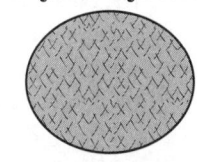

Microcrystalline aggregate

Color: Various, banded or layered **Mohs' Hardness:** $6\frac{1}{2}$ - 7 **Cleavage:** None **Refractive Index:** 1.544 - 1.553 **Dispersion:** None	**Streak:** White **Specific Gravity:** 2.55 - 2.65 **Fracture:** Uneven **Birefringence:** 0.009 **Pleochroism:** None

Crystal System: Hexagonal (trigonal) microcrystaline aggregates
Chemical Composition: SiO_2 silicon dioxide
Transparency: Translucent to opaque
Fluorescence: Varies with band; partly strong, yellow, blue-white

The term onyx is properly applied to chalcedony with straight parallel bands. Commercially, the name is improperly used to refer to grayish chalcedony or agate dyed to produce a solid colored material. Black onyx, blue onyx, green onyx, etc., are all examples of the use of this misnomer. If the bands contain sard mixed with either white or black, the stone is called sardonyx; with bands of carnelian, it is called carnelian onyx.

The name onyx is Roman and originally referred to the fingernails. The name was applied to the stone since it appeared to share a similar luster to polished nails. In Hebrew the same word signifies "shining or gleaming stones."[1]

Historically onyx has been associated with negative attributes. Lapidaries of the Middle Ages stated that wearing the stone exposed its owner to demons, ugly dreams, and lawsuits. Chevalier Jean de Mandeville advises in his fourteenth century *Lapidaire*:

> Onyx is a stone with white or red veins. It has bad
> virtues because he who wears it at the neck or on

the back will see devils; it causes many frivolities, it brings anger and disagreement, but it gives hardiness. If one hangs it at the neck of an infant, it will increase its saliva. It is found in India, Libya, and in Arabia.[2]

To counteract these ill effects sard was always to be worn with any onyx setting. It was said that if worn on the neck it would cool desire and passion in lovers. The sixteenth century Italian writer Girolamo Cardano reported that the stone was used for that purpose in India.[3] It was also said to be used in other parts of the world to provoke discord and separate lovers. The juxtaposition of opposites in one substance seemed to suggest separation within a unit. Brides and grooms were cautioned against wearing this gem as a divorce would soon follow.

The Italian physician Camillus Leonardus gave both positive and negative attributes for the gem in his sixteenth century lapidary:

This stone represents many horrible things in sleep. He who carries it about him stirs up quarrels and contentions. It increases spittle in children, and hastens a birth... Being hung about the neck of one who has the epilepsy, it prevents his falling.[4]

One fabled onyx was used as a protective and healing stone. This fist-sized gem is said to have been part of a the now lost silver reliquary in the Abbey of St. Alban. The shrine was commissioned by the Abbot Geoffrey de Gorham in 1010 to hold the remains of St. Alban. This large onyx was left unset so it could be taken to the homes of women in labor. If held by them, it would relieve the pains of childbirth. The stone is described as being engraved with the figure of Esculapiu, ancient god of healing, on one side and the figure of "a boy bearing a buckler" on the other.[5] This kind of engraving was unknown in tenth century Europe and suggests the stone was an earlier pagan amulet placed in the shrine and given Christian meaning. The stone was also engraved with the word *Kaadman*. This may have been a corruption of the word *cadmeus* or *cameus*, earlier Latin forms of the modern word cameo.

Contemporary authors list very different attributes for onyx. The stone is reported to bring increased vigor, strength, stamina, andself control. It is also used to alleviate worry, tension, and nervousness, and it acts to eliminate confusion and nightmares. It "can be used to provide glimpses of that which is 'beyond' and activate memories of one's roots and reality."[6] Claimed medicinal applications include the treatment of bone marrow diseases, foot problems, and soft tissue disorders.

The Myth of The Origin of Onyx: Greek

The goddess Venus was resting on the banks of the Indus river. As she slept, Cupid used the point of one of his enchanted arrows to give her a manicure. The parings of her nails fell into the waters of the sacred river. Since the nails were of heavenly origin, they sank to the river bottom and were metamorphosed into onyx.

"When nature had finished painting flowers, coloring the rainbows, and dyeing the plumage of the birds, she swept the colors from her palette and molded them into opals." by the artist Du Ble

Crystal System

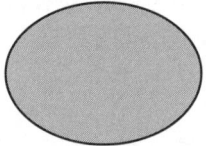

Amorphous

Opal

Color: All colors
Mohs' Hardness: $5\frac{1}{2} - 6\frac{1}{2}$
Cleavage: None
Refractive Index: 1.37 - 1.47
Dispersion: None

Streak: White
Specific Gravity: 1.25 - 2.23
Fracture: Conchoidal, splintery, brittle
Birefringence: None
Pleochroism: None

Crystal System: Amorphous; kidney or grape-shaped aggregates
Chemical Composition: $SiO_2 \cdot nH_2O$ hydrous silicon dioxide
Transparency: Translucent to opaque
Fluorescence: White: white, bluish. brownish, greenish, Black; usually none, Fire; greenish to brown

Opal is primarily silicon oxide and is therefore closely related to chalcedony. While chalcedony is only silica, opal also contains a small amount of water. What makes the gem unique is its play of color known as "fire." Most opal is a dull gray or yellowish material which is without fire, called "common opal." Common opal is often waxy in appearance and has little or no gem value.
Gemologists distinguish four types of gem opal:
White opal— opaque white material with flashes and speckles of rainbow colors. This is the opal most commonly seen in jewelry.

> Black opal— the same flashes of color, but the body color is dark gray to black. This gem material is rare and costly.
> Fire opal— a transparent or translucent stone which displays a red or orange body color, and may or may not display fire.
> Water opal— contains brilliant flashes of color suspended in a transparent and colorless stone.

The birthstone of October, opal displays all the colors of autumn. The name originated with the Latin *opalus*, from the Greek word *opalios*, which was derived from the Sanskrit *upala*, meaning precious stone. According to the author Orpheus, "On Olympus the opalios, the delight of Immortals, so fair to view that it charmed the strong eye and strengthened the weak."[1] He continued in his praise, it "fills the hearts of the gods with delight."

Known as the "cupid stone" to the ancient Romans, the *cupid paederos* or "child beautiful as love," served as a symbol of hope and purity. The historian Pliny the Elder confined almost all of his descriptions of gems to form, color and data regarding popularity. In the case of the *opalus* he writes,

> There exists today a gem of this kind, on account of which the senator Nonius was proscribed by Antony. Seeking safety in flight, he took with him of all his possessions this ring alone, which it is certain, was valued at 2,000,000 sesterces (valued at $80,000 in 1913 by G. F. Kunz)." He further states that this opal was "as large as a hazel-nut.[2]

An opal of such size would be a truly historic stone. Many have doubted Pliny's reference is to opal, but it is unlikely his description could be of any other gem.

> Of all the precious stones, it is the opal that presents the greatest difficulty of description, in displaying at once the piercing hue of carbunculus, the purple brilliancy of amethystos, and the sea green of smaragdos, the whole blended together and refulgent with a brightness that is incredible.[3]

The only other stone which could qualify is an iris quartz with internal fractures, which also produces a rainbow of color. The places listed by Pliny as sources of opal show no sign of the stone today but do produce quartz gems.

An ancient legend relates a story of the creation of opal. The God of Storms became jealous of the Rainbow God and, in a rage, shattered the rainbow. As the pieces fell to Earth, they were petrified and became opals. Arab tradition says that opals are the remnants of lightning strikes to the ground and the flashes in the stone are the captured lightning. The gem was also used by Native Americans and Australian aboriginal shamen to invoke visions and allow them to travel in dream quests.

The twelfth century scholar Albertus Magnus provided a description of a stone which was most certainly an opal. He speaks of an *orphanus*, a gem set in the imperial crown of the Holy Roman Empire.

> The orphanus is a stone which is in the crown of the Roman Emperor, and none like it has ever been seen; for this very reason it is called orphanus. It is of a subtle vinous tinge, and its hue is as though pure white snow flashed and sparkled with the color of bright, ruddy wine, and was overcome by this radiance. It is a translucent stone, and there is a tradition that formerly it shone in the night-time; but now, in our age, it does not sparkle in the dark. It is said to guard the regal honor.[4]

The opal has long been associated with eyes or vision. Lapidaries of the Middle Ages listed the gem as an *ophthalmios*, or "eye stone." If a Russian saw an opal mixed with other goods for sale, they would buy nothing the rest of the day, as opal was considered a form of the "evil eye." A Norse myth refers to a sacred stone of opal-like description called the *yarkastein*. This gem was formed by a clever smith Volömer (the Scandinavian name for Vulcan) in a gruesome way. Volömer would form the gems from children's eyes he had stolen. Opal was often given as a cure for diseases of the eye.

Marbodus writes that the gem "conferred the gift of invisibility on the wearer, enabling him to steal by day without risk of exposure to the baneful dews of night."[5] Albertus Magnus writes of the *ophthalmus lapis* and calls it the *patronus furum* or "patron of thieves." This attribute may have originated with the story of Gyges, originally told by Plato in his *Republic*. The fable tells of a shepherd who discovers a magical ring which makes the wearer invisible. The original telling does not say what stone was set in the ring, but later historians claim it was an opal. Stories of invisibility may be based on the fact that a variety of opal which displays no play of color, *hydrophane*, actually disappears in water. When dry it is a milky white, but when immersed, it slowly takes on water and gradually fades from sight. Its refractive index is so close to water's that it cannot be seen in the liquid.

Later writers repeated these early virtues and added others. The tradition of color determining attribute continued. Since opal contains so many colors it was often given the characteristics of many other gems. Cardano recorded his purchase of an opal and the pleasure it brought him. He wrote that a diamond cost thirty times the price of his opal, but they were equal in value to him. One of the earliest references to the stone in English writing was made by Dr. Stephen Batman in the late sixteenth century:

> Optallio is called Oppalus also and is a stone distinguished with colors of diverse precious stones, as *Isid* saith. . . . This stone breedeth only in *Inde* and is deemed to have as many virtues, as hiewes and colours. Of this *Optallius* it is said in Lapidario that this *Optallius* keepeth and saveth his eyen that beareth it, cleere and sharp and without griefe, and dimmeth other men's eyen that be about, with a maner clowde, and smiteth them with a maner blindnesse, that is called *Amentia*, so that they may not see neither—take heede what is done before their eyen. Therefore it is said that it is the most sure patron of theeves.[6]

The writings of the sixteenth century Italian author Giovanni Porta list additional traits for opal. He states, "It stimulates the heart, preserves from contagious and infected airs, drives away despondency, prevents fainting, heart disease, and malignant affections."[7]

> *October's child is full of woe,*
> *And life's vicissitudes must know;*
> *But lay an Opal on her breast,*
> *And hope will lull her woes to rest.*[8]
>
> An alternative for the last two lines reads,
>
> *When fair October to her brings the Opal*
> *No longer need she fear misfortune's perils.*[9]

Recent history has seen the opal stigmatized. It has been labeled a stone of "ill omen" or "bad luck." Some of this nineteenth century belief may have been fostered by the jewelry industry. It was about this time that public began holding jewelers liable for damage done to goods left for repair or manufacture. Opals can fracture easily and without warning, even with great care. Jewelers may have decided it was better to discourage opal wear than to risk the liability of a damaged stone.

Another origin of the superstition may have been the publication of Sir Walter Scott's early nineteenth century romantic novel *Anne of Geierstein*. The story begins in 1474 with the story of an enchanted princess, Hermoine. She had been created from a magic lamp which vanished when she appeared. She was said to be the granddaughter of a Persian shaman, and the daughter of a magician. The story tells of her marriage to a wealthy baron and the birth of their daughter. At the christening of the child, Hermoine wore a brilliant opal in her hair. The stone was said to mimic the mood of its owner. It would sparkle with happiness and shoot red fire when the wearer was angry. In the course of the christening, holy water splashed on the gem and it turned to a lifeless colorless pebble. At the same moment Hermoine fell to the floor in a swoon and was taken to her room to recover. All that night a storm raged and lightning shown about the castle. The next day her husband came to wake her and found a pile of ashes on the bed. All that was left in the ashes was the lifeless opal. From this simple tale the rumor started that opals brought death and tragedy to their owners. It is believed Scott chose an opal simply because of its supposed ability to project the mood of its owner. His novel may also inspired the following passages from the nineteenth century artist and critic John Ruskin,

> The opal in its native rock presents the most lovely colors that can be seen in the world, except in the clouds. . .
> If fire is applied to the stone it immediately disappears, leaving only a puff of smoke. . .
> Health conditions of those who wear the stone will affect its color, and brilliancy.[10]

Queen Victoria attempted to revive interest in the opal. She wore many pieces of jewelry set with beautiful opals as a young woman. Her efforts may have backfired due to an embarrassing accident at her coronation. The gown she wore was fastened in the back with an opal-set broach. During the processional, the pin unfastened, revealing more of the new Queen than was considered acceptable. Many said it was only further proof that the gem was unlucky.

The end of the nineteenth century brought what seemed to be more evidence of royalty plagued by the misfortune of an opal. Alphonso XII of Spain gave his queen an opal on their wedding day. Soon after their betrothal, the queen mother, the king's sister, and his sister-in-law each died. The King then wore the "fatal gem" and met the same fate as the female members of his family. After all these disastrous deaths, the widowed queen Christina attached the stone to a gold necklace and placed it on the statue of the Virgin of Alumbera, patron saint of Madrid. The necklace is reported to still be around the statue's neck. (It should be noted that all these deaths occurred during the summer and autumn of 1885, the same time as the great cholera plague of Spain that killed over 100,000.)

Miners in the Lightning Ridge region of New South Wales, Australia probably added to the superstition. They told a tale of 'night stones.' These were reportedly opals which glowed in the dark with such intensity that they could make an impression on photographic film. To collect the gems the miners would descend into treacherous mines, prone to cave-in. The vast majority of these fatal accidents would occur between midnight and one in the morning. New prospectors were warned to be out before the witching hour. The curse of the mines in the dead of night was blamed on a spell cast by the glowing gems.

The English gem authority Edwin Streeter never visited the opal mines of Australia, but he reported a curious fact about their harvesting. He states, "When first taken out of the earth, it (opal) is soft, but it hardens by exposure to the air."[11]

Additional stories based in Queensland, Australia are told in the book *Opals and Agates*. It chronicles the local beliefs surrounding the gem, one of which is," An opal of mammoth size rules the stars in their courses, presides over destiny of the gold in the mine, guides the love of women, and dictates the shedding of human blood."[12]

The Mystic Ring of Gyges: Greek

Gyges was a poor shepherd in the service of the King of Lydia. He found his romantic life unsatisfying at home so he left in search of a woman. During his journey he found himself in the midst of a violent earthquake. Gyges came to the brink of a newly created chasm of considerable depth. Being an inquisitive man, he entered the chasm's depths and was amazed to find a hollow brazen horse. It had small openings in its side which allowed Gyges to peer inside. Through one of the openings, he saw the body of a man and on the man's finger was a glittering opal ring. He entered and removed the ring, slipped it on his finger, hurried out of the chasm, and returned home.

It was the custom in Lydia that once a month the king's shepherds would report to their monarch. The shepherds assembled to prepare their reports and tell stories to one another. Gyges, anxious to relate his good fortune, had readied his reports ahead of time. Waiting for the others to finish their work and listen to his story, Gyges sat and nervously twisted his newly found ring on his finger. As he turned the setting of the ring toward his hand a strange thing happened. The other shepherds looked in his direction and exclaimed that he had disappeared. Surprised, he reversed the position of the ring and was welcomed back by his colleagues.

Returning home, Gyges experimented with his new possession and became convinced that he now owned a wondrous ring. He knew that he could carry out his romantic ambitions. At the next gathering of the shepherds, he asked to be the one to bring the reports to the king. When he arrived at the palace, he ignored the guard's directions to the king's offices and went instead to the queen's rooms. Before he knocked on the lady's door, he turned the setting inward and became invisible. He entered the queen's chambers, made love to her, and then proceeded to murder the king. Impressed by his romantic prowess, the queen married Gyges and placed him on the throne as the new king. From that time on, the two reigned happily ever after.

The Opal: a dialogue from *Gems of Beauty* by the English writer Countess of Blessington, 1836

Mother: Come, let me place a charm upon thy brow, and may good spirits grant, that never care approach, to trace a single furrow there!
Daughter: Thy love, my mother, better far than charm, shall shield thy child–and yet this wondrous gem looks as though some strange influence it had won from the bright skies–for every rainbow hue shoots quivering through its depth in changeful gleams, like the mild lightnings of a summer eve.
Mother: Even so doth love pervade a mother's heart; thus, ever active, looks through her fond eyes.[13]

Pearl

Crystal System

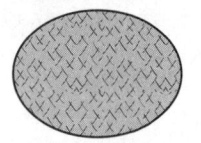

Microcrystalline aggregates

Color: Pink, silver, cream, blue, golden, green, black with possible overtones of the same
Mohs' Hardness: 3 - 4
Cleavage: None
Refractive Index: 1.52 - 1.66 (Black 1.53 - 1.69)

Streak: White
Specific Gravity: 2.66 - 2.85
Fracture: Uneven or scaly
Birefringence: Weak -.156
Pleochroism: None
Dispersion: None

Crystal System: Microcrystalline
Chemical Composition: 84-92% calcium-carbonate, 4-13% organic substances, 3-4% water
Transparency: Translucent to opaque
Fluorescence: Weak; unusable, Natural Black; red-reddish

A pearl is the response of a mollusk to irritating foreign material within its shell. Mollusks secrete mineral deposits around the offending particle to protect their delicate tissues. If this concretion includes an outer layer which contains nacre, the pearl will exhibit the luster which makes it so highly valued. Saltwater pearls are generally found in an inedible oyster known as the *Pinctada*. Freshwater pearls are found in mussels called *Unio*. Many other mollusks produce protective secretions, but few display the iridescence associated with gem pearls. This unique spot of reflected light is known as the pearl's "orient." The exterior or nacre will show a color known as overtone, displayed on the color of the body or background. The overtone is seen in reflected light as a luster coming from the surface of the pearl.

> Body color is divided into white, black, and colored, including red, purple, violet, blue, green, or yellow. The overtones exhibited include green, yellow, orange, pink, and purple. Black pearls display grays, bronze, dark blue, blue-green, and green, as well as a range of metallics. White pearls show creams, light rose, cream-rose, or so-called "fancy" colors. Fancies will display three colors: cream, rose, and blue or green overtones.

Birthstone of June and valued as the stone of serenity, virtue, and purity, the pearl has maintained its status as a prized gem throughout history. The Egyptians, Persians, Jews, and Hindus all prized the pearl and wrote of its value and lore.

Pearls were one of the gems most revered by the Romans. They were a favorite of both men and women. Mysterious in origin, yet readily available, pearls were displayed by Romans in a variety of ways. Ladies' gowns and head-wear, men's robes, and even pieces of furniture would be covered in their entirety to create as obvious a display of wealth as possible. The Emperor Caligula is said to have worn slippers made of pearls. It was said that noble ladies wore pearls to bed so they would be reminded of their wealth when they awoke. The pearl was believed to be the favorite jewel of Venus and was thought to be necessary for love potions. Dissolved in acid it would be added to wine and presented to one's object of affection. Romans also bedecked their hair with the gem to ensure success in love.

Pliny the Elder relates a legend of the King of Pearls. This monarch was distinguished by his size, age, and ability to protect his spherical subjects. The king was protected by a bodyguard of sharks in his kingdom of rocky shoals. If a diver should happen to capture a pearl, the king would kill any future divers who invaded the same location. Pliny concluded the story by lamenting that despite the noble's efforts, his charges continued to be sacrificed to adorn ladies' ears.

Pliny passed along the contemporary belief of how pearls were formed. He states that the "dews of heaven" would fall into the sea and be captured by oysters at their breeding time. He further explains that the quality of pearls varied, depending on the quantity and quality of the dew collected and the weather conditions present. If the dew was pure, a lustrous pearl resulted; if the dew was fouled, the product would be dull. Cloudy skies during this rain of dew would spoil the color; lightning would retard pearl formation, and thunder caused the mollusks to miscarry these treasures.[1] His *Natural History* also contains medicinal attributes for the gem. The affinity between pearls and their owners was said to be demonstrated as an indicator of health. The clouding of pearls showed the owner's ill health, while a loss of luster would be seen when the owner died. If ingested at the first sign of illness, they would purify the system and drive away all maladies.[2]

Tales of the conversion of dew or rain into pearls is found in many successive pieces of literature. Oriental poets state that on the twenty-fourth of the month, the day known as *Nisan*, pearl oysters rise from the sea and open their shells to receive a precious rain. The poet Sadi in his *Bostau* states,

> A drop of water fell one day from a cloud into the sea. Ashamed and confused at finding itself in such an immensity of water, it exclaimed, "What am I in comparison of this vast ocean? My existence is less than nothing in this boundless abyss!" While it thus discoursed of itself, a pearl-shell received it in its bosom, and fortune so favoured it that it became a magnificent and precious pearl, worthy of adorning the diadem of kings. Thus was its humility the cause of its elevation, and by annihilating itself, it merited exaltation.[3]

The scholar Benjamin of Tudela adds to the story that on the sixteenth of the next month, the day known as *Tisri*, "two men being let down into the sea by ropes onto the bottom, bring up certain creeping worms, which they have gathered, into the open air, out of which these stones are taken."[4]

Soldiers of the Middle Ages, returning from the Crusades, spread an appreciation for pearls throughout Europe. The thirteenth century writer Alfonso X added medicinal attributes to its value.

> The pearl is most excellent in the medical art, for it is a great help in palpitation of the heart, and those

---

Then too the pearl from out its shell unsightly, in the sunless sea,
(As 't were a spirit, forced to dwell in form unlovely) was set free,
And round the neck of woman threw a light it lent and borrowed too.[5]
Loves of the Angels; Second Angel's Story: *the poet Sir Thomas Moore*

who are sad or timid are helped in every sickness which is caused by melancholia, because it purifies the blood, clears it, and removes all impurities.[6]

The *Lapidaire* of Chevalier Jean de Mandeville makes reference to the pearl's structure. "Pearls are a stone called onion, because they are of many layers, they have several surfaces, one after the other, like an onion." He also explains the color variations found.

> It is condensed into gravel on sea shores and rivers, in dew from the sky, as is the diamond and it is made from this dew, if it is condensed in the morning and it comes out pure and light, then therefore it is a white pearl, but if the dew is condensed at vespers (in the evening) the pearls are not pure and they are murky and badly colored.[7]

The Italian physician Camillus Leonardus repeated the story of the pearl's creation from dew in his *Speculum Lapidum* of 1502:

> Margarita, or Pearl, has the first place among white gems, generated by celestial dew in some sea shell-fish, as is held by authors. The shell-fish, its reported, early in a morning, at a certain season of the year, leaves the bottom of the sea to draw in the air, of which pearls are generated; and according to the clearness of the air taken in, pearls are either lucid or muddy.[8]

Leonardus also gives an explanation of why some pearls have holes and some not.

> Perforated and unperforated and perforated by art. That the pearls which are perforated, are these which have lain a long while in the shells, and being quite ripe are spew'd out into the sea, where by a long stay and a perfect ripeness, they are perforated. Those

> *See these pearls that long have slept;*
> *These were tears by Naiads wept.*
> *For the loss of Marinel.*
> *Tritons in the silver shell*
> *Treasured them, till hard and white*
> *As the teeth a Amphitrite.—*[9]
> The Chorus Song of the Second Maiden
> Bridal of Triermain: *Sir Walter Scott*

which are not perforated are more styptick than the perforated." Adding to medical lore he continues, "When boiled in meat they cure the quarten ague [a four day flue]; bruised with milk they heal putrid ulcers; and being so taken wonderfully clear the voice. . . If taken with sugar, they yield help in pestilential fevers; and render him who carries them chaste.[10]

The Renaissance saw a lavish use of the gem. They became so popular that one period of European history is referred to as the Pearl Age. Members of royalty tried to out do each other in their display of the gem. Some royal families even tried to monopolize their use by making decrees that the wearing of pearls was to be exclusive to their family. In 1530 the Augsburg family permitted each noble lady to own four silk dresses, but banned the use of pearls by anyone but their family members. In 1612, a law from the Duke of Saxony read:

> The nobility are not allowed to wear any dresses of gold or silver or those garnished with pearls; neither shall professors and doctors of the universities nor their wives wear any gold, silver or pearls for fringes, or any chains of pearls, or caps, neck ornaments, shoes, slipper, shawls, pins, etc., embellished with gold or silver or pearls.[11]

The use of the pearl to represent conspicuous consumption has a literal history dating to ancient Egypt. Cleopatra is said to have drunk a toast to Mark Anthony containing a dissolved pearl. It is said her intention was to demonstrate the wealth of Egypt to her Roman suitor. The pearl was said to be more costly than the rest of the entire feast held in his honor. Sir Thomas Gresham repeated the gesture centuries later by toasting Queen Elizabeth I with a dissolved pearl at a city banquet. An outraged Thomas Heywood commented, "Here £15,000 in one clap goes. Instead of sugar, Gresham drinks the pearl unto his queen and mistress."[12]

The virtues of pearls have been enumerated in many cultures. The Chinese believe that pearls are liquid globules expelled from the mouth of the sky dragon or rain god which petrify as they reach the sea. They hold that pearls lose their luster with time by a factor of one hundredth with each passing year. Buddhist tradition states that the third eye of Buddha is a pearl and the essence of wisdom. Buddhist reliquaries in India must be adorned with highly esteemed red pearls as they are one of the seven precious objects in Buddhism. Hindus value a special class of pearls as a powerful amulet, these are the "pearls" retrieved from the forehead, brain, or stomach of elephants.

Monarchs in the ancient East were anointed with gold-dust and seed pearls at their coronations to ensure wealth and wisdom. Moslems are told in the Koran that the faithful will spend a blissful eternity in the arms of a "houri," secreted within a pearl. Christian lore contains a number of symbolisms in the gem. As a precious gem which is found in a rough outer shell, it symbolizes the Christian sole residing within a rude human body. As a symbol of purity, it may represent Christ or particularly the Virgin Mary. A name which has been closely connected with the pearl is Margaret (or Margarita, as found in Leonardus' writings), a named derived from the Persian *Murwari*, meaning pearl or "child of light." The name was popular with early Christians and was used to suggest purity, simplicity and beauty. The virgin martyr, St. Margaret of Antioch. . . .who, before the fifth century, was the embodiment of feminine innocence and faith overcoming evil, and also is often represented wearing a string of pearls,"[13] is held up as a living example of the purity with which pearls are associated. References to the pearl in scriptures has served to reinforce its Christian symbolism. The Apostle John writes in the *Book of Revelations* that the gates of the New Jerusalem consist of mammoth pearls. From this reference we find the common notion that heaven is guarded by "pearly gates."

Mystical properties have been assigned to the pearl throughout the centuries. As a birthstone it gives the wearer the ability to overcome irritations and affords a cure for fevers. Astrologically, the pearl governs the sign of Cancer and does harm if worn by those of the signs Aries and Virgo. To dream of pearls is symbolic of sorrow and disillusionment. A dream which contains the image of a broken pearl necklace foretells a broken and sorrowful life. Ladies in mourning in the nineteenth century wore pearls in their mourning rings. In Asia and Europe a gift of pearls will bring sorrow and tears unless received as part of an ancestral inheritance. Pope Adrian wore many amulets to help keep him virtuous,

> *And those pearls of dew she wears*
> *Prove to be persaging tears*
> *Which the sad morn and let fall*
> *On her hast'ning funerall.*[14]
> An Epitaph on the Marchioness of Winchester: *John Milton*

including a sun-baked toad with pearls. The thirteenth century explorer Marco Polo states that on the island of Chipanga, "The inhabitants have pearls in abundance, which are of a rose color. When a dead body is burnt, they put one of these pearls in the mouth."[15] Mandeville writes in his chapter on the "marguerite" of the many uses of the gem to treat emotional and bodily ailments.

According to Diderot's Encyclopedia this symbol stood for pearl to chemists of the eighteenth century.

> Pearls pacify anger and melancholy. They give joy and serve to comfort the heart, the liver, swellings of the stomach and against inflammations of the blood and other humors. Pearls help the sight and obtain peace and agreement and they bring thought and good memory to the person who wears them.[16]

Modern mystics believe the pearl signifies faith, charity and innocence, that it enhances integrity and helps focus attention. Powdered pearls are used to treat digestive disorders, relieve biliousness and bloating, treat soft tissue injuries, and increase fertility.

The Origin of Pearls and Their Virtues: India

A story of the origin of the pearl is given in Indian Sanskrit texts. Vala, a demon god, was slain and dismembered by the demigods. The parts of his body were strewn about the earth and universe to create the various gemstones we know today. The teeth of the great demon were said to have been scattered throughout the celestial regions. They fell like stars into the oceans below and became the seeds for the various species of pearls. Possessing the luster of moonbeams, the seeds entered into the shells of oysters lying within the oceans depths and grew to be pearls.

These same Vedic texts give extensive information concerning the source of pearls and their varied powers. Besides oyster pearls, seven

> *Fish wept their eyes of pearl quite out,*
> *And swam blind after.*[17]
> *anonamous English commentary on the extravagant use of pearls in Elizabethan times.*

other types of pearls are listed; including those found in conch shells, recovered from boar's heads, elephant heads, king cobras' hoods, bamboo stems, clouds, and fish heads.

Pearls from the cobra's hood are highly valued. Anyone possessing such a treasure attains piety, good fortune and, eventually, becomes an illustrious leader of men. If a man obtains a snake pearl, he should have the rite of installation performed by a priest to ritually install the gem into

A pearl dealer sorts and grades pearls for wear and medicinal use in an illustration from the Ortus Sanitatis *of 1483 by Johannis de Cuba.*

the owner's home. This man shall never be troubled by snakes, demonic beings, diseases, or any form of disturbance.

Cloud pearls are considered to be a natural by-product of the sun that can illuminate the sky in all directions and dispel the darkness of cloudy days. A cloud pearl is considered so priceless that the entire earth would not be equal in value. Cloud pearls rarely reach the earth, because they are usually taken away by the demigods. A man who possess a cloud pearl would bring good fortune to the entire earth, and no evil could touch the land within 8000 miles of his birthplace.

Ningyo and The Origin of Pearls: Japanese

A myth of a sea-goddess, whose tears are the source of pearls, is related in *The Mythology of All Races*:

> A denizen of the sea, is Ningyo, The Fisher-woman. Her head is that of a woman with long hair, but her body is that of a fish. This mermaid-like creature often appears to human beings in order to give them advice or warning. Pearls are said to be her tears, and according to one tale, a fisherman who caught her in his net but set her free, received her tears as a reward which filled a casket with pearls.[18]

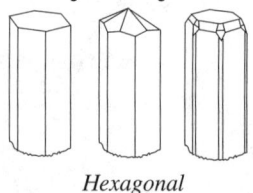

Peridot

Crystal System

Hexagonal

Color: Yellow-green, olive-green, brownish
Mohs' Hardness: $6\frac{1}{2}$ - 7
Cleavage: Imperfect
Refractive Index: 1.654 - 1.690
Dispersion: 0.020
Pleochroism: Very weak; color-less to pale green, lively green, olive green

Streak: White
Specific Gravity: 3.27 - 3.48
Fracture: Brittle, small, conchoidal
Birefringence: 0.036

Lapidarium:

Crystal System: Orthorhombic; short, compact prisms, vertically striated

Chemical Composition: $(Mg, Fe)_2 SiO_4$ magnesium iron silicate

Transparency: Transparent

Fluorescence: None

Called olivine, chrysolite, peridot, "evening emerald," forsterite, and fayalite, the names given this stone have caused confusion through the years. Peridot (pronounced PEAR-uh-doe) is the term used by gemologists to refer to this gem. "Evening emerald" stems from the days when any less expensive gem was given a fanciful name based on the "major" gems emerald, ruby, sapphire, or diamond. The name chrysolite was used by many scholars and jewelers in the past. Misleading combinations of this term have caused confusion for jewelers and the buying public. The gem chrysoberyl has been called "Oriental-chrysolite" or "Brazilian chrysoberyl," beryl called "aquamarine-chrysolite," and topaz referred to as "Saxony-chrysolite." Today peridot is the name preferred by gemologists, believing chrysolite is confusing given the other chryso-prefixed gems. Mineralogists use the term "olivine" for this gem, but call olivine rich rock "peridotite." Forsterite (pure magnesium silicate) and fayalite (pure iron silicate) are mineralogical terms for close relatives to peridot.

Called the "gem of the sun" by the ancients, peridot has been known by many names. The term peridot comes from the old French word *peritot* which itself is of uncertain origin. *Peridota* was the name for this gem in England beginning in the thirteenth century and later was shortened to peridot. The older term "chrysolite" is from the Greek word meaning golden and stone. Pliny the Elder used chrysolite to refer to golden-green stones which came from an island in the Red Sea. The island was called Topazos in ancient times and so he also called these same gems topazius. The ancient author Agatharcides called this island the "Serpent Isle," the Egyptians knew it as Zebirget, and in modern geography it is called St.

John's Island. This same location, off the coast of Egypt, is still a major source of rough peridot.

Ancient papyri record the mining of these yellow-green stones as early as 1500 B.C. The stones were highly prized by the Egytian kings. It is reported the island's inhabitants were forced to collect the gems for the pharaoh's treasury. Legend says the entire island was guarded by jealous watchers who had orders to put to death any trespassers. The story continues that the gems could be found after nightfall due to their radiance. The miners would mark the spot for retrieval the following day.[1] Ptolemy II was said to have had a statue of his wife Arsinoë carved from a single block of the gem. It is likely this was actually a carving from a massive piece of fluor-spar.

The finest examples of peridot were brought to Europe from Egypt during the Crusades. They were called emeralds and presented to the Church treasuries. Many of the "emeralds" held in church inventories today are actually deep colored peridots.

Christian tradition says chrysolite is the stone of the apostle Bartholomew. This is based on a treatise by the tenth century writer Andreas, Bishop of Caesarea. He is one of the earliest writers to associate the Foundation Stones of Revelations with the twelve apostles. He states, "The chrysolite, gleaming with the splendor of gold, may symbolize Bartholomew, since he was illustrious for his divine preaching and his store of virtues."[2]

The stone was associated with the sun, and therefore, was said to dissolve enchantments and evil spirits. It was believed that the powers of darkness would always be subdued by the sun or anything associated with it. Damigeron in *De Virutibus Lapidum* states, "Perforated, threaded on a stiff hair .. and worn 'round the left arm, it overcomes any demon."[3] Similar powers have been recorded in a number of texts. The *Lapidaire* of Chevalier Jean de Mandeville states,

> Chrysolite is the color of green sea water which glitters like gold; when put in the sun, it will sparkle like fire. It chases away bad thought and spirits. This stone is good for those who dabble in necromancy. It is found in Ethiopia.[4]

He continues in a separate entry,

> It is good to wear on oneself, against natural fear. The man who wears it is never sespected of evil doings, and it greatly helps him enter wherever he

wants to go, because he will be rendered gracious and friendly. If one finds a pierced crysolite and puts it in a donkey's jaw, the devil will be chased away. It should be encased in gold and worn on the left side.[5]

Marbod in his *De Lapidibus* of 1531 prescribes the stone to relieve "vague terrors of the night." He continues to say that full protection is gained if the stone is pierced and strung on the hair of an ass and attached to the left arm.[6] William Jones advised in the Nineteenth Century that, "The chrysolite expelled phantoms, and, what was more serviceable, rid people of their follies; bound round with gold and carried in the left hand, it dispersed night hags."[7]

Peridot continues to enjoy a following as a healing stone in the late twentieth century. The stone is reported to enhance healing and spiritual growth. It is a highly recommended stone for healers and psychic counselors. Metaphysical healers prescribe it for the heart, lungs, spleen, stomach ulcers and intestinal tract. As with most green gems, peridot is also said to strengthen eyesight, particularly in conditions of astigmatism and nearsightedness. The stone is also used by midwives as a facilitator in birthing. These practitioners claim it stimulates contractions and causes dilation of the birth canal.

Quartz: Rock Crystal

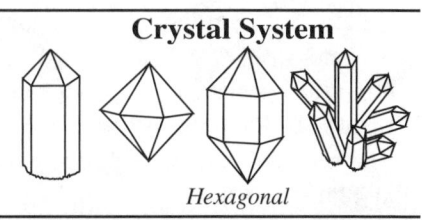

Crystal System

Hexagonal

Color: Colorless	**Streak:** Colorless
Mohs' Hardness: 7	**Specific Gravity:** 2.65 - 2.66
Cleavage: None	**Fracture:** Conchoidal, very brittle
Refractive Index: 1.544 -1.553	
Dispersion: 0.013	**Birefringence:** 0.009
	Pleochroism: None

Crystal System: Hexagonal (trigonal) hexagonal prisms
Chemical Composition: SiO_2 silicon dioxide
Transparency: Transparent
Flourescence: None

Quartz is one of the most abundant minerals, found in nearly every exposed rock on the earth's surface. It is a compound of the two most common elements in the earth's crust, silica and oxygen. If formed in an open space, quartz grows into large magnificent crystals. Deposited by super heated alluvial waters, or solidified within molten rock, its crystals may be microscopic. This cryptocrystaline quartz acts as a geological cement or filler. Quartz is the "glue" which bonds together sandstone. Many gem forms are examples of quartz replacing other material. Quartz may replace the cell structure of wood to form petrified wood, or it may fill the spaces between asbestos fibers to create tiger-eye.

Pure quartz—also known as rock-crystal, is colorless.

Smoky quartz— a golden to brownish or gray variety, possibly colored by radiation. Very dark smoky quartz is called *morion*. In Scotland it is called *cairngorm* (named for the Cairngorm Mountains).

Citrine (see page 69)— yellow quartz, colored by iron. It mimics the colors of topaz and is marketed under the misnomers "citrine topax" or "quartz topaz."

Rose quartz— ranges from a pale pink to a deep rose and may be transparent or cloudy. It is less common than the other gem varieties listed.

Amethyst (see page 42)— is the highly prized color. It ranges from a pale lilac to a deep royal purple.

The text which follows refers to the history and tradition of rock-crystal.

Most of the beliefs connected with quartz center around its use as a communications device or an object which enhances vision or brings "visions" to its possessor. It is interesting that one of its modern uses is in the controlling of radio frequencies.

Quartz was discovered to be piezoelectric in 1921, meaning it will generate an electrical charge. A plate of quartz cut parallel to the direction of its prism faces produces a charge when stressed. When vibrated by the introduction of an electrical current it has been used to stabilize the frequency of radio transmissions. If alternating current is used the quartz slice oscillates dimensionally. This predictable and regular oscillation is used in controlling radio frequencies and in the regulation of time pieces. Quartz is also pyroelectric and will generate a charge when heated or rubbed vigorously.

The magical powers of quartz have been recorded since earliest times. Large crystals of fine quality were found in the eight thousand-year-old Egyptian Temple of Hathor. This ubiquitous gem has been held in reverence by nearly every ancient culture. The Greek priest Onomacritis, founder of the Hellenic mysteries, gave the following advice in the fifth century B.C. regarding this transparent crystal. "Who so goes into the temple with this in his hand may be quite sure of having his prayer granted, as the gods cannot withstand its power." He further states, "that when this stone is laid upon dry wood, so that the sun's rays may shine upon it, there will soon be seen smoke, then fire, then bright flame."[1] The flames where known to the ancients as "holy fire" and thought to be the most pleasing way to burn offerings to the gods.

The word quartz is an old German mining term of unknown origin which has been in general use since the sixteenth century. Also known as 'rock-crystal,' it was known to the Greeks as *krystallos,* meaning clear ice. They held that krystallos was ice which had petrified and was held in a permanent solid state. This belief continued for centuries as is recorded in the eleventh century *Book of Various Acts* by the monastic writer Theophilus. "Crystal, which is water hardened into ice and ice of great age is hardened into stone."[2] Today's common term 'crystal' is derived from this same Greek root and is used to refer to many naturally occurring mineral forms. It is also commonly used to distinguish high quality flint glass from other forms of clear glass.

The ancient Romans also connected colorless quartz with ice. Pliny the Elder related the story that it was congealed water, found in dark mountain caverns under extremely cold conditions. Crystal balls were often held in the hand to reduce the effects of summer heat. Romans also used quartz crystals for glandular swelling, to reduce fevers and relieve pain. The first century mystic Apollinus of Tyana used quartz to "transport" himself. It is said that Apollinus dematerialized and materialized in the presence of Caesar Domitian. He then used the gem to disappear and reappear at the foot of Mt. Vesuvius.[3] The Renaissance Viennese scientist Pribill claimed to have recreated this feat through extensive experimentation. He stated the secret lay in the cut of the stone. First exposed to the tropical sun, then held in the mouth and accompanied by the proper incantation, Pribill claimed to have disappeared and reappeared on numerous occasions.

The double "chi" served as a sign for crystallus (rock crystal) in eighteenth-century chemistry. It probably also served as a symbol for the general process of crystallization.

XX The belief that quartz is super frozen water has led to its use in rainmaking and divining water. The Maya used quartz crystals in the Yucatan to divine water and to ensure a bountiful harvest. The Ta-ta-thi of New South Wales, Australia use rock crystal in an elaborate ritual to bring rain. Tribal wizards break off a fragment of the crystal and toss it to the heavens. The remaining part is wrapped in feathers and immersed in water. After it has soaked for an appropriate time, it is buried underground or otherwise hidden in a secure place. Rains are certain to follow if the ceremony is performed properly. Similarly, rock crystal pebbles are buried in fields by Irish farmers to bring rain and assure a good crop.

A story of the origin of quartz is given in Indian Sanskrit texts. The demon god Vala was slain by the demigods and dismembered. His body parts were scattered across universe, creating the gemstones we know today. The Vedic texts say the potent semen of Vala were transformed into the seeds of quartz. These seeds germinated in the Himalayas and the "lands to their north." The virility of Vala's seed caused quartz crystals to spread across the globe. The texts state that wearing pure crystals of quartz, when set in gold, will bring good fortune in life and protect one from dangerous animals, including tigers wolves, leopards, elephants and lions. Quartz also brings the wearer extraordinary sexual prowess and protects from drowning, burning and theft. Finally, these Indian texts advise that wearing pure quartz while drinking a toast to one's ancestors will bring them lasting happiness.

A number of beliefs about quartz relate to fire or sparks. This may have originated in its use as a fire starter, as related by Onomacritis, or because of the bright sparkles produced from particularly clean surfaces of well formed crystals. It is interesting that a gem so closely associated with water and ice should also be associated with fire or fire gods in many cultures. In the first century medical men of Rome used rock crystal balls to heal wounds. They prescribed allowing the sun's rays to pass through the ball and onto the wound as the best method of cauterizing and promoting healing. The use of a crystal ball as a source of fire has been repeated throughout history. The early sixteenth century *Mirror of Stones* contains a passage which confirms its applicability, but adds to the ancient prescription. "A ball made out of crystal, and exposed to the sun, inflames any combustible matter that is put under it, but not before the ball is heated."⁴ Legends say that the stone emits sparks when it is struck and, if buried with the deceased, may serve to light the way in the after-life.

The alchemist's symbol for rock-crystal is also a sign for pebbles.

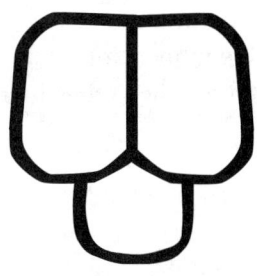

The use of quartz as a burial stone is widespread. Christian mythology includes references to burial with rock-crystal and cites a passage in Revelations to give validity to this belief, Rev 2: 17 "To him that overcometh... I will give a white stone, and in the stone a new name written, which no man knoweth save he that receiveth it." In their studies of traditions in the British Isles, Janet and Colin Bord found, "Until recently, crystals of quartz and white stones (called Godstones) were placed in Irish graves, and the fisher-folk of Inveraray (Strathclyde Argyll) followed the custom of placing white pebbles on the graves of their friends."⁵ The Celts used quartz as an essential part of their burial rituals, burying pieces of the gem in the spiral barrows they constructed for the dead. It is interesting to note that the structure of quartz is a helix or spiral, a property officially discovered in 1926. Since earliest recorded history the spiral has served as a symbol of life and rebirth in many cultures.⁶ White quartz pebbles are also placed in the mouths of the deceased in Scottish tradition to allow them to communicate more efficiently in the after-life. North American mound builders included unworked white quartz pebbles as part of the sacred objects buried in ceremonial mounds and Pueblo graves featured white quartz pendants and carved quartz fetishes. Their widespread use in burial may suggest a common belief that these white or clear stones possess purity and were associated with moral cleanliness.

White quartz pebbles and transparent crystals find significance in many cultures for a variety of purposes. Many native American peoples believed spirits dwelled in rock crystal and would speak to their medicine men. Apache Indians used rock crystal as "good medicine" in many rituals and considered the stone to bear the divine light.

The book *Love is in the Earth* details an example of the steps performed by Native Americans to bring about healing using quartz crystals. The healer would first breath on the crystal to remove any negative forces. He would then initiate a mental connection to the crystal and to the healing spirits working through him or her. The person requesting healing would place his or her hand of preference between, but not touching, the hands of healer. The crystal would be held in the hand of preference of the healer and pointed toward the palm of the preferred hand of the subject. The healer would rotate the crystal counter-clockwise toward the palm of the subject's hand of preference and ask him or her to feel the influencing energy from it and report when it is strong. The healer would then rotate the crystal clockwise and ask the subject to determine which position of the crystal conveys the most compatible energy. When the crystal is in the most compatible position, the healer would allow the energy present to flow through it until the subject feels its energy level change or feels the treatment has had its effect.[7] This same ceremony is still prescribed by modern day crystal healers.

In the Scottish Highlands, a crystal set in silver and worn about the back was thought to be effective for diseases of the kidneys. In the Shetland Islands quartz pebbles were, "believed to cure barrenness when dropped into a pool in which women wash their feet.[8] Tribes in Burma revered quartz and "fed" the stones by rubbing blood over them at regular intervals. Tasmanian aboriginals believe the gem allows communication with others, living or dead. A practice in China is to hold a quartz pebble in one's mouth to avoid thirst during a journey. The Buddhists count quartz as one of the seven precious substances. The belief that rock-crystal is congealed water may have also led to its use as a remedy for dehydration. In Japan, a small crystal ball is worn as a cure for dropsy and other "wasting" diseases.

Medical practitioners of the Middle Ages included quartz in the preparation of healing elixirs. One method of preparing such a potion was to heat rock-crystal to a high temperature and plunge it into cold water to shatter it into small fragments. The resulting powder was mixed

> Marbodus, on the use of crystal for scrying
>
> *If e'er thou seek, where deep rivers flow,*
> *To force the water-sprites the fates to show,*
> *Take the* Diadochus *upon thine hand;*
> *No gem more potent o'er the fiendish band.*
> *Within its orb to thine affrighted eyes*
> *Shall myriad shapes of summon'd demons rise.*
> *But mark! if brought in contact with a corse,*
> *Forthwith the gem shall lose its native force.*
> *Like to the Beryl shows the wondrous stone,*
> *That dreads the touch of one by death o'erthrown.*[9]

with tartaric salts and heated to create a liquid solution. This was cooled, put into a distilling glass with the "best spirit of wine" and then placed in a bath of luke warm water. After several successive distillings, the red liquid was reduced to a tincture of quartz. The prescription advised administering this tincture, forty drops at a time, in a "proper" drink as a cure for dropsy, scrofula or hypochondriac melancholia.[10] Another recipe results in a "magisterium of rock-crystal." The crystal was heated and dipped into spirits of vitriol to fracture it. This was repeated ten times and the resulting liquid was filtered through blotter paper. The remaining particles were ground on a marble slab and given internally as a cure for gout or the formation of organic stones in various organs. The filtered vitriol liquid was then advised as a diuretic if added to meat broth in a dose of seven to ten drops per serving.

The fourteenth century *Lapidaire* of Chevalier Jean de Mandeville contains a number of uses for quartz, including a cure for ineffective gems. Mandeville states,

> Crystal is white and makes man cold. If one makes a crystal powder and gives it to nursing women to drink, their milk will increase. If a stone which seems to have lost its properties is held to crystal, the properties will be recovered if one confesses the sin which caused the stone to lose its properties. It should be worn on the right side of the body.[11]

The use of quartz as a remedy continued throughout European history. The sixteenth century German name for rock crystal was *schwindelstein*, meaning vertigo stone, alluding to its use as a charm that protects the wearer from dizziness. The Italian physician Camillus Leonardus writes, "Crystal hung about those who are asleep, keeps off bad dreams; dissolves spells; being held in the mouth, it assuages thirst; and when bruised with honey, fills the breast with milk."[12] Other medical uses of the gem have included powdering the stone and drinking it in wine as a cure for dysentery and placing it in contact with a wound to stop bleeding.

> *And if they demand wherefore your*
> *wares and merchandise agree,*
> *You must say, jet will take up a straw;*
> *amber will take on fat;*
> *Coral will look pale when you be sick,*
> *and crystal staunch blood.*[13]
> Robert Wilson (d. 1600)

The carving of clear quartz has been practiced for thousands of years. One of the most popular forms has been that of a sphere or "crystal ball." As mentioned earlier, these spheres have been used since ancient times. This practice has continued into modern times in the form of "scrying," or the use of a crystal ball as a way to see visions of the past or future. The practice of scrying is such a vast topic that an entire text on that subject alone would serve only as an introduction. For this reason the topic is only mentioned here as an example. Crystal carving has brought its own legends. The monk Theophilus writes in his *Book of Various Acts* of a method to prepare quartz and allow ease in its carving:

> But should you wish to sculpt crystal, take a goat of two or three years and, binding his feet, cut an opening between his breast and stomach in the position of the heart, and lay in the crystal, so that it may lie in its blood until it grows warm. Taking it out directly, cut what you please in it as long as the heat lasts, and when it has begun to grow cold and to harden, replace it in the blood of the goat...[14]

Given this prescription, it is a wonder the many quartz carvings which exist where ever completed.

Students of crystal power and healing refer to rock crystal as the "Stone of Power." It is used to amplify psychic thoughts and bodily energy, and it provides a source of purification. Clear quartz crystals are commonly used as amplifiers of energy fields and may be affixed to a wand to aide in this application.

Modern practitioners also divide members of the quartz family by color or appearance and list different attributes for each. Those most often mentioned are listed.

Blue quartz— improves understanding and the ability to "tune in" while in devotion or meditation.

Chlorite quartz— aids in one's connection to the earth, particularly plant life; it also strengthens the entire digestive system.

Clear quartz— holds and amplifies thought and can be "programmed" with conscious intent or direction of thought to serve as a healing tool with any particular issue.

Elestials quartz— a deep healing stone that helps one uncover deep and/or unconscious memories. Caution is advised in this application.

Rose quartz— known as the "love stone" it is said to open and balance the heart center, especially the emotional heart in matters of romantic love. It also strengthens the ability to give and receive romantic or platonic love and to heal anger and disappointment.

Rutilated quartz—increases longevity and slows the aging process. As an elixir it strengthens the immune system, enhances thought forms, absorbs negativity and reduces background radiation.

Tourmalinated quartz—the combination of these stones (tourmaline and quartz) naturally helps balance masculine and feminine energies, alleviates fear and promotes emotional peace.

The Amazing Crystal Skull

Without a doubt the most impressive rock-crystal carving found to date is a Mayan artifact in the form of a 475.5 ounce piece, known as the Mitchell-Hedges skull. In 1927 F.A. Mitchell-Hedges was clearing debris at the ancient Mayan city of Lubaantum, located in modern-day Belize, when his daughter saw a bright object in the dust. It was a beautifully carved and polished skull made of rock-crystal and missing the jaw. The jaw was located three months later in an excavation approximately twenty-five feet from the original find.

The skull was fashioned from a single block of clear quartz and measured five inches high, seven inches long and five inches wide. These dimensions make it nearly the same size as an actual human skull. The detail has been called amazing, to the point that it could be identified as a rendering of a male skull, lacking the globular prominence or superciliary ridges characteristic of female craniums.

The carving has been subjected to a series of tests, revealing a number of curious findings. Conservator and restorer Frank Dorland was given permission by the Mitchell-Hedges estate to test the skull in cooperation with the Hewlett-Packard Laboratories in Santa Clara, California. A test of the sculpture's crystal structure showed that the skull and jaw were carved from the same single block. Another amazing thing was that the natural axis of the crystal had been disregarded by the carver(s). The first thing modern crystal carvers do is determine the axis of a crystal to prevent fracturing and chipping during the cutting process. It was found that the unknown artist(s) did not use any metal tools or leave a single tool mark on the surface. Most metals would have little use in working with a stone that has a Mohs hardness of seven.

Dorland detirmined tiny marks near the carved surfaces revealed the use of diamond point chisels. He further speculated that the finish was achieved by successive grinding and polishing with silicon-crystal sand. The problem then arises that if these were the methods used, Dorland calculated a total of over 300 man-years of labor went into fashioning the skull. These facts mean that either this single carving was the longest lasting artistic project in history or some lost technology was employed by these ancient peoples.

The puzzle of the skull does not end with its making. The zygomatic arches (the bones extending across the brow) form light pipes which act optically to channel light from the base of the skull to the eye sockets. The eye sockets also act as miniature concave lenses to bring light from the base into the upper cranium. The interior of the piece is a ribbon prism which serves to magnify and illuminate objects held beneath it. These various optical properties combine to bring about a dazzling effect. Placed over a light source, the entire skull becomes illuminated, and the sockets become glowing eyes. Dorland reported that the skull "lights up like it was on fire."

Further analysis of the skull shows an early understanding of weights and balance. The jaw fits precisely into two polished sockets that allow it to open and close. The entire carving balances exactly where two tiny holes are drilled into the base. These openings undoubtedly served as support points for the sculpture. The entire arrangement is so perfectly designed that the skull will nod up and down in the slightest breeze with the jaw opening and closing as a counter-weight. The effect created is that of a living person, talking and carrying on a conversation.

Added to these measurable observations, many strange phenomena have been reported that defy explanation. It has been said that the skull changed color. The frontal lobe area clouds, while at other times it appears perfectly clear. A dark spot often develops on the right side and slowly blackens the entire skull in a matter of five to six minutes. The skull then clears in as short a time as it took to darken. Other observers have reported a glow surrounding the carving for as long as six minutes, like an aura. Ringing noises have been heard coming from within the skull, odors have emanated, and reflections of buildings and other objects have been seen in the eyes, despite the fact that the skull is resting against a dark background. It also gives sensations of heat and cold to the touch, even though it is kept at a constant seventy degrees Fahrenheit.

Dorland has speculated that the crystal stimulates the opening of a psychic door to unknown parts of the observer's brains. He notes that quartz gives off electrical vibrations which may interact with natural brain waves, like radio waves or transmissions, a property which quartz is known to exhibit. Other writers have stated that the skull, in the hands of a skilled meditator and mental focalizer, may serve as a tool to communicate to a higher mental plane, to amplify psychic and earth energy forces, and bring about healing in a person's mind or body.

What ever the intended use, or possible future applications, it is certain that the technology employed to create the rock crystal skull is beyond that which we understand or expect to find in an ancient culture. Future studies may yield answers but are hampered due to its inaccesability. It is now held in the possission of an anonymous private collector.

> ***Epigrams*: The fifth century Roman poet Claudian on Crystal's Attributes**
>
> *Pass not the shapeless lump of Crystal by,*
> *Nor view the icy mass wtih careless eye;*
> *All royal pomp its value far exceeds,*
> *And all the Pearls the Red Seas's bosom breeds.*
> *This rough and unform'd stone, without a grace,*
> *Midst rarest treasures holds the chiefest place.*
>
> *With th' Alpine ice, frost-harden'd into stone,*
> *First braved the sun, and as a jewel shone,*
> *Not all its substance could the gem assume—*
> *Some tell-tale drops still linger'd in its womb.*
> *Hene with augmented fame its wonders grow,*
> *And charms the soul the stone's mysterious flow:*
> *Whilst stored within it, from Creation's birth,*
> *The treasured waters add a doubled worth.*
>
> *Mark where extended a translucent vein*
> *Of brighter Crystal tracks the glistening plain:*
> *No Boreas fierce, no nipping winter, knows*
> *The hidden spring, but ever ebbs and flows:*
> *No frosts congeal it, and do Dog-star dries;*
> *E'en all-consuming Time its youth defies.*

A stream unfetter'd pent in Crystal round,
A truant fount by harden'd waters bound:
Mark how the gem with native sources foams!
How the live spring in refluent eddies roams!
How the live rainbow paints the opposing ray,
As with the imprison'd winter fights the day!
Starange nymph! above all Naiads' fame supreme—
Gem, yet no gem— a stone, yet flowing stream!

Erst while the boy, peased with its polish clear,
With gentle finger twirl'd th icy sphere,
He mark'd the drops pent in its stony hold,
Spared by the rigour of the wintry cold.
With thirsty lips the unmoisten'd ball he tries,
And the lov'd draught with fruitless kisses plies.

Streams which a tream in kindred prison chain,
Which water were, and water still remain,
What art hath bound ye, by what wondrous force
Hath ice to stone congeal'd the limpid source?
What heat the captive sames from winter hoar,
Or what worm zephyr thaws the frozen core?
Say in what hid recess of inmost earth,
Prison of floating tides, thou hadst thy birth?
What power thy substance fix'd by icy spell,
Then loosed the prisoner in his lucid cell?[15]

"The loveliest precious stone of which I have any knowledge—an uncut ruby."
John Ruskin

Ruby

Crystal System

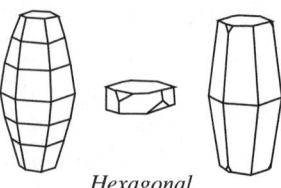

Hexagonal

Lapidarium:

Color: Varying red
Mohs' Hardness: 9
Cleavage: None
Refractive Index: 1.762 - 1.770
Dispersion: 0.018

Streak: White
Specific Gravity: 3.97 - 4.05
Fracture: Small conchoidal, uneven, splintery, brittle
Birefringence: 0.008
Pleochroism: Strong; yellow-red, deep ruby-red

Crystal System: Hexagonal (trigonal) ; hexagonal prisms or tablets
Chemical Composition: Al_2O_3, aluminum oxide
Transparency: Opaque, translucent, transparent
Fluorescence: Strong; ruby-red

The name ruby is reserved for the red variety of corundum, a mineral which may be found in nearly every color. Corundum occurs throughout the world in many kinds of rocks, but large transparent crystals of deep color are rare. A gem may only be properly called ruby if it is corundum of a red to purple-red color and a medium to dark shade. The color red in corundum is due to trace amounts of chromium found within the crystal's structure.

The finest examples come from Burma, although Burma also produces inferior quality stones. The top colored Burmese gems are called "pigeon's-blood;" however, most of the rubies in use today came from other localities and are not as fine as these top grade stones. Sri Lanka also produces rubies, but these are paler than Burmese gems. Rubies from Thailand tend to be dark with a purplish cast; they often resemble deep red garnets.

Some rubies may contain inclusions of long rutile crystals. This secondary mineral orients itself in the six-sided symmetrical pattern of the host crystal. When a cabochon is cut to take advantage of this phenomenon, it exhibits a six-rayed star and is called a "star ruby." Large flawless rubies and emeralds are the most valuable of all gems as compared with any other flawless stone of similar size.

July's birthstone is a rare and valued gem of worldwide legend. Its present name first appeared in print in the 1310 *Oxford Dictionary* and in western literature when Chaucer wrote in 1380, ". . . lyke ruby ben your chekys rounde."[1] The English word comes from the Latin *ruber*, meaning red. The stone has many names in Sanskrit, owing to its valued status to the Hindu. It is called *ratnaraj* (king of precious stones) *ratnanâyaka* (leader of precious stones) or, for a particular shade, *padmarâga* (red as the lotus).[2] Another Sanskrit word for ruby, *kuruvinda*, and the Tamil word *kurandam* are the sources of our name for the mineral class corundum.

The Romans included the ruby with other red stones under the name *carbunculus*. The historian Pliny warned in his writings about gems, to be wary of imitations and enhanced stones. Possibly speaking of rubies that appeared too pure to be garnet he writes, "Some say that the Aethiopians steep their dusky and dark carbunculi in vinegar. As a result, in fourteen days, they become pure and lively and remain so for fourteen months."[3] Pliny also states that the counterfeits may be distinguished "by grinding on a mill" and by weight, "for glass imitations are the lighter of the two." Attempting to grind a ruby would have proved an arduous task, while comparing weights may have proved a reliable means of separation.

Much of the early lore regarding this stone originates in India. Legends say the Lord created this stone and then man to own it. Dark red and star stones were called male and the pink and pigeon blood called female. The color red was considered to signify stimulation, heat, life and power. Statues of Buddha in this country are usually adorned on the forehead with a small ruby, since red is the symbol of the reincarnation of this god. Burmese gem miners held that pale rubies, if buried, would ripen to a deep color. Ruby was said to be the ripened member of the corundum gems. Jan Heyghan von Linschoten traveling in India from 1582 to 1593,

> *Two versions of a natal stone poem*
>
> *The gleaming ruby should adorn*
> *All those who in July are born,*
> *For thus they'll be exempt and free*
> *From lover's doubts and anxiety.*
>
> *The glowing Ruby should adorn*
> *Those who in warm July are born;*
> *Thus will they be exempt and free*
> *From love's doubts and anxiety.*[4]

wrote, "The cause whereof is because that in the rocks and hills where they grow their first colour is white and by force of the Sunne, are in time brought to perfection and ripeness" as rubies, but "wanting somewhat of their perfection and being digged out before that time they are of divers colours, as I said before and how much paler they are and lesse red."[5] To Ceylonese miners, flawed stones were considered overripe.

Hindus believed the gem contained an inextinguishable fire which would glow through clothing and had the ability to make water boil. The light of a ruby is thought to be the chosen dwelling place of their gods. As an offering to Krishna, the ruby would bring reincarnation as an emperor. If the stone was considered small, the worshiper may be reborn as only a king. As a medicine, it was said to preserve bodily health and to serve as a cure for all diseases.[6] An ancient Indian legend tells the source of the ruby mines. Three marvelous eggs were laid by a serpent in a nest. The first egg hatched out the Mogul of Pag, the second the Emperor of China, and the third the ruby mines of Burma.

Hindus divided the ruby into four castes (like diamonds); true Oriental ruby is the *Brahmin*, rubicelle a *Kshatriya*, the spinel a *Vaisya* and the balas ruby a *Sudra*.[7] The most highly prized, the *padmarâga* or Brahmin ruby, gave perfect safety to the owner and shielded him or her from harm and misfortune. One must be careful not to allow these purest of rubies to come in contact with inferior stones as they would be contaminated and their power diminished. "Balas rubies" (the lowest caste) are a misnomer, They are actually one of two kinds of gem quality red spinel. Spinel is often found in close proximity to corundum, but it is an oxide of aluminum and magnesium unrelated to ruby. The balas ruby gained its name from the *Balascia*, ancient name for *Badakhshan*, the region that served as the source for the best stones imported into Europe in the Middle Ages. Many balas stones found their way into major collections and have only been recently separated from true ruby (see Spinel p. 202).

Other beliefs in India are contained in ancient Vedic texts. These relate a story of the origin of the ruby and its powers. The texts tell of Vala, a demon god, who was slain and dismembered by the demigods. The parts of his body were scattered about the earth to create the various gemstones we know today. The blood of the demon was taken by Surya, the sun-god, who then fled to the vastness of space. Ravana, great king of Sri Lanka, attempted to block the sun-god's flight like the power of a solar eclipse. Ravana's power frightened Surya, causing him to drop the blood, which fell into the deep pools of Bharata.[8] From then on these pools became as sacred as the Ganges River and were known as Ravanda-Ganga. The banks of these perfumed pools became covered with beautiful gemstones, rubies and all other colors of corundum. The story concludes by advising of the existence of two grades of rubies, *padmarâga*, or "top crystal" and *kuruvinda*, or stones lacking deep color, clarity or luster.

Thirteenth century Sanskrit literature by the physician Naharari of Kashmire notes ruby as a cure for flatulence and biliousness.[9] Another of the powers is to make one invulnerable to wounds from sword, spear or gun. To be most effective the stone was to be inserted into the skin to become one with the body.

Rubies also have been given a place in Christian lore The fourth century writer Saint Epiphanius told of rubies' ability to shine in the darkness. As in Indian lore, he contended they could even be seen through one's robes. Marbodus writes in his *Lapidaire en Vers* that the ruby is, "the most precious of the twelve stones God created when He created all creatures." He continues to state that by Christ's command a ruby was placed on Aaron's neck, "the ruby, called the lord of gems; the highly prized, the dearly loved ruby, so fair with its gay color."[10] In many accounts the stone of Judah, called nophek, in the Breastplate of Aaron may have been a ruby or garnet. Possibly due to this reference, and the fact that Judah was the source of Israel's royalty, the ruby has long served as a favorite stone of nobility and leaders of the church. It is reported that the betrothal ring of Martin Luther was a gold ring set with a ruby and engraved with scriptural passages.

Most of western lore regarding the ruby was imported into Europe along with stones brought back from travels to the East and the Crusades. Marco Polo wrote in the thirteenth century that a Sinhalese monarch owned a red gem "four inches long and as thick as a finger,"[11] for which the Kublai Khan offered the value of a city. The value and virtues of the stone were repeated by many writers in history. Phillipe de Valois says in his *Lapidaire*, "the books tell us the beautiful clear and fine ruby is the lord of stones; it is the gem of gems, and surpasses all other precious stones in virtue."[12] The ruby is assigned strong protective powers by many authorities. The fourteenth century writer Chevaliere Jean de Mandeville included many attributes in his *Lapidaire*.

> *They say, through patience, chalk*
> *Becomes a ruby stone;*[13]
> Translations : *Ralph Waldo Emerson,*

The ruby is called Epitest in Greek: it is red, sparkling and dazzling. It comes from parts of India, Libya, and Tourniche and is found on the banks of the rivers of Paradise, near Alexandria. This stone has several properties: it acquires and maintains power and rulers, it procures peace and agreements, it makes Man devoted towards God, it appeases anger and maintains seductions, it makes the person wearing it safe in all dangers and if one throws it into boiling water, the water will stop boiling and will lose heat, and if this stone is in the sun for a period of time, it will emit red rays like fire, it will preserve fruits on trees, vines and on the earth, and it will save houses from tempests. This stone should be worn on the left part of the body. There are also rubies of Alexandria and from Thir [Probably the Greek island Santorini], but they don't have as many properties as those of the Orient.[14]

The *Old Dream Book* says, "the ruby indicates joy and good fortune and the more rubies the more joy. The owner of the ruby is feared by his enemies, when he wears the stone."[15]

Besides these general attributes the gem was said to be an aid medicinally. Jones gives the following prescription in his nineteenth century writings, "The ruby, bruised in water, relieved infirmities of the eyes, and helped disorders of the liver."[16] The gem was said to be a cure for all diseases, but the treatment may have seemed severe to many. Fobes gives the following description of a cure:

> The ancient method of applying the ruby as a cure was to place it on the tongue, which was rendered at once cold and heavy, so that only incoherent sounds could be emitted. The fingers and toes also became cold, and a violent shivering followed. Thus, the bad symptoms disappeared and a sense of elasticity and well being followed—the cure was complete.[17]

The ruby has also been said to demonstrate clairvoyant powers. The gem reportedly changes color when misfortune is about to over take its owner. Two such tales are given in references on gems. One story tells of a ruby owned by Katherine of Aragon which changed its hue when Henry VIII was considering divorcing her. Wolfgang Gabelchover included the following account in his commentary on the sixth book of the treatise *De Gemmis* by Andrea Baccio:

> It is worthy of note that the true Oriental ruby, by frequent changes of color and by growing obscurity, announces to the wearer some impending misfortune or calamity; and the obscurity and opacity is greater or less according to the extent of the coming ill-fortune. Alas! that what I had often heard proclaimed by learned men, I should myself experience; for as, on the fifth of December, 1600, I was travelling from Stuttgart to Calw with my beloved wife Catherine Adelmann of pious memory, I plainly observed in the course of the journey that a very beautiful ruby which she had given me, and which I wore on my hand, set in a gold ring, once and again lost its splendid coloring and became obscure, changing its brightness for a dark hue. This dark hue continued not for one or two days only, but so long that I was greatly terrified, and, removing the ring from my finger, concealed it in a case. Wherefore, I repeatedly warned my wife that some great calamity was impending either for her or for myself, this which I inferred from the change and variation of the ruby. Nor was I deceived, for within a few days she was seized with a dangerous illness, which resulted in her death.[18]

Modern practitioners of crystal therapy include ruby as a gem capable of numerous remedies. The stone is said to promote physical and emotional health and stability. The ruby also helps to clear negative thoughts and emotions and enhances the ability to feel and express unconditional love. The owner of this gem also demonstrates a willingness to serve others while serving as a leader.

This king of stones has found its place in many pieces of literature. The first allusion to ruby cheeks made by Chaucer has been followed by other analogies. Ralph Waldo Emerson spoke in glowing terms about the gem:

> They brought me rubies from the mine,
>   And held them to the sun;
> I said, 'They are drops of frozen wine

> From Eden's vats that run.'
> I look'd again—I thought them hearts
> 	Of friends, to friends unknown;
> Tides that should warm each neighbouring life
> 	Are lock'd in sparkling stone.
>
> But fire to thaw that ruddy snow,
> 	To break enchanted ice,
> And give love's scarlet tides to flow,—
> 	When shall that sun arise?[19]

Shakespeare makes reference to ruby cheeks in *Macbeth*, ruby lips in *Julius Caesar*, and a ruby kiss in *Cymbeline*. Rubies have also served as the "fruit" of wealth. In Hawe's *Pleasure and Pastyme* he describes a grand hall as having its roof overspread with a golden vine, whose grapes are represented by rubies. Mandeville describes a vine "that hath many bunches of grapes, some white, all the red being rubies."[20]

Sapphire

Crystal System

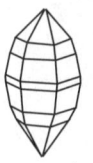

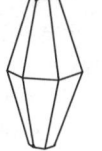

Hexagonal

Color: Blue in various hues, colorless, pink, orange, yellow, green, purple, black
Mohs' Hardness: 9
Cleavage: None
Refractive Index: 1.762 - 1.770
Dispersion: 0.018
Streak: White
Specific Gravity: 3.95 - 4.03
Fracture: Small conchoidal, uneven, splintery, brittle
Birefringence: 0.008
Pleochroism: Blue: definite; dark-blue, green-blue

Crystal System: Hexagonal (trigonal) ; dipyramidal, barrel-shaped, tabloid-shaped
Chemical Composition: Al_2O_3, aluminum oxide
Transparency: Opaque, transparent
Fluorescence: Blues; (purple) none, Yellows from Sri Lanka; weak; orange, Colorless; orange-yellow or purple

Sapphire is the term used for all corundum gems that are either colorless or any color except red. A corundum that is a red or purple-red and medium to dark in hue is called ruby. Pink or light red corundum is called pink sapphire. Corundum occurs throughout the world in many kinds of rocks, but large transparent crystals of deep color are rare. Pure corundum is colorless, but trace elements found within the crystal are the cause of the rainbow of colors available.

The finest sapphires are found in Ceylon in shades of pale blue, violet, deep blue, yellow, white, green, greenish-blue, brown, and pink. The lovely pinkish-orange shade is known as padparadschah. Other areas known for sapphire, particularly greenish-yellow and grayish-blue stones, are the Chantabun area of Thailand and Battambang region of Cambodia. Sri Lanka is known for its color variety of corundum mined in the Ratnapura region, Ratnapura meaning "city of gems" in Sinhalese. Some of the most famous sapphires come from India's Vale of Kashmir in the Himalayas. Fine colored gems have been found in this area in the past hundred years. Indian sources are ancient and have been known since the time of Marco Polo.

Sapphires may contain inclusions of long rutile crystals. This secondary mineral orients itself in the six sided symmetrical pattern of the host crystal. When a cabochon is cut to take advantage of this phenomenon, it exhibits a six-rayed star and is called a "star sapphire."

Legends and lore of the sapphire speak almost exclusively of blue gems. When a stone is referred to only by the name sapphire in the following text, it may be assumed that blue sapphire is the stone discussed.

The modern word sapphire derived from the Latin *sapphirus*, Greek *sappheiros* and the Hebrew *sappir*. All these were possibly derived from the Sanskrit *sanipriya*, literally meaning "dear to the planet Saturn." The word originally referred to lapis-lazuli and possibly all other opaque blue stones. For this reason the lore of sapphire is often uncertain as to what stone it truely speaks. Many of the attributes given sapphire mimic those of other blue stones. Ancient writings regarding the stone mention that it was often engraved with some passage. The hardness of the gem would preclude nearly any attempt to carve it, so most of these probably were engraved lapis. Tradition says the Ten Commandments were engraved on great sapphire tablets. Knowing the technology of the time and the hardness of sapphire, it is certain that God alone could have engraved the Laws into a corundum.

The second century B.C. Greek historian Damigeron wrote extensively about the virtues of the gem:

> Sapphire is indeed a stone made highly honorable by God. Some also call it Ormiseum or Hormesion. Kings are accustomed to wear this stone about their neck, for it is a great protector. It also has the powers of preserving a man from envy, and rendering him agreeable to God, and keeping the body whole and at the right temperature; tied on it soothes heavy flows of sweat and cools the internal parts of him who wears it. If ground with milk and placed on old ulcers it is a great remedy, and cures epiphoras [perhaps an abscess?] of the eyes if rubbed on the forehead. And if anyone has gripings in his guts, it will cure them if taken in milk. And if anyone who has a wound in his intestine, grind the stone with milk, smear it on the intestine and it will be healed. But if anyone has a sore tongue, rub it with the ground stone and it will be cured. It cures all wounds whether recent or old in the same way, that is, if ground in milk and rubbed in. It is very effective against envy and preserves the body. It renders the holy God propitious to him who asks clearly and honestly. It also allows one to understand the holy responses to divination by water. It is useful to diviners because he who has

one carved in the shape of a scarab will divine all things truly. And he who is chaste while wearing it will be pleasing to God and man.[1]

His prescriptions are the basis for many of the attributes given by later western writers; however, most believe he was writing about lapis-lazuli.

The lore of India is most likely based on the true corundum sapphire we know today. Hindu legends say the sapphire is bitter to taste and lukewarm in temperature. Medicinally it is said to relieve a build up of phlegm, flatulence, and excess bile. A potion made with sapphire was prescribed as a cure for scorpion bite, ulcers, boils, pustules, and ruptured membranes.

Vedic texts of India give two stories of the origin of sapphires, yellow and blue. The stories begin when Vala, a demon god, was slain and dismembered by the demigods. The parts of his body were strewn about the earth to create the various gemstones we know today. These writings treat these two colors of sapphire as separate stones. Yellow sapphires were said to originate when the skin of Vala was transformed into mystic seeds. These seeds fell on the lands crowned by the Himalayas and formed the mines still being worked today. This tale says the eyes of Vala were the color and shape of the blue lily flower's petals. His eyes, torn from his head, became the seeds of blue sapphires which fell to the sacred lands of Sri Lanka and the surrounding areas of Southeast Asia. The seeds were so numerous that these lands are said to glow with their beauty.

Another story of the origin of sapphires on the island of Ceylon was expressed by an ancient Indian poet. "When the young Cingalese [Sinhalese] maidens sway with the tips of their fingers, the stems of the lavali blossoms, then do the two dark blue eyes of the Daitya fall, eyes with a sheen like that of the lotus in full bloom!"[2] To express how sapphires abound in Ceylon another poet describes a river as, "That lovely stream, the Kalnquaga, which meandered, as a sapphire chain, over the shoulders of the maiden Laubea. [Laubea is the Sinhalese name for Ceylon]".[3]

Throughout the centuries the primary medicinal use of sapphire has been as an "eyestone." An instrument for the removal of foreign objects and the treatment of eye diseases. This tradition may have a basis in the practices of ancient Egyptian physicians. They used an oxide of copper mixed with boric acid as an eye wash. Known as *lapis armenus*, this astringent has been shown by modern medical research to be effective and is still prescribed. Later translations of Egyptian texts substituted lapis-lazuli, assuming the more valuable material would be more effective. Medieval writers assigned the attributes of all blue gems to "sapphirus." Marbod's classic eleventh century poem includes sapphire as an eye elixir, "Dissolved in milk it clears the cloud away, from the dimmed eye and pours the perfect day"[4] A century later Albertus Magnus declared the stone to be an effective means of removing foreign objects from the eye. The stone was dipped in cold water before and after the operation. He also warned that not every sapphire was useful; only certain selected stones should be set aside for medical use. Other medieval writers asserted that in experiments the sapphire removed particles of sand and dust if it was warmed while directly over the eye. The action must have been strong as it was claimed the eye "ejected all foreign substances" with this treatment.[5] The stone also shared the attribute of emerald in its ability to soothe the sight. Intent gazing into a sapphire was said to protect the eye from all injury. An inventory of gifts presented by Richard Preston in 1391 at the shrine of St. Erkinwald in Old Saint Paul's includes a sapphire placed there for the cure of eye diseases.

Marbod passed along many of the claims of Damigeron in his writings. These became the basis of claims by later authors, writers who added to the legend. St. Jerome declared the sapphire, "will procure the wearer the favor with princes and all others. pacify enemies, free him from enchantments, bond and imprisonments and it looseth men out of prison and assuageth the wrath of God." The gem was also credited with detecting poisons. The medieval writer Wolfgang Gabelchover claimed he had conducted experiments to find the most useful stones. A test of sapphire's quality was to place a spider in a vessel, suspend the stone above the spider and swing it back and forth in a pendulum fashion. If the stone possessed sufficient power, the spider would soon die. Later authors specified oriental sapphire due to its transparency and supposed purity and effectiveness. Bartolomæus Anglicus continued the claim that it is an antidote for poison in a writing of 1495, as translated by John of Trevisa.

> His vertue is contrary to venym and quencheth it every deale. And yf you put an attercoppe [old English for spider] in a boxe and hold a very saphyre of Inde at the mouth of the boxe ony whyle, by vertue thereof the attercoppe is overcome & dyeth as it were

sodenly, as Dyasc. sayth [pseudonym of Dioscorides].
And this same I have assayed oft in many and dyvers
places. His vertue kepeth and savyth the syght, &
clearyth eyen of fylthe wythout ony greyf.

He continued that it was a favorite in necromancy. "Also wytches love well this stone, for they wene that they may werke certen wondres by vertue of this stone."[6]

The scholar Chevalier Jean de Mandeville lists many applications of the sapphire as a curative touching stone in his fourteenth century *Le Lapidaire*.

> Sapphires are worthy of God, kings and counts. The first kind comes from an Eastern River and is found in the gravel. There is a kind a sapphire which comes from deepest Turkey. They are dark and do not glimmer, but one has nevertheless found them to have great virtues. Daniel loved the sapphire stone much that he called it a sainted stone, or gem of gems. When one looks at it, it elevates the thoughts to the celestial realm, it comforts the limbs of the body and impedes the body from being imprisoned; or if one has it on oneself, it will help one to be delivered: the prisoner should touch his irons and the for sides of the prison with it. It is very good for bringing people together. It is marvelously good at removing all the bumps inside the body. If mixed or watered down with milk, it will cure the bumps with the virtue and the force that God has given and bestowed upon the sapphire. It will cool off a man who has too much heat in his body, and will make the hot sickness leave and will put the person in good health. It will take away dirt and filth from the eyes and will clean and purify them. It is good for pain of the head, and for a person who has bad breath. It gives good advice to he who

---

A maiden born when Autumn's leaves
Are rustling in September's breeze,
A Sapphire on her brow should bind
'Twill cure diseases of the mind.'

A sapphire touchstone is used to remove a foreign object from the eye. The illustration is a woodcut from the Ortus Sanitatis *of Johannis de Cuba, published in 1483.*

wears it as well as making him sure in all business. He who wants to feel the sapphire and to know its properties and its virtues should be chaste, pure and clean, without any filth on him when he wears it. The sapphire should be encased in gold because its properties and virtues will therefore increase.[8]

Other applications of the gem are listed in the *Speculum Lapidum* of 1502 by the Italian physician Camillus Leonardus .

> Some call it the jewel of jewels for its beauty. It checks the ardor of lust, and makes a person chaste and virtuous, and restrains too much sweat. It takes away the filth of the eyes and the pains of the head. It discovers frauds; expels terror. It is of great service in magic arts, and is said to be of prodigious efficacy in the work of necromancy. It discharges a carbuncle with a single touch. The eyes being touch'd with it, it preserves them from being injur'd by the small pox.[9]

The use of the stone to cure carbuncles and boils was given in many medical texts. Joh. Bapt. Von Helmont provides a detailed cure for boils caused by the plague in his *A Ternary of Paradoxes* of 1650. He recommended rubbing a stone of deep fine color gently and slowly around the boil. The stone then would be removed gently. He advised that the patient would feel little relief at first, but gradually he or she would feel relief. The theory seems to be that a magnetic force exists in sapphire that continued to draw away, "the pestilential virulency and contagious poyson from the infected part."[10]

To increase a sapphire's effectiveness engraved images were recommended. The advice of the thirteenth century writer, Rabbi Ragiel, included in The *Book of Wings,* states, "The figure of a ram or a bearded man engraved on a sapphire has the power to cure a person from many ailments and to free him from poison and demons."[11]

The Roman Catholic church has given the gem a place of honor. It has been considered a charm which promotes chastity and guards against evil influences. It is said that for this reason Pope Innocent II had all the bishops wear rings with sapphire settings. Marbod, Bishop of Rennes, chose sapphire as the greatest of all stones, "By nature with superior honours graced, As gem of gems above all others placed," gave it celestial powers, "E'en Heaven is moved by its force divine"[12] and thought it particularly appropriate for ecclesiastical rings. Along with the powers given to the clergy, sapphire was also said to be appropriate for royalty. However, Elver Burton warns in his *Anatomy of Melancholy* that, "if a King saw a blue gem in his crown while dreaming, it foreshadowed the loss of part of his kingdom, if not all of it."[13]

Another Christian symbolism is the assignment of the Foundation Stones, listed in Revelations XXI, to each of the twelve apostles. One of the earliest writers to make such an association was the tenth century writer Andreas, Bishop of Caesarea. He states that sapphire is the stone of St. Paul and gives the following reasoning.

> As the sapphire is likened to the heavens (from this stone is made a color called lazur), I conceive it to mean St. Paul, since he was caught up to the third heaven, where his soul was firmly fixed. Thither he seeks to draw all those who may be obedient to him.[14]

Star stones have been awarded special attributes. In Ceylon they are considered to give protection against witchcraft or the evil eye. Known as the "Stone of Destiny" as the crossed lines represent faith, hope and charity. The seventeenth century writer de Boot uses the term *siegstein* to speak of the star sapphire, a word which translates to 'victory-stone' in English.[15] Star stones are also said to possess the language of the genii. Fobes writes, "In many sapphires are found the cuneiform characters, like feathers and twin crystals. The cuneiform characters found on Babylonian tablets were without doubt suggested by the dorms discovered in a certain kind of uncommon sapphire crystal found and venerated as supernatural by the Indian natives."[16]

Contemporary crystal therapists have added a few medicinal uses for sapphire. The stone is used to treat disorders of the blood, combat excess bleeding and strengthen the walls of veins. It is often prescribed for hemophilia. Additional universal healing powers are claimed. One modern text states, "They are now being used to assist in the intense energy transfer necessary for healing the Earth, for ameliorating those forces which can be detrimental to the environment and for the purification of the realm."[17]

Sard

Crystal System

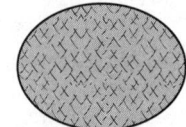

Microcrystalline aggregate

Color: Red-brown variety of chalcedony **Mohs' Hardness:** 6 1/2 - 7 **Cleavage:** None **Refractive Index:** 1.530 - 1.539 **Dispersion:** None	**Streak:** White **Specific Gravity:** 2.65 **Fracture:** Uneven, shell-like **Birefringence:** up to 0.006 **Pleochroism:** None

Crystal System: Hexagonal (trigonal) fibrous aggregates
Chemical Composition: SiO_2 silicon dioxide
Transparency: Dull, translucent
Fluorescence: None

The opaque brownish form of chalcedony is known as sard. It is similar to carnelian in make up and appearance but darker in color. It is one of many varieties of cryptocrystalline (microscopically crystallized) quartz.

The name sard is a modern shortening of the Latin *sardius* and based on the Greek city name *Sardis*, capital of Lydia in Asia Minor. The stone is mentioned in many ancient writings, but few attributes are ever listed. The second century B.C. Greek historian Damigeron writes only a few lines concerning the stone. "This has a light color, or is spotted; worn by a woman it makes them lovable. Carve the vine and the ivy intertwined on it."[1] The only use listed in Marbod's classic poem is as a balance to the powers of onyx. He also volunteers a commentary on its relative value, "Cheapest of gems, it may no showe of fame, for any virtue save its beauty claim;"[2] The fourteenth century Lapidaire of Chevalier Jean de Mandeville contains one of the most extensive descriptions and lists of attributes of any late medieval literature. Mandeville does not, however, make a distinction between sard and sardonyx when he states:

> Sard is a white stone; there are five kinds. Some are divided like an onyx, with white part and a black part. Other kinds are topped with redness, and these are called sards. The darkest kinds are the most valuable. The sard gives a beautiful color to the face, makes a person chaste and humble, gives happiness and comforts the sight, destroys spells and enchant-

ments. It also works against a woman's flowing and any other flowing of the blood. They are found in India, Libya, Arabia and at Sodom.[3]

The *Speculum Lapidum* of Camillus Leonardus adds gender to the stone's description, "The males shine brighter than the females; for the females are the fattest and glitter more obscurely."[4]

One bit of Christian lore is the use of Sard as a symbol for the apostle Philip. This is based on the writings of the tenth century Bishop of Caesarea, Andreas, one of the earliest writers to associate the Foundation Stones of Revelations with the twelve apostles. He proposes sard's connection with St. Philip and includes healing virtues in the following passage.

The sardius with its tawny and translucent coloring suggests fire, and it possesses the virtue of healing tumors and wounds inflicted by iron; hence I consider that it designates the beauty of virtue characterizing the apostle Philip, for his virtue, animated by the fire of the Holy Spirit, cured the soul of the wounds inflicted by the wiles of the devil, and revived it.[5]

Sardonyx

Crystal System

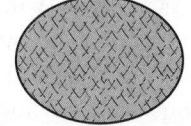

Microcrystalline aggregates

Color: Red-brown alternating with bands of white or brown	**Streak:** White
Mohs' Hardness: $6\,^1/_2 - 7$	**Specific Gravity:** 2.60 - 2.65
Cleavage: None	**Fracture:** Uneven
Refractive Index: 1.544 - 1.553	**Birefringence:** 0.009
Dispersion: None	**Pleochroism:** None

Crystal System: Hexagonal (trigonal) microcrystaline aggregates
Chemical Composition: SiO_2 silicon dioxide
Transparency: Translucent to opaque
Fluorescence: Varies with band: partly strong; yellow, blue-white

> Sardonyx consists of alternating bands of sard and white and/or black onyx. It is another variety of cryptocrystalline (microscopically crystallized) quartz.

This gem, listed as the birthstone of August in many texts, is almost unknown by the general public and rarely found in modern jewelry. References to sardonyx in ancient literature are also limited. Early Greek and Roman writers make no mention of this marriage of stones. However, the medieval author Marbod starts his entry with a clear description,

> The Sard and Onyx in one name unite,
> And from their union spring three colours bright.
> O'er jetty black the brilliant white is spread
> And o'er the white diffused a fiery red:
> If clear the colours, if distinct the line,
> Where still unmixed the various layers join,
> Such we for beauty and for value prize,
> Rarest of all that teeming earth supplies.[1]

He also states with regret that the stone is powerless. The power given to sard is said to act as a counter-agent to onyx's negative attributes. For this reason many writers may have assumed that sardonyx was then neutral. Camillus Leonardus disagrees, "Its virtue is to put a restraint on lascivious motions, and make men merry and agreeable."[2]

Christian symbolism assigns each of the Foundation Stones, listed in Revelations XXI, to one of the twelve apostles. Andreas, Bishop of Caesarea states that sardonyx is the stone of James and gives the following reasoning.

> By the sardonyx, showing with a certain transparency and purity the color of the human nail, we believe that James is denoted, seeing that he bore death for Christ before all others. This the nail by it color indicates, for it may be cut off without any sensible pain.[3]

Modern practitioners of crystal therapy list more attributes than the ancients. It is said to bring balance and happiness to marriages, give courage to its owner, attract quality friends and cure insomnia. It is also listed as a stone which, by its presence, works to reduce crime in the immediate area.

> Wear a Sardonyx, or for thee
> No conjugal felicity.
> The August born without this stone
> 'Tis said must live unloved alone.[4]

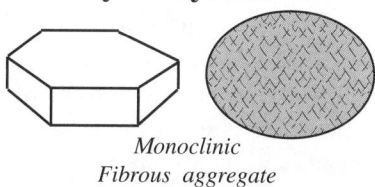

Serpentine

Crystal System

Monoclinic
Fibrous aggregate

Color: Dark green to apple-green, variegated or mottled with white
Mohs' Hardness: $2\,^1/_2$ - 4 bowenite and williamsite varieties 5 - $5\,^1/_2$
Cleavage: None
Refractive Index: 1.560 - 1.570

Streak: White
Specific Gravity: 2.50 - 2.60 bowenite up to 2.80
Fracture: Spintery to conchoidal
Birefringence: up to 0.006
Pleochroism: None
Dispersion: None

Crystal System: Monoclinic; fine grain fibrous aggregate
Chemical Composition: SiO_2 silicon dioxide
Transparency: Translucent to opaque
Fluorescence: None

The stone most often confused with jade is serpentine. The colors are like jade, green and yellowish green, but it is softer and easily distinguished by refractive index. Many Chinese carvings, assumed by their owners to be nephrite jade, are actually serpentine. The mineral crystallizes in the form of fibers, thin layers, masses, and sometimes single crystals. It may be green, very dark green, brownish red, brownish yellow, yellowish green or white. It may also be mottled with red or show a net-like webbing of magnetite.

The name serpentine derives from its resemblance to snake skin, particularly when interlaced with magnetite. Italian peasants use the stone as an amulet to protect against the bite of venomous creatures. It is also said to be capable of drawing the poison out of a bite by direct application. Although it is easily carved or cut en cabochon, its powers are neutralized if it is not left in its natural shape. Some say touching the stone with an iron tool renders it useless, but this is not a widely held belief.[1]

Modern beliefs include the use of serpentine medicinally. The stone is used as a healing tool in the treatment of heart and lung disease. It is proposed that by touching a subject the stone is an effective treatment for diabetes and hypoglycemia, and as a talisman, it aides in the absorption of calcium and magnesium. As with the belief in its ability to draw out poisons, it is recommended as a way of eliminating parasitic infestations of the body.

Spinel

Crystal System

Isometric (cubic)

Color: All, ruby red favored
Mohs' Hardness: 8
Cleavage: Imperfect
Refractive Index: 1.715 - 1.735
Dispersion: 0.020

Streak: White
Specific Gravity: 3.57 - 3.90
Fracture: Conchoidal, uneven
Birefringence: None
Pleochroism: None

Crystal System: Isometric; octahedron, twins, rhombic, dodecahedron
Chemical Composition: $Mg(Al_2O_4)$ magnesium aluminum oxide
Transparency: Transparent
Fluorescence: Red; strong red to pink

Spinel is found in a wide range of colors due to various impurities, including deep-red, brownish-red, rose-red, carmine-red, orange blue, blue-violet, purple, green, and black. Pure spinel is colorless. Spinel tends to have fewer flaws than comparably sized corundum. Large stones, those over ten carats, are usually reserved for museum display. The source of the world's finest gem quality stones is the gem gravels of Sri Lanka and Burma. Coincidentally, these are the

> same localities noted for ruby and sapphire. Most of the spinel found in commercial jewelry is synthetic. The stone is produced to mimic any of the birthstones used today. Synthetic spinels are often labeled as the stones they are meant to represent.
>
> The misnaming of this stone has been the rule, rather than the exception, for this gem. It has most commonly been known as balas ruby, ruby spinel, rubicelle and sapphirine. Most of the lore written concerning spinel has been as its misnomer, balas ruby. The gem referred to in ancient literature is the deep-red or carmine-red stone.

The lore of the spinel is found in the lore of the ruby or carbuncle. As a deep red stone it is given the attributes, by color association, of these two stones. Hindu tradition divides ruby into four castes. One type of "ruby" is the spinel, called the Vaisya, another is the Sudra or "balas ruby." How red spinel was divided into two types is not explained in Indian literature. The balas ruby gained its name from the *Balascia*, ancient name for *Badakhshan*, the region that served as the source for the best stones imported into Europe in the Middle Ages. Many balas stones found their way into major collections and have only recently been separated from true ruby.

One of the most famous balas rubies is included in the British crown jewels, the Black Prince's Ruby. The stone is set in the Imperial State Crown of 1937, the best known of all the British State Regalia. It is an uncut, but highly polished, water worn natural crystal of about 170 carats. Its actual weight is unknown as it has never been removed from its setting to be weighed. The Black Prince, son of Edward III, received the jewel from Don Pedro the Cruel, King of Castille as reward for his valuable service at the Battle of Najera in Northern Spain, 1367. It is said to be one of the balases worn by Henry V on his helmet at the Battle of Agincourt.[1] Legend says the stone deflected what would have certainly been a fatal blow during the battle.

Another famous spinel, weighing more than 400 carats, is set at the apex of the crown made for Empress Catherine II of Russia. It resides in the Diamond Treasure of Russia in the Kremlin. A fabulous spinel, although said to be very dark and lacking brilliancy, is the 361 carat Timur Ruby. It is not cut in a traditional form but the natural facets of the crystal have been polished. The stone was given to Queen Victoria and remains in the India Room at Buckingham Palace. It was listed in the official catalog of the Crown Jewels in 1851 as a very large "spinelle ruby."

Modern gem therapists divide spinel by color and recommend its use for various personal therapies according to the stone's hue.

Black spinel— serves as a protection from external harm.
Blue spinel— is used to reduce sexual drive and promote chastity.
Brown spinel— works as a filter to clarify thoughts.
Dark blue spinel— enhances psychic powers.
Green spinel— stimulates kindness, love, and devotion.
Orange spinel— brings fertility and reduces frigidity (a counter stone to blue spinel) and it adds to one's creativity.
Red spinel— provides physical vitality.
Violet spinel— encourages spiritual development.
White spinel— aides in communication and enhances mystic contacts.
Yellow spinel— stimulates intellectual powers.

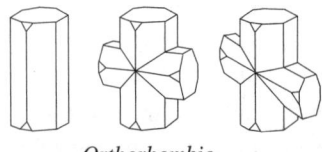

Staurolite

Crystal System

Orthorhombic

Color: Tan to brown to black
Mohs' Hardness: 7 - 7 $^1/_2$
Cleavage: Distinct pinacoidal
Refractive Index: 1.736 - 1.746
Dispersion: 0.021

Streak: White to gray
Specific Gravity: 3.65 - 3.77
Fracture: Conchoidal, uneven
Birefringence: .010
Pleochroism: Distinct

Crystal System: Orthorhombic, interpenetrating twins common, prismatic
Chemical Composition: $HFeAl_4Si_2O_{13}$ hydrous silicate of aluminum and iron
Transparency: Transparent to opaque
Fluorescence: None

> Staurolite is a reddish brown crystal that often occurs twinned, forming a natural cross. These twins may cross at a ninety degree angle to form a Greek cross or at a sixty degree angle as in a St. Andrew's cross. The twin crystals may be marketed under one or more of the following names: "fairy-stone," "cross-stone," "twin stone," or "lapis-crucifer." Clear crystals are found but rarely sold as cut stones.

The name staurolite comes from the Greek word σταυροζ, or *stauroz*, meaning cross. It is also sometimes called Baseler taufstein (baptismal stone) from its use in baptisms in the area of Basel, Switzerland. The "lapis crucifer" referred to by de Boot was probably not staurolite, but likely chiastolite. These stones are found in New Mexico, Switzerland, Britain, and France. Another common source is Fairy Stone State Park in Patrick County, Virginia. The origin of the stones is the subject of a local legend.

Once upon a time, very long ago, there was a perfect place that you may have seen in your dreams: a woodland glen of peace and serenity, of misty mornings and long, warm afternoons. Fairies lived in this valley and on surrounding hillsides. Their home was idyllically suited to them, with mountains, lakes, streams, and shady woods.

> On day a messenger arrived to tell the fairy king and queen news of the crucifixion of Jesus Christ in a far-off city. When the fairies, wood nymphs, and elves heard the sad news, they wept. Their tears fell to Earth and crystallized to form small stone crosses. As the centuries passed, the fairies gradually disappeared from this glen, but the little stone crosses still remain.[1]

Another North American legend says Pocahontas wore a necklace full of small fairy crosses and that she gave one to Captain John Smith as a token of good luck. A display card which accompanies staurolites sold in some Appalachian souvenir shops contains the following story:

> The rain god sent down millions of little rain drops, which were in the form of crosses. A voice out of the firmament accompanied the stones and said, "I am the soul in all that lives, time without end am I, and the life of things to be, the Spirit Celestial and

Supreme. All was, all is, and all ever shall be." The all voice spake, and motion was. "By virtue of my presence all the living are brought forth into life. I am the quickener, The mover, the creator. In obedience to my will, man named me, not after anything on earth, but by the sounds the wind uttereth; E O I H, Elohim, Jehovah, Creator, The Great Spirit, Father, The Almighty, The Ever Present."[2]

The European stones have long been used as amulets for protection from witchcraft, illness, accidents and other disasters. Another legend of England says the cruciform configuration was used to heal Richard the Lionhearted of a malarial-type disorder during the Crusades. In the British Isles the stones are worn as charms and, as in North American legend, were thought to have dropped from heaven.

Contemporary crystal therapists say, when worn as an amulet, "fairy-stones" provide comfort in times of extreme stress, eliminate depression, ease addictive traits, and reduce one's tendency to over-extend personal time. It is said that President Theodore Roosevelt wore a staurolite charm for these reasons on his watch chain.

Firm friendship is November's and she bears
True love beneath the Topaz she wears.
Anonymous

Topaz

Crystal System

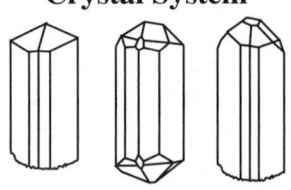

Orthorhombic

Color: Colorless, yellow, red-brown, light blue, pinkish red, pale green
Mohs' Hardness: 8
Cleavage: Perfect
Refractive Index: 1.610 - 1.637
Streak: White
Specific Gravity: 3.49 - 3.57
Fracture: Conchoidal, uneven
Birefringence: 0.008
Dispersion: 0.014
Pleochroism: Yellow; definite, lemon-honey-straw yellow, Blue; weak, light blue, pink, colorless, Red; definite, dark red, yellow, pink red, Pink; weak, brown, Red; weak; yellow-brown, Yellow; weak; orange-yellow

Crystal System: Orthorhombic, prisms with multi-faceted ends, often 8-sided in cross section
Chemical Composition: $Al_2(SiO_4)(F, OH)_2$ fluor containing aluminum silicate
Transparency: Transparent
Fluorescence: Weak; unusable, Natural Black; red-reddish

Topaz forms late in the crystallization of rocks and may reach enormous size. Crystals of fine color may reach a foot in length. One hundred-pound crystals have been found in the mines of Brazil. It is very hard with a hardness of eight but has well developed cleavage. It resists wear well but may be easily chipped or split by a slight blow. The gem's color may be any shade of yellow to brown, light to medium blue, or colorless but is rarely pink. Many of the popular colors; blue, or greenish-blue, sherry, and pink are often artificially enhanced by heat or radiation.

The jewelry trade has attached the term topaz to a variety of stones. "Smoky topaz" is actually smoky quartz; "citrine topaz," "Bohemian topaz," and "occidental topaz" are all names for citrine; "Oriental topaz" is a name used for yellow corundum. Genuine topaz may only have its name properly modified with the prefixes "precious" and "Imperial." The name Imperial topaz is reserved for fine, warm golden-brown stones from Brazil.

The name of this birthstone for November may have originated from one of two possible sources. According to Pliny the Elder, the name comes from Topazios, an island in the Red Sea which was the location of ancient peridot mines. The island's name is from the Greek word meaning "to guess," since it is often shrouded in fog. A second possible source is the Sanskrit word topas, meaning "fire." The reference to fire was appropriate because many early writers were only familiar with yellow to orange colored stones. Early texts mention yellow-green topaz, a possible confusion with the peridot harvested from its namesake island.

Earliest writers only mention topaz and do not list any virtue for the stone. It is known that the stone served in ancient times as an amulet to ward off the evil eye. It was tied to the left arm to protect its owners from any curse, as well as to guard against liver trouble.

Marbod, eleventh century Bishop of Rennes, describes two varieties of the gem. He probably is referring to citrine and/or peridot in some of his verses.

> From the seas remote the yellow Topaz came,
> Found in the island of the self-same name; . . . .
> And but two kinds to eager merchants known.
> One vies with purest gold, or orange bright;
> The other glimmers with a fainter light:

He does list two attributes in his classic poem,

> Its yielding nature to the file gives way[1]
> Yet bids the bubbling caldron cease to play. . . .
> One only virtue Nature grants the stone,
> Those to relieve who under hemorrhoids groan.[2]

The gem's ability to cool boiling water is the one virtue passed down to most of the later authors of lapidaries.

Early Christian lore credits the stone with curing all eye troubles. This is based on the writings of the tenth century scholar Andreas, Bishop of Caesarea, one of the earliest writers to associate the Foundation Stones of Revelations with the twelve apostles.

> The topaz, which is of a ruddy color, resembling somewhat the carbuncle, stops the discharge of the milky fluid with which those having eye-diseases suffer. This seems to denote Matthew, for his was animated by a divine zeal, and, his blood being fired because of Christ, he was found worthy to enlighten by his Gospel those whose heart was blinded, that they might, like new-born children, drink of the milk of faith.[3]

Harriet Fobes claims in her book that St. Matthew himself was cured of eye diseases by the use of a topaz.[4]

The virtues of the stone have grown through the centuries. Saint Hildegarde, wife of Theodoric, claims the stone emitted so brilliant a light that prayers could be read in the darkened chapel without any additional light. She therefore recommends the stone as a cure for "dimness of vision." Her treatment consisted of placing the gem in wine for three days and three nights, rubbing the eye with the stone before going to sleep so the liquid just touches the eye and then using the wine as an eye wash for five days.[5]

The fourteenth century Lapidaire of Chevalier Jean de Mandeville restates the attributes listed by earlier writers and adds some not previously noted. Mandeville also reiterates some of the misconceptions that topaz and peridot are like stones.

> Topaz is very light citrine, [meaning lemon-colored, The use of this color name has added to the misuse of citrine quartz as a form of topaz] and sometimes the color of emeralds. Topaz is found in many places, especially on an island of Arabia called Topaz, whcih explains the origin of the name of the stone. Sometimes topaz is found in such large sizes that large pictures are made of it. This stone cools things off greatly because when one is put in boiling water, the water will stop boiling. Topaz will cure gall; if one has the gall, it will not spread as much. Topaz will also check one's anger and will lessen lechery.[6]

Other references list many applications for the stone. Topaz was said to become invisible in contact with poisons. Powdered and placed in wine, it was claimed to act as a cure for asthma, insomnia, burns and hemorrhage. As an amulet it is used to drive away sadness, strengthen intellect, and grant courage. All these powers were said to increase and decrease with the phases of the moon and be even more powerful if used in moonlight. In the thirteenth century *Book of Wings* author Rabbi Ragiel states, "The figure of a falcon on a topaz helps to acquire the good will of Kings."[7] The topaz is also considered precious by African bushmen, it is used in ceremonies for healing and contacting spirits.

The thirteenth century Hindu physician Naharare states topaz is an, "appetizer of excellence, it tastes sour and is cold. It is a remedy for flatulence. Any man wearing this stone will have long life, beauty, and intelligence."[8] A Roman physician of the fifteenth century prescribed a particular topaz in his possession as a cure for the plague. He claimed

> *Who first comes to this world below*
> *With dull November's fog and snow,*
> *Should prize the Topaz's amber hue.*
> *Emblems of friends and lovers true.*[9]
> An alternate version of this poem reads;
> *Who first comes to this world below*
> *Under Sagittarius should know*
> *That their true gem should ever show a topaz.*

the stone had been owned by Pope Clement VI and Pope Gregory II. The stone was used to touch the plague sores and this cured the disease. The fact that the stone had been owned by two pontiffs added to the power of the cure.

Modern practitioners of crystal therapy say topaz is a "stone of true love and success in all endeavors."[10] The gem is used to relieve insomnia, nightmares, fears, and anxiety. Medicinally it is used as a treatment for loss of the sense of taste and in healing wounds and skin eruptions. It is also recommended that topaz be used in conjunction with amethyst to give a soothing and stabilizing effect to white magic and healing procedures.

Tourmaline

Crystal System

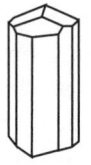

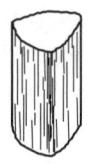

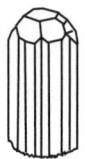

Hexagonal (trigonal)

Color: Colorless, pink, red, yellow, brown, green, blue, violet, black, multicolored
Mohs' Hardness: $7 - 7\frac{1}{2}$
Cleavage: None
Refractive Index: 1.624 - 1.657
Streak: White
Specific Gravity: 3.02 - 3.20
Fracture: Uneven, small conchoidal, brittle
Birefringence: 0.020
Dispersion: 0.017
Pleochroism: Red; dark red - light red, Yellow; definite, dark yellow-light yellow, Brown; definite, dark brown-light brown, Green; strong, dark green-light green, Blue; strong, dark blue-light blue

Crystal System: Hexagonal (trigonal); usually long crystals with triangular cross section and rounded sides, definite striation parallel to main axis, often several prisms grow together.
Chemical Composition: $(NaLiCa)(Fe_{11}MgMnAl)_3Al_6((OH)_4(BO_3)_3Si_6O_{18})$ aluminum borate silicate (varied composition due to trace elements)
Transparency: Transparent, opaque
Fluorescence: Colorless; weak, green-blue, Red; weak, red-violet, Pink, Brown, Green, and Blue; none

Tourmaline is known for its dazzling array of colors. Some stones display two or three color bands along the length of a single crystal; some show concentric zoning in cross section. The most delicate and varied colors in the mineral kingdom are found in the tourmaline group.

The finest green tourmaline resembles emerald, but these stone are rare. Most green examples exhibit tinges of brown, blue or yellow. The blue, colorless, orange, yellow and red stones are rare. The reddish-purple or violet stones are the most valuable. Color zoned stones known as "bicolor" or "multicolor" stones are gaining popularity in the jewelry trade. A most popular type is the "watermelon tourmaline," a stone that is pink in the center and green around the outside.

Tourmaline is a general group of stones that share similar atomic structures and chemical make up. Many color varieties carry different names, leading to some confusion; rubellite is the name for pink or red stones, indicolite the name for blue tourmaline, reddish-violet stones are called siberite, brown stones are dravite, and achroite are colorless. Adding to this collection of terms are mineral species found within the tourmaline group, such as elbite, schorl.

The gem is a very practical stone for everyday wear. Its color mimics many more expensive stones, it has no cleavage (reducing breakage and chipping), and its hardness of seven to seven and one-half resists dulling.

The vast range of colors that tourmaline shows may have led to its name. The root word for the name, *turmali*, is Sinhalese for "mixed precious stones." This indicates an inability of ancient gem dealers to differentiate tourmaline from other colored stones found in the gem gravels of Sri Lanka. Pliny the Elder wrote of the *lychnis*, a stone from India. He states that some resemble the carbuncle or ruby, and another type is violet and called the *iona* (from the Greek *ion*). He also tells of the lynchnis' peculiar ability to attract small pieces of straw or paper if heated by the sun or the friction of rubbing. His description is probably in reference to tourmaline's pyroelectric quality. If the gem is heated or rubbed it becomes static charged, attracting dust, lint and small pieces of paper.

It is said that the electrical properties of the stone were first noticed by Dutch school children on a tour of a display of tourmaline in Amsterdam. They noticed small bits of straw from the original packing material were still attached to the stones. Tests conducted by Dutch lapidaries confirmed the static electric charge stored in the stones. The story of the children's discovery is charming, but Dutch jewelers probably knew of this phenomena well before the school children. They were in the habit of testing stones for durability by placing them in a fire. The heated tourmaline would collect ashes from the fire as it was removed. Whatever the source, the stone was known for a time in Holland by the name *aschentrekker* or "ash-attractor." The first record of scientific proof of the pyroelectric property of tourmaline is found in the work of eighteenth century Swedish botanist Carl Von Linné. He called the tourmaline the "electric stone."

Tourmaline is a gem that has often been confused with other gems. Theophrastus speculated that true emerald may have been evolved from another gem. As evidence, he tells of a Cyprian emerald which is half emerald and half jasper. The conclusion was that the metamorphosis must have been incomplete in this stone. If he used the word jasper to refer to a red stone, it is evident by his description that he is referring to a color zoned tourmaline crystal. The Dutch mineralist Johann de Laet also tells of interesting "emerald" crystals he found in the Spiritus Sanctus mines of Brazil in 1647. The crystals were said to be "cylindrical in form, striated, and vitreous in luster."[1] The color was compared to a transparent prase. Laet's description is characteristic of the typical tourmaline crystals found in Brazil.

The tourmaline is used by tribes in Africa, native Americans, and aboriginal group of Australia as a talisman that protects against all dangers. It is thought to bring healing powers to a shaman or medicine man. Ancient ceremonies in India included the use of the gem as a tool to bring insight and help in the discovery of that which is good. It would also serve to make known who or what was the cause of troubles or evil deeds.

European lore does not list attributes of the tourmaline, as the stone was not known as a separate gem. A few medicinal uses are given in the English tale of *Reynard the Fox*, derived from the eleventh century *Roman de Renard* by William Caxton, published in 1481. A magic ring is said to contain what may be a tri-colored tourmaline. The ring was set with a stone "of three maner colours," red, white and green. Red had the attributes given to the ruby, "The shynyng of the stone made and gaf as grete a lyghte as it had been mydday." White was remedy for diseases of the eye, headache and most all ills, "sauf only the very deth." This portion of the stone would be stroked on the affected parts, or the stone would be soaked in liquid and drunk to relieve internal maladies. The third part was "grene lyke glas," with small purple spots. This portion brought love and friendship, as well as victory in battle.[2]

Those working with modern crystal therapy recommend the use of a large crystal as a tip for wands. Such a tool is said to clear auras and energize the spirit. Tourmaline is also used "to attract inspiration, to diminish fear by promoting understanding and to encourage self-confidence."[3]

My Tourmaline: an American tale

The electric properties of the gem led to an interesting tale. The story *My Tourmaline* by Saxe Holme, published in 1886, is based on the supposed sensitivity of children to the emanations of stones. The story also commemorated the discovery of a fine specimen taken from Mount Mica in the state of Maine.

One day while on a walk in the country, a little girl's attention was drawn to a spark of color from the gnarled roots of an old tree. Upon closer inspection the girl found a beautiful crystal. As she struggled to retrieve the gem, her leg became tangled in the tree's roots causing a severe sprain. The injury to her leg resulted in a six week stay in bed. Her only consolation was the presence of her newly found crystal during her dreary days of convalescence.

The little girl was convinced the stone had a life of its own. She knew this by the prickling and tingling sensation felt when she held the crystal in her hand. The stone was a perfectly shaped crystal, as if cut by a lapidary. It was red at one end and green at the other with a clear border between. She feels the special sympathy of the stone for her condition and used it as a healing aid. Held in a silken bag and placed against her cheek, it acted to calm her restlessness and bring her comfortable sleep.

Word of this fine specimen reached beyond the girl's home. Upon her recovery, she consented to placing the stone in the collection of a distant museum. She was soon heartbroken to learn it was carelessly lost in transport. Convinced she would never see her "Stonie" again, she decided she must go on with her life.

The little girl grew to womanhood, married and traveled to Europe as a young bride. She and her groom were invited to view the mineral specimens of a noted collector. As soon as she entered the collector's home, she felt an odd presence. When she was allowed to view his prized tourmalines, she recognized her lost crystal and renewed their connection.[4]

The heav'n-blue turquois should adorn
All those who in July are born;
For those they'll be exempt and free
From love's doubts and anxiety
Anonymous

Turquois

Crystal System

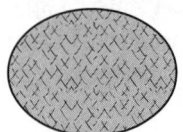

Aggregate

Color: Sky blue, blue-green, apple green
Mohs' Hardness: 5 - 6
Cleavage: None
Refractive Index: 1.61 - 1.65
Dispersion: None
Pleochroism: Weak

Streak: White or greenish with brown or black spots
Specific Gravity: 2.31 - 2.84
Fracture: Conchoidal, uneven
Birefringence: Not detectable with common instruments

Crystal System: Triclinic; seldom and small, usually grape or kidney-shaped aggregate
Chemical Composition: $CuAl_6((OH)_2/PO_4)_4 \cdot 4H_2O$; a copper containing basic aluminum phosphate
Transparency: Opaque
Fluorescence: Weak; green-yellow, light blue

Turquois is a secondary mineral, leached from surrounding earlier rocks. It is deposited in veins and as rounded masses, called *nodules*, by ground waters. The gem most commonly exists in thin irregular layers which require the neighboring minerals, called *matrix*, to be included in cut stones. Occasionally, nodules are found in alluvial deposits, but these deteriorate as weathering breaks down the gem. The highest quality turquois is found in veins thick enough to allow cut stones which are free of matrix. Turquois is almost always found as an opaque micro-crystalline mineral. The blue color of the stone is attributed to its copper content. Greenish tinges are often seen and are caused by traces of iron.

The gem is often identified and/or graded in the jewelry trade by place names. These names do not necessarily denote the place of origin.

Persian turquois is an intense medium blue material that is slightly harder, less porous, takes a high polish and has a slightly higher specific gravity than other varieties.

American or Mexican turquois is pale to light blue, or greenish to bluish green. It is often quite porous.

Egyptian turquois has a poorer color than American or Mexican due to a higher iron level. This is often seen as a strong greenish cast or yellowish green body color. Egyptian is also characterized by its higher density and lower porosity.

Matrix turquois contains portions of the surrounding rock. This may be in masses, spots, bands or as a fine overall pattern known as *spiderweb*. Evenly distributed and distinct spiderweb matrix is the most desirable of this type.

Subdivisions of the matrix variety may be named for a particular mining location which is know for a specific color or patterning of stones. This again is no guarantee of a gem's actual origin. As an opaque gem, most turquois is cut en cabochon. Beads, carvings and slabs for inlay are also favored forms.

Before World War I the primary source of this gem was Persia, called Iran today. America has become the most important producer of turquois in the late part of the twentieth century. Much of this contemporarily mined material is porous and of inferior color. The introduction of waxes, oils, silicones and plastics is a common method of enhancing colors and the quality of polish. All of these methods are unstable and will deteriorate with age and use.

Although often spelled with an 'e' at the end, mineralogists in the United States have agreed to a standard spelling without the ending letter. This same standard will be used in the text of this book.

The oldest pieces of wrought jewelry known to exist are four gold and turquois bracelets. They were found on the arm of the 7500 year-old mummy of Egyptian Queen Zer. These artifacts were unearthed at the turn of the twentieth century. Turquois, birthstone of December, is one of the few stones that has maintained its popularity from earliest antiquity to the modern era. Like lapis-lazuli, turquois was prized by many early cultures. It was valued by the Egyptians, Babylonians, Aztec, Maya and Incas. Tibetans and Native Americans have treasured the stone from their earliest history to the present day.

The name of the stone has taken many forms throughout the centuries. Our modern word *turquois* is derived from the French *pierre turquois*, meaning "Turkish stone." The gem probably first entered Europe by way of the eastern trade routes through Turkey, and was thought to originate there. The first written account to use this name is credited to the early thirteenth century author Arnoldus Saxo in his works on the properties of stones. He writes:

> The *turcois* is a stone of a yellow color, verging on white. It is so called from the regions of Turkey, whence it originates; it has the quality of preserving the eyesight from external injuries when superimposed on the eyes, and it induces hilarity.[1]

Some of the earliest written references list the stone as a form of lapis-lazuli. Pliny the Elder writes of a stone he calls *callais*. He states it "is like *sapphiros* (lapis-lazuli) in color, only that it is paler and more closely resembles the tint of the water near the seashore in appearance."[2] Using the color description as a key, scholars believe he is speaking of turquois. Another stone in Pliny's *Natural History* is called the *callaina*, also thought to be a turquois.

The English scientist Thomas Nicols attempted an inventory of names for the gem in his *Arcula Gemmea* of 1653. He claims:

> In Latin- *Turchus, Turchicus, Thurchina, Turchesia, Turchoys*
> Pliny called it *Boreas*
> Martinus Rulandus said it is the sixth jasper *Turcica*
> Dutch- *ein Orientisher Turckise*
> Greek- *Jaspis Aerizusa*
> Mesue called it *Feruzegi*
> Baccius called it *Turcicus*

It may be seen that most of his list is based on the same root as our French derived name. The term credited to Mesue is obviously a form of the Persian names for the stone, *ferozah* and *firohah*, which both translate as "victorious." Fobes states in her work *Mystic Gems* that turquois is the "national stone of Persia, called piruzeh," yet another name for the same gem.[3] In Tibet it is called *gyu*, similar to the Chinese word for jade, *yu*, their most precious gem.

The lore of turquois originates in the most ancient of writings. The Egyptians describe the stone in numerous papyri and associated the color of the gem with many of their gods. The god Amun is depicted as blue in murals, Isis is sometimes shrouded in a blue veil, and the Nile takes its name from the Sanskrit *nila,* meaning blue. Both lapis-lazuli and turquois enjoyed a high status due to their color. This "sympathy of color," and the granting of power to objects of the same color, is common in many cultures. Besides its connection with various deities, Egyptian physicians prescribed the stone, set in silver, as an amulet against diseases of the eye. It was thought to be particularly effective in treating cataracts. Ancient Egyptian physicians used an oxide of copper mixed with boric acid as an eye wash. Known as *lapis Armenus*, this astringent has been shown by modern medical research to be effective and is still prescribed. Later translations of Egyptian texts substitute turquois in the recipe. We may assume they knew of the copper content of the gem.

The papyrus of King Snefru, dated 4700 B.C., tells the story of a lost turquois recovered. Included in Petrie's *Egyptian Tales*, published in 1895, it is called *Baufra's Tale*.

> The King, on one occasion, being disconsolate, sought diversion at the suggestion of Zazamankh on the lake of the palace. He was rowed in a magnificent boat by 20 fair maidens, and was made glad by the sight of their rowing. But one of them dropped a jewel of new turquois from her hair and refused to be comforted for its loss. Then Zazamankh, seeing the King chagrined by this untoward event, spoke a magic speech, whereby the waters were divided and the jewel recovered. Then was the King much pleased over the skill of his chief reciter Zazamankh[4].

In western lore the turquois was granted the attributes of other blue or blue-green stones. The Roman historian Pliny gave little distinction to the callais, giving it the same attributes as the sapphire or lapis-lazuli. The sixth century Greek physician Alexander Trallianos wrote in a treatise on epilepsy:

> Wear on your finger a jasper shining blue-green like the turquois, and you will be cured from the disease (epilepsy); it is of great value.... Wholesome, if worn on the finger, are also the chrysolite and the jasper, which shine blue like the atmosphere or green blue like the turquois.[5]

The jasper referred to is the same stone named by Pliny, but is translated by scholars as turquois.

It was not until the gem was given its connection with Turkey, and its modern name, that it gained its unique attribute. Stories of its use as a protective stone for horses and their riders passed along the trade routes to Europe. Turquois has been used for centuries on the bridles of Arab horsemen and may have originally been the "horse amulet" of Persia. Horses were thought to pull the sun through the heavens and capable of keeping a sure course. The blue of the stone replicated the sky and added sure-footedness to the animals while on earth. The same gem was also said to protect horses from thirst, overheating and exertion. The Persians also had a strong belief in the evil eye and believed the color blue blocked its effect. Camels, mules and horses in Turkey and Iran still have turquois beads strung on their tail or mane for protection.

Many works by Arab authors have included detailed passages on the virtues of turquois. The Arabic botanist Ibn-el-Beithar (1197-1248) claimed if taken internally, the gem made one immune to the scorpion's sting. He also said that if ground and taken in a drink it would cure ulcers, stop growths in the eyes and repair any resulting damage. These same virtues had been listed by the Roman physician Dioscorides in the first century A.D. but given to lapis-lazuli. This may be one of many cases where turquois was seen as a variety of lapis. Dioscorides also states in his *Book of Stones*, "It shines when air is pure, and gets pale when air is dim."[6] Another Arabian authority, Ahmed Teifascite (b. 1253) wrote the following about the gem:

> The turquois is a stone of the same nature and quality as copper and is produced from the vapors of that metal, the turquois possesses the quality of becoming

> *If cold December gave you birth,*
> *The month of snow and ice and mirth,*
> *Place on your hand the Turquois blue,*
> *Success will bless you if you do.*[7]

> clarified or bright in time of serene and clear weather, and vice versa, becomes dull and obscure when the weather is dark and cloudy.....The stone brightens and refreshes the vision when it is looked at fixedly, and it is similarly beneficial to the eyes when it is used together with eye salves.[8]

The use of turquois as an eye stone is expanded upon by the twelfth century writer Muhammed Ibn Mansur

> The eye is strengthened by looking at a turquois. If one sees a turquois early in the morning, then he will pass a fortunate day. One should view a turquois at the time of the new moon.[9]

The works of these eastern writers either influenced European thought or were drawn from sources familiar on both continents. Some authors credited authorities of the Near East in their own writings, and some did not. The early thirteenth century scholar Albertus Magnus, referring to these Arab writers, states, "Moreover, they say it preserves sight and protects those wearing it from harmful accidents."[10] The late fourteenth century scientist Chevalier Jean de Mandeville states in his *Le Grand Lapidaire;*

> Turquois is three kinds of stone: green and opaque; it is found in parts of Cimbe [Location unknown]; the other is green mixed with white, like milk mixed with green; and it comes from the Orient and is better than the other; it comforts sore eyes and protects against the causes of them; it makes one bold and the horse that wears it does not get colic from drinking cold water; and Indians say it serves better in battle and better protects a man than a diamond, and that this has been tested and proven countless times; moreover, it has such potency that the man who acquires it, cannot make children and a woman will not conceive.[11]

The concept of turquois as a stone of sympathy became fashionable during the sixteenth and seventeenth century. Many tales have been related of the stone's ability to predict illness and even demonstrate emotion. Bernardus Casesius, writing in 1636, retells a story of such a stone and uses as documentation the writings of previous scholars.

A certain man is said to have possessed a turquois of great value and exceeding beauty, and when he had paid his debt to the laws o' fate, the gem seems to have wept over its owner's death, its glorious luster becoming dimmed, and it began to appear obscured for a long time. Caussinus relates this and says that this stone is the symbol of the most devoted friendship. The same story is told by Ruëus, who adds that he knew the owner of the stone in question, and that it had become noticeably obscured after his death; and he also adds that the same gem was subsequently bought at a low price by another citizen, but as soon as it had been transferred to a new owner, it recovered its pristine brilliancy. But, as we have said, after the death of the first owner, it appeared as if dimmed for a long time and exhibited a crack in its median line.[12]

Turquois is known to change color with wear and exposure due to its porosity. It will lighten in color if washed with soap or an astringent like alcohol. The absorption of oils will cause it to darken or turn greener in hue. It is also easily chipped and fractured along natural lines of separation established by matrix impurities. The instability of the stone was noted by many historians, but attributed to emotional influences or the stone's ability to absorb misfortune.

An example of the turquois' ability to protect a horse and rider, and to protect the wearer from injury, is given in the seventeenth century writings of Anselmus de Boot. This court physician to Emperor Rudolph II related the story of an outstanding turquois owned for thirty years by a Spanish nobleman. The gem was eventually sold as part of the nobleman's estate. All who saw the stone soon after the sale were amazed by its loss of color. The new owner of the stone was de Boot's father, who bought it for a fraction of it original value. Ashamed by the lack of color the stone possessed, he decided to give it to his son, saying, "Son, as the virtues of the turquois are said to exist only when the stone is given, I will try its efficacy be bestowing it upon thee."[13] The son did not respect the original

> "A compasionate turquoise which doth tell
> By looking pale, the wearer is not well."
> An Anatomy of the World, *John Donne* [14]

value of the stone and had his coat of arms engraved on it, an act usually reserved for common jasper to be worn in a signet ring. After wearing the engraved gem for only a month, the stone's color and luster returned.

The long held belief that turquois protects a horse and rider seems to have been confirmed by another incident reported by de Boot. While on a later trip, and wearing the same stone, he was thrown from his horse in a violent manner. He was surprised to find that neither he nor his horse were injured in any way, but one quarter of his stone was broken away. Convinced the stone had protected him, de Boot continued to wear the damaged gem. He reported that he was again saved by the stone from serious injury. Lifting a heavy load, de Boot was certain he had broken his ribs. He was "saved" by the turquois signet breaking in two and limiting his injury to a mild muscle strain. De Boot also stated in 1609 that no man was considered well dressed without a turquois on his hand. He also reported that few women could be found wearing the stone. This adornment of the male hand with turquois was a fashion statement continued by many Englishmen into the late nineteenth century. A contemporary of de Boot's wrote in general terms of the stone's protective properties. Scientist and author of a treatise on gems, Joh. Baptist Von Helmont, states, "Whoever wears a turquoise, so that it or its gold setting touches the skin, 'vel non, perinde est,' may fall from any height; and the stone attracts to itself the whole force of the blow, so that it cracks, and the person is safe."[15] The court jester of Emperor Charles the Fifth did not seem to share majority opinion when he offered this advice to the monarch, "If you should happen to fall from a high tower whilst wearing a turquois, the stone will remain unbroken, but for you—I cannot say!"[16]

The virtues attributed to turquois have not been limited to its protection in falls. The lore of many western countries include turquois as an amulet. The English have used turquois as a love charm for centuries. Shakespeare writes of such a use in the *Merchant of Venice*; Leah gives a turquois ring to Shylock to win his love and induce a proposal. Turquois settings are also given by English maidens to their lovers as forget-me-not rings. In Germany, it serves as a barometer of one's affection. Worn by lovers as a talisman of mutual good luck, the stone is said to lose its color when love fades. Russian couples are fond of turquois rings as wedding bands, and into the early twentieth century soldiers in the Czar's army carried the stone to guard against wounds.

Turquois has maintained its lore and mystery throughout history. Even as many other stones fell from fashion or were dismissed as lifeless minerals, the attributes of this gem continued to fascinate. A late nineteenth century sceptic lists turquois as the only truly magical gem.

> Other precious stones have lost all the marvelous powers that belonged to them for centuries; the emerald no longer relieves the fatigued eyesight; the diamond can not now dispel fear; the sapphire, though still cold to the touch, has ceased to be able to extinguish fire. In these perverse days the hailstorm comes down even upon the wearer of an amethyst, and bright red coral attracts rather than repels robbers. But the turquois still retains one of its mysterious properties and flaunts it in the face of modern science. Sometimes slowly, sometimes suddenly, it unaccountably turns pale, becomes spotted, or changes from blue to white; and specimens that behave in the capricious manner are found more commonly than those whose color is distinctly permanent.[17]

Many cultures in Asia and the Near East prize turquois. From the deserts of Persia to the high mountains of Tibet this gem serves as a religious ornament, protective amulet and source of cures. The many attributes of the stone are summarized eloquently in a passage from the *Mani-Mala*, or *Chain of Gems* by Tagore. This compilation of gem wisdom from Arabian and Persian writers was published in 1879 and has served as a primary source for many later scholars.

> The turquoise possesses the virtues of the *Bish* stone. It cures all diseases of the head and the heart. By application over the eyes in the shape of Surmá, it increases lustre, prevents the fall of fluid therefrom, brings back the color of the pupils if they get white, and restores natural vision to those who are almost blind at night. It is a sovereign remedy for hernia, swellings, flatulence, dyspepsia, insanity, and ulcers inside the stomach or abdomen. In combination with other ingredients, it would relieve and cure the pains and swelling of the body caused by assault. Whether taken with other drugs or simply with honey, it has the power of curing epilepsy, spleen, stricture, etc. In cases of poisoning or snake bite, a durm or a quarter tola weight of turquoise should be given with wine;

for scorpion bites a third of the quantity would suffice. But as the above prescription may cause harm to the stomach, it should always have added to it a quantity of katilá. Hakim Aristatalis (possibly Aristotle) has limited the dose to one-eighth of a tola. Worn on the fingers as a ring, the turquoise brings about happiness of mind, dispels fear, ensures victory over enemies. He who after looking at the moon on the Pratipada (the first day of the new moon) casts his eyes over this stone becomes the master of fabulous wealth.[18]

The gem is used extensively in Tibet for ornamentation and as a symbol. Legend says the blue ceramic tile work seen throughout the region was formed by melting turquois over the pieces of clay. The color of the stone is common in Tibetan myths. It is the color of the hair of a goddess, the eyebrows of supernaturally born children, horses manes, tadpoles, flowers, bees and the tips of blessed arrows. Marriage vows are sealed by the symbol of a turquois affixed to an arrow and carried on the bride's back. Tibetan shamen revere the stone as a protector of things earthly and spiritual. A protection against the evil eye is given when a turquois is set in the forehead of the statue of Buddha and kept on prominent display. Another association of Buddha with the stone is a legend which says Buddha used a turquois pebble to destroy a monster. The tale is often compared to the David and Goliath story of the Bible.

The Chinese think of turquois as a common stone and do not use the gem in their myths and legends. The dominance of jade does not allow much reverence for other green gems. Of interest are two stories of the gem's origin. Thought to be a transformation of fir branches, the name for the stone translates to "green fir-tree stone." Another account of the formation of the gem is the belief that moss which grows on a rock for many years is consumed by the rock, assuming the color and resulting in turquois.

The greatest volume of myth and lore comes from the Amerindian people. Turquois is found in abundance in Mexico, Central America and the Southwestern part of America. The gem is also found in other areas of the Americas due to ancient trade carried on between tribes. It serves many functions to native groups but is generally regarded as a sacred stone which acts as a protective agent and bringer of powers. The color of the gem results in a close association with water, the sky or rain. Individual native cultures have in the past, and still do, revere the stone in their own legends and traditions.

The Aztecs held turquois in high regard. It shared first rank with emerald and jade. To them, blue was the color of supreme authority. The burial rights of the Aztec included placing a turquois bead in the mouth of the deceased to present to the gods upon arrival in heaven. The stone decorated many sacred objects and religious sites. Cortez was presented with a turquois encrusted insignia of Quetzalcoatl, their highest deity, when he first encountered the natives of Mexico. Early chroniclers of Aztec history describe the entire interior of temple rooms covered with turquois. One of these early writers, Sehagun, tells of a temple with four rooms, one of which was called the "apartment of emerald and turquois, because inside, instead of the lace work or the plates of gold, it was ornamented with mosiac work of emerald and turquoises, in a most beautiful manner."[19] Sehagun also reports a unique method of finding the gem used by the Toltecs. He claims they looked for the stone in wet places as the water level was rising. The "miners" would look for a puff of smoke to rise from the water, and at that point the stones were found in the ground below the water. The gem was also used by these ancient people as a medicine. The *xiuhtomoltetl* (turquois stone) was ground or crushed, mixed with water and administered as a cure for feebleness or nausea.

The Pima believe the loss of a turquois is due to bad magic. This loss will bring to the former owner a mysterious ailment which may only be cured by a knowledgeable medicine man. The healer can restore the sufferer's health by placing a blessed turquois, piece of slate or clear crystal in water. This newly consecrated stone is then presented to the patient to complete the cure.

The color of turquois gives the stone its power in the Zuni culture. Blue is the color of heaven and the color of the West. Turquois colored masks and body paint are an important part of ceremonial garb. Blue turquois is a male stone, any other color is female. The cradle boards of male children are set with perfectly round stones directly behind the child's head. This is said to insure masculinity in later life. The positive or upper world is symbolized by the sun, eagle and turquois, and the negative or underworld by the rattlesnake and the toad. A sacred meal "fed" to the gods as an offering consisting of ground corn, ground turquois and bits of shell. Fetish use is a common practice to the Zuni. A fetish is able to capture the spirit of living things and allow its possessor to hold the qualities of what it represents. They are often kept in fetish bowls and fed ground turquois, corn pollen and other bits of nourishment. A hole in the side of the bowl lets the spirit exit and return. A fetish form of great power is a seashell with the figure of a toad made of turquois set in black pitch. The blue coyote fetish represents the West and the place where the sacred sun resides when not in sight. A sheep with turquois eyes is used to insure the shepherd a fertile flock. Both of these objects also guard against disease, wild animals and accidental death. A protective use of the stone is in the form of turquois set prayer sticks. These sticks are placed in the walls of a house during construction as an offering to the dwelling god. This gesture will bring luck to the house and protect it from falling. A medicinal application is the placing of a turquois and herbs in the mouth to treat mouth sores. A legend of the Zuni tells of a young boy who is carried by an eagle to the Mountain of Turquois. The mountain shines so brightly in the sun that its reflected light serves as the source of the sky's blue color.

The Pueblo people say the turquois stole its color from the sky. Used as a charm, the gem brings good luck. The Chaco Cañon ruins at the Pueblo Bonito in northeastern New Mexico are rich with turquois in burial mounds. Early archaeologist George H. Pepper recorded finding 24,932 beads in one dig during 1896. Nine thousand beads and pendants were found on or around one skeletal remain. The Santo Domingo Pueblo hold *Keres* as a primary protective fetish. It is the figure of a gypsum prairie dog with turquois eyes. The people of this same Pueblo are known to have expressed their disapproval of the mining of the stone by early explorers. When Major Hyde opened mines in the region in 1880, he was told the stone should not be possessed by anyone whose Savior was not Montezuma.

The turquois is also considered a vital stone for the Apache. Called *duklij*, meaning blue or green stone, it is considered the most important stone a medicine man can possess. Without the gem his medicine is without power. Useful in bringing rain, it may be found by a diligent searcher at the end of the rainbow in moist ground. The Apache warrior believes a turquois affixed to a gun or bow makes it more accurate. The stone is also worn as a protection against all diseases.

Talismanic powers are also given to turquois by the Navajo. It has a role in nearly every aspect of daily life. The gem thrown into a river, with a prayer to the rain god, is a way to summon rains. Carrying a carving of a horse will bring, and help retain, many horses for the owner. A turquois attached to a gun stock assures an accurate shot. A bead fastened to the hair protects against lightning strikes and sudden thunderstorms. The same hair ornament also acts as a protection from snake bite. Shepherds carry fetishes made of the gem to insure abundant lambs from their ewes. Each household keeps a bag of herbs, shell and turquois to protect the family from unforeseen events. The gem serves many important functions in the Navajo religious life. Tied to prayer sticks, it is left as an offering at shrines. An offering of the stone to the wind spirit allows control over the winds. The Navajo believe the wind blows because it is constantly searching for turquois. The medicine man uses the stone to gain knowledge of the cause of diseases, places to dig for water, the location of lost items and the fidelity of one's partner in marriage. Milestones in a Navajo's life are marked with the gem. A child's ears are pierced at two hours of age to accept turquois earrings. Additional earrings may be added at puberty, marriage, initiations and other important events throughout life. The stone also is used as a form of currency, the oldest and purest of color having the greatest value.

To the Hopi, turquois is the petrified excrement of lizards. It is their most highly valued stone and hardly any ceremony can be performed without it. Mosaics of shell, jet and turquois are used to adorn fetishes used in ceremonies. A particularly strong fetish consists of bundles containing staffs of clothing, shell, turquois, quartz crystals and horse hair. The bundles are wrapped into one shaft by thread with a bundle of hair protruding from the top. No one is to speak of this fetish or touch it except for the owner. This bundle protects its owner and acts as a protective charm for life. The gem also serves as a love potion. The stone is held in front of a magician and a chant is sung which includes the name of a man and woman. The spell cast excites the woman, causes her to choose the man named in the song and drives her to his home. The magic of a young boy's first hunt is marked with beads of turquois. After his first rabbit kill, the sight of this event of passage is marked by spreading turquois beads on the ground. The stone also helps to insure success for the young hunter in the future. Fetishes adorned with the gem are used to bring luck in the chase and hunt. The purity of turquois is critical to the Hopi. It is believed that to wear a stone that contains matrix while gathering additional stones will cause any gems found to crack.

The Acoma believe turquois will bring all good things to those who wear it. It is said, "the same way that men are attracted to women, so are friends, game, or anything good attracted to turquoise."[20]

Birth of the Hopi and the Rescue From the Flood: Hopi

The Hopi people are said to have come from the underworld. They are said to have originated from a great skeletal creature in the underworld. This creature was very handsome and beautifully adorned with gems. Turquois hung from his neck, and his ears were decorated with earrings of the same stone. From this common ancestor the Hopi emerged from the earth and traveled to the East, source of the sun and all life. In their travels they encountered many trials and tribulations. The last of these was a great flood of water. To triumph over the flood, the chiefs of the tribe met in council. They prepared a gift of two large spheres made of powdered shell and turquois. This gift was presented to the evil water serpent, the cause of the flood. With this offering of highest value, the land immediately became dry. The Hopi where saved and have forever lived on the dry land.

How Turquois Were Obtained From Chief Morning Green: Hopi

The children of the Hopi village Casa Grande were playing an ancient game called *toka*. As they took a break from their play, they noticed a blue-tailed lizard descend into the ground. Where the lizard entered the earth the ground was green. They reported this strange event to the chief of the village, Chief Morning Green. Upon hearing the story, the chief ordered the place the children pointed out be excavated. Chief Morning Green and his people were pleased to discover an abundance of turquois in the diggings. The chief ordered the production of many fabulous items, each to be encrusted with the newly found gems. He also distributed many stones to each of the members of his tribe.

News of the beauty of the turquois reached the ears of the Sun as it rested in the East. The Sun sent a brightly colored parrot to go to the village and gather all the stones. A daughter of the chief met the bird as it approached the village and returned to her father to announce the arrival of this colorful visitor from the Sun. The chief told his daughter to take a charmed stick he prepared to the parrot, let the bird pick it up, and lead him back to the chief. The girl returned to the parrot and tried the charm, but the bird was not fooled. The girl went to her father and reported the failure of the charm. The chief gave her a list of other ways to trick the parrot into following her. She once again went to the bird and, as her father had instructed, offered pumpkin seeds, watermelon seeds, seed of cat's claw, kernels of corn and charcoal granules. None of these tempted the bird, but the girl noticed it ate a small blue bead that was near her offerings. When she saw this, the chief's daughter brought more blue beads and then small bits of turquois. The bird's interest caused the girl to bring more and more turquois. When she thought it had eaten its fill, she tried to capture her prey. The parrot escaped and flew back to the home of the Sun in the East. The bird vomited the turquois to the Sun-god who distributed it to the people who reside around his house of rising. This is why the people to the East of the Hopi have turquois.

Chief Morning Green became enraged by the loss of part of his treasure. He used his power to cause a violent rain to fall on the East. His power over the rain was so great that he was even able to extinguish the Sun. Alarmed by this, the old priests gathered to decide what to do. They sent Man-Fox to the East to re-establish the fire of the Sun, but he failed. Rapid Runner was then sent to visit the Thunder, the only one who owned fire, and steal his torch of fire. As he fled, Thunder shot at him with a flaming arrow. The arrow's sparks scattered as it flew and set fire to the Sun and spread fire to all things. This is how the energy of fire came to dwell in all things.

An Origin Legend: Navajo

The First Man and Woman decided their first chore was to create the sun and the moon to light their world. Legend says the sun they created was a disk of clear stone set apart from the sky by a rim of turquois set in its edge. After the world was lit, the Man and Woman saw the earth was covered with water. To create dry land, the Man and Woman drained parts of the earth by digging vast trenches. The only tools capable of such enormous work were spades made of turquois tipped with coral. With these sacred spades, the earth was separated into vast seas, dry land and the trenches which drain the land into the sea.

Why the Buffalo's Horns are Black: Navajo

In the beginning, the earliest buffalo had horns of brilliant turquois. Two boys admired the beautiful horns and devised a way to claim them. They used sunbeams to construct a web and ensnare the great beast. The boys offered to set the buffalo free in exchange for the marvelous horns. In exchange for the horns they offered all the jet the Navajo possessed and a promise the Navajo would never use jet again. The buffalo agreed and made the trade. This is why the Navajo have so much turquois and the buffalo have jet black horns.

In a variation, the turquois horned buffalo convinces the boys that he has a headache. He also tells the boys that if his horns are replaced by jet he will be cured of the pain. He also promises to bless the Navajo with many buffalo if he were set free with his new jet horns. The swap was made and the animal set free to display his new jet horns.

Zircon

Crystal System

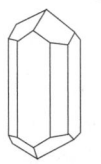

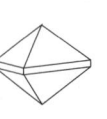

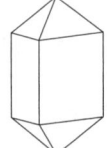

Tetragonal

Color: Colorless, yellow, brown, orange, red, violet, blue, green (hyacinth; yellow-red to red-brown)
Mohs' Hardness: $6 \frac{1}{2} - 7 \frac{1}{2}$
Cleavage: None
Refractive Index: 1.780 - 2.01
Dispersion: 0.030 - 0.040

Streak: White
Specific Gravity: 3.90 - 4.71
Fracture: Conchoidal, very brittle
Birefringence: 0.002 - 0.059 (none in green)
Pleochroism: Yellow; very weak, honey, yellow, yellow-brown, Red; very weak, red, light brown Blue; definite; blue, yellow, gray

Crystal System: Tetragonal; four-sided prisms with pyramidal ends
Chemical Composition: $Zr(SiO_4)$ zirconium silicate
Transparency: Transparent
Fluorescence: Blue; very weak, light orange, Red and Brown; weak, dark yellow

The gems with a chemical composition which is basically zirconium silicate are called zircons in the jewelry trade. There are actually three types, known as high, medium and low. The distinctions are made regarding their refractive indices. The majority of gem

> zircons are of the high variety; however, medium and low stones are also used. They occur naturally in almost all colors, but most are reddish to yellowish brown. Although they do exist naturally, most blue and colorless stones available in the retail trade are the result of heat treatment. References to zircon may be found in past literature under the names jacinth, hyacinth, or jargoon.

Zircon, the modern birthstone of December, may be one of the first crystals separated from common rock by early man. In its natural state it stands out as a bright and distinct material. References to the gem exist in biblical writings as well as those of the ancient Persians, Greeks and Romans. Centuries of doubt about its chemical composition led to its moniker as the "Mystery Stone." Psellus, a contemporary of Marbodus, numbers jacinth "amongst the stones about which nothing was then definitely known."[1]

This common mineral has been known by different names in different lands, but descriptions of the gem reveal that these varied peoples were speaking of the same stone. The derivation of the modern name is unknown. It may have come from the Persian word *zargun*, meaning gold-colored, or from the Arabic word *zarkun*, meaning vermilion. Zircon crystals may occur in many colors. The contemporary gemologist Robert Webster states that the word may have come from the Italian word *gaicone* which he speculates was a corruption of the Persian zargun. The first use of the word zircon is found in the writings of Weiner in 1783. He reserved the term for crystals from Ceylon. The word *jargoon*, a popular name for the stone in French writings, may have come from the same Arabic or Persian roots. The Greeks named the gem *huakinthos*, the same name given to a particular blue or violet wild flower. From this the name the modern name *hyacinth* is derived. The name is also shared with a mythical youth. It is said that the boy Hyacinth was slain and wild hyacinth flowers sprang from his spilled blood. It is believed that the Greeks used the name to separate the blue and violet varieties of the gem from other crystals. Jacinth is a simple variation on the original Greek name. The Roman historian Pliny compared the hyacinth with amethysts while other authors may have used the name to described blue or violet sapphires. Even as late as the eighteenth century colorless stones found in Matara, Sri Lanka were thought to be inferior grade diamonds and were called "Matara diamonds."

The lapidaries of the Middle Ages gave reference to zircon as a protective amulet. It was worn by travelers to ward off plague, wounds and injuries. The stone would dull if the wearer had been infected by plague or in close proximity to the disease. The gem was said to enhance the finances of the owner and make him prudent in business. The wearer of zircon was also immune to lightening. It was thought that this power was so strong that even the impression of the stone made in wax would give the same protection if the wax seal was carried. It was reported that in regions which experienced many lightening strikes, no report existed of a bearer of the stone being struck.

St. Hildegard, the Abbes of Bingen (d. 1179), gives this account of the proper use of *jachant* or jacinth.

> If anyone is bewitched by phantoms or by magical spells, so that he has lost his wits, take a hot loaf of pure wheaten bread and cut the upper crust in the form of a cross,–not, however, cutting it quite through,–and then pass the stone along the cutting, reciting these words: 'May God, who cast away all precious stones from the devil . . . cast away from thee, (name), all phantoms and all magic spells, and free thee from the pain of madness.'[2]

The patient was to then eat the bread. If the stomach was considered too feeble, unleavened bread was to be used. All other solid food was to be treated in the same way. If the sickness was in the heart it will be relieved if the sign of the cross is made over the heart while the above mentioned ceremony was performed.

Christian symbolism assigns each of the Foundation Stones, listed in Revelations XXI, to one of the twelve apostles. One of the earliest writers to make such an association was the tenth century Bishop of Caesarea, Andreas. He states his conclusion in the following passage, "The jacinth, which is of a celstial hue, signifies Simon Zelotes, zealous for the gifts and grace of Christ and endowed with a celestial prudence."[3]

Camillus Leonardus expands the attributes of jacinth when he writes, "They invigorate animal life, especially the heart. They disperse sorrow and imaginary suspicions. They increase ingenuity, glory and riches."[4] Cardano states that he was in the habit of wearing a large jacinth to induce sleep. His account reveals that he did not consider his stone to be of high quality because of its golden color.

The fourteenth century *Lapidaire* of Chevalier Jean de Mandeville gives an extensive description of the gem's colors and kind. He also restates some of the common attributes listed for the stone by other authors but adds virtues not included in other writings.

> Hyacinth is a stone which has three colors, citrine red, violet or blue. They are found in many places, all give joy and happiness. The red kinds are the best and are similar to garnets, but there is a difference, because they are not as pure in color, but rather are paler. The violet Hyacinth is precious because it changes color, from light to murky, depending on the weather. Some are part red and white and are hard stones and all have the same qualities: he who wears this stone can go safely wherever he wishes, this stone brings peace to the corrupt, pestilence cannot destroy he who wears this stone, nor can he be injured by any weapon. He can go safely in the countryside because he will be received in a proper manner in hotels wherever he goes. The Hyacinth will bring about fair and reasonable petitions or requests. Many are found in Ethiopia: they are effective against snakes and venom and render man gracious unto God and to others. Hyacinths and Garnets have the same properties.[5]

His text gives the hyacinth a prominent place and was the inspiration for many to carry the stone as a protection during travels. Hyacinth is also later described as a poor man's substitute for high quality garnets.

The medical virtues of the stone have been extoled since Roman times. Pliny states in his *Natural History* that, "if drunk in wine, or even worn, would expel the stone in the bladder and cure the jaundice."[6] Den Sina (Avicenna) claimed that the zircon possesses a peculiar tonic effect and is an antidote for poisons. The stone was also said to prevent any deadly effects that may accompany a wound. The sixteenth century Swiss physician and alchemist, Paracelsus the Great, recommended powdered zircon mixed with laudanum as a cure for fevers caused by bad air or water. The nineteenth century author William Jones writes that the jacinth drives away fever and dropsy, clears the sight and expels noxious fancies. He also included in his account the recommendation that the stone be set in gold to increase its powers.[7] Modern gem healing practitioners call zircon the "stone of virtue." It is used to treat the sciatic nerve and nerve structures which lead away from the spine. It is used to increase bone stability, quicken the healing of broken bones, help torn muscles heal and correct problems with vertigo.

Testing gems by submitting them to fire was a common practice in the fifteenth century. Stones that shattered or lost color were considered to be of inferior quality. This illustration shows such a "trial by fire" for precious stones as prescribed in the Ortus Sanitatis *by de Cuba.*

Mythical Gems

The lore of gems includes many "stones" that are not recognized by modern gemologists. These are generally classified as "mythical gems" or "stones of imagined origin." Many of these stones are said to have been recovered from animals, both real and mythical. It is generally accepted that some of these gems are in fact calcifications that are common within the organs of many animals. It is known that humans generate "stones" in various organs; gall stones and kidney stones are two examples. The same calcifications are found in many other mammals. Pearls are examples of calcifications that are commonly harvested from mollusks and revered for their beauty.

Other creatures carry stones within their systems. Chickens and other foraging birds often ingest small pebbles along with the insects and grains they find on the ground. It is also known that the digestive systems of many animals generate coatings around foreign objects that are swallowed and would otherwise cause ulcerations or poisoning if left undigested. Other creatures do not hold stones within their bodies but may collect stones in the course of their daily routine. Rodents and birds both collect stones as objects they include in their nests.

The stories of animals as a source of gems may have originated from the disemboweling of animals as they were butchered. Early people that found gems in an animal's nest may have believed the animal generated the stone. Ancient scientists performing dissections may have struggled to explain hardened material found in the organs of their subjects. The rarity of such discoveries must have added to their mystery. From such occurrences myths arose that granted powers to these unusual finds. Some of the stones thought to be of magical origin are now understood to be common natural occurrences.

Some of the gems cataloged in ancient literature are truly obscure and often carry names that add little to our understanding of their source. Other stones are now recognized as commonly occurring hardened growths or calcifications. Mythical gems have become the subject of many authors and are even written of in contemporary texts on magic and mysticism. As recently as the nineteenth century, these stones were include in gemology texts. The list given here is not intended to be complete. The names included are those given by numerous authors throughout the centuries.

Aetites or Lapis Aquilaris

The aetites, lapis aquilaris, or eagle-stone is said to be found in the nests of eagles. It is believed to help eagles in reproduction. The famed gem engraver Dioscorides tells of the following use for the stone. The aetites should be powdered and mixed with flour to form balls of dough the size of hens' eggs. Persons suspected of theft are brought to a court

An eagle placing the aetites in its nest. The illustration is from the Ortus Sanitatis *by Johannis de Cuba, published in 1483.*

and given one of the balls with a small drink. The guilty among them will be unable to swallow the dough and will choke if they try. Seventeenth century medical texts list the aetites as a stone that should be applied to pregnant women to prevent miscarriages. Many texts describe the stone as scarlet in color and able to protect the owners from harm as well as making them amiable, sober and rich. The twentieth century gem scholar G. F.. Kunz concludes aetites were quartz pebbles collected by the birds during nest construction.

The lengthy poem regarding gems *De Lapidibus by* Marbode, Bishop of Rennes' (1035-1123) contains the following verse regarding this gem,

> Chief amongst gems the Aetites stands
> Borne by the bird of Jove from farthest lands:
> As safeguard to his nest, and influence good
> To ward off danger from the callow brood.
>
> Shut in the pregnant stone another lies —
> Hence pregnant women its protection prize;
> With this gem duly round her left arm tied
> Need no mischance affright the teeming bride.
>
> Sober the wearer too shall ever prove,
> Shall wealth amass, and reap his people's love:
> Victory shall crown his brows; his offspring dear,
> Shall healthy live nor fate untimely fear.
>
> The epileptic wretch, saved by its worth,
> No more shall fall and writhe upon the earth.
> Should'st thou suspect they friend of treason foul,
> The privy prisoner lurking in the bowl,
>
> Thus prove his mind: him to thy banquet bid
> And let this stone beneath the dish be hid,
> When, if he harbour treachery in his thought,
> Whilst there the stone lies he can swallow nought:
>
> Remove the gem, delivered from its power
> The tasted meats he'll greedily devour.

The stone they say is found, with scarlet dyed,
Hid on the margin of old ocean's tide.[1]

Alectorius or Allectory

A gem said to be found in the gizzard of a capon was the alectorius. The bird was to be sterile for at least three years before the clear crystal could be harvested. The gem is said to relieve any thirst of its owner. Marbode's *De Lapidibus* contains a passage relating the lore of this stone.

Extracting an alectorius from the head of a capon. Most authorities placed the gem in the capon's gizzard. Johannis de Cuba ignored this in his illustration from the Ortus Sanitatis, *first illustrated text about drugs, published in 1483.*

> Not least the glory of the gem renowned
> Within the belly of the capon found,
> Which, made an eunuch when three years have flown,
> Though twice two more in swelling bulk has grown;
>
> Its utmost size no larger than a bean,
> Like purest water or the crystal's sheen;
> Hence Alectorius is the jewel hight,
> For gifts of strength extolled, and matchless might.
>
> If parched with thirst place this within thy mouth,
> 'T will in a moment quench thy burning drouth;
> Aided by this on many a well-fought day
> Crotonian Milo bore the palm away:[2]

Aspilates

A stone said to be found in the nests of Arabian birds is called the aspilates. The attributes are usually given as being similar to those of the aetites. It may be assumed that the reference to Arabian birds denotes where they are found and not their national origin.

Bezoar

Bezoar, besuar, and beza are all terms used to refer to a potent stone. The name is derived from the Persian word *pad-zahr* or *bad-zahr,* meaning counter-poison. It is believed the stone is harvested from the Cervicabra, a wild animal listed in various texts as a deer, goat, pig or monkey from Persia. The consensus of most authors is that the animal is a wild goat. Some writers say the stone is found in the stomach, but most agree it is a kidney stone with a kernel center surrounded by hardened layers. The value of the stone increases with its size. Medieval apothecaries soaked the stone in wine and administered the resulting tincture as an antidote for poisons. Other prescriptions suggest grinding the stone to a powder and applying it directly to wounds, particularly smallpox lesions. Since such a stone is mostly calcium carbonate, it works as an astringent to draw out infection. Camillus Leonardus speaks highly of the stone in his *Speculum Lapidum*:

> (Bezoar) is a red, dusty light, and brittle stone; by some it is described as of a citron colour.... All agree

it obtains the first place in remedies against poisons. ... any thing which frees the body of that ailment, is called the Bezoar of the ailment. And thus its name is become general.[3]

A recipe to make "oriental bezoar" is included in the old *Meteria Medica*. It calls for the mixing of equal parts white amber, red coral, crab's eyes, powdered hartshorn, pearls and black crab's claws.[4] The stone is also recommended as a cure for scorpion stings. It is said to be most effective if the image of a scorpion is engraved on the stone, this should be done when Scorpio is in ascendancy.[5]

Cabot Stone or Cimedia

A stone of prognostication that is found in a particular fish (the species is never revealed), is the cabot stone. It is said to be able to forecast weather and is considered of great value to mariners. It becomes clear and shinny when the weather will be clear, and fogged if storms are approaching. Leonardus writes of a stone of similar attributes, but calls it a cimedia. He claims it, "is taken out of the brain of a fish of the same name."[6]

Corvia

A stone recovered from the nest of a crow and given great value is called the corvia. Leonardus gives an explanation of how the stone is obtained and its supposed virtues:

> On the calend of April, boil the eggs taken out of a crow's nest until they are hard, and being cold, let them be placed in the nest, as they were before. When the crow knows this, she flies a long way to find the stone, and returns with it to her nest, and the eggs being touched with it, they become fresh and prolific. The stone must immediately be snatched out of the nest. Its virtue is to increase riches, to bestow honours and foretell future events.[7]

Cock-stone

A crystal formed stone found in a particular cock is called a cock-stone. The test of which cock contains such a stone is found in its eating habits. A cock that never drinks as it eats, unlike most fowl, is a likely candidate to contain a cock-stone. This gem was revered by the Romans as a stone that could make the owner invisible. In the Middle Ages the gem was said to keep the owner free of thirst.

Crab's Eye

The name crab's eye is a misnomer. This gem is actually a calcium carbonate deposit found in the abdomen of the crayfish. An ancient anonymous writer gives a good description of the stone:

> in the crayfish, just beneath the head near the stomach grow two small pea-sized stones which are in the shape of flattened spheres hollowed out on one side and look rather like eyes. Nevertheless, this they are not since the creature has tow perfectly good eyes in the usual place. In Latin they are called *Lapides cancri, Oculis cancri*. They ease out the teeth, dry weeping sores, neutralize acidity, are good for sweetening inveterate ulcers and blisters and for checking diarrhoea, bleeding and vomiting.[8]

The crayfish is an animal of the planet Pluto and is used as a treatment for cancer. A prescription from the old herbalist Brother Aloysius consists of mashing living crayfish with garlic in an earthen pot. The resulting paste is applied to cancerous tumors and left in place for twenty-four hours while the patient is kept awake. The disease is supposed to enter the crayfish paste and cure the cancer patient. It should also be noted that the symbol for the zodiac sign, Cancer the crab is often drawn resembling a crayfish rather than a crab. The crayfish in nature is known as an animal that removes unpleasant things from its surroundings.

Demoniu

A stone mentioned in various texts but never described is the demoniu. This mysterious gem is said to make wearers invisible and keep them safe from illness.

Diadochus

Described as having the color of beryl, the diadochus is a gem of value to conjurors. It is said to give answers to all questions and display the images of devils if cast into water while a secret incantation is repeated. The gem is used to disturb devils and is used as a gem of exorcism. Held in the mouth, spirits from beyond are summoned. The owner of such a gem is cautioned that its powers will dissolve if it is touched to a dead body.

Doriatides

A gem retrieved from the head of a cat, recently cut off in a swift manner, is known as the doriatides. The stone is described as shinny and black with the power to grant any wish made by the owner.

Draconius

Albertus Magnus described a black stone of pyramidal form called the draconius. It comes from the East where it is harvested from the heads of dragons. The stone looses all its powers if the dragon has not been recently killed. Magnus states that the dragon must still be panting while the stone is removed. Leonardus describes the method of capturing the

Killing a dragon to retrieve its precious stone. An illustration from the Ortus Sanitatis *by Johannis de Cuba, 1483.*

dragon and extracting the stone:

> Some bold fellows, in those eastern parts, search out the dens of the dragons, and in them they throw grass mixed with sopo-riferous medicaments, which the dragons, when they return to their dens, eat, and are thrown into a sleep, and in that condition their heads are cut off, and the stone extracted. It has a rare virtue of absorbing all poisons, especially that of serpents. It also renders the possessor bold and invincible, for which reason the kings of the East boast of having such a stone.[9]

Echites

A stone that is most likely a form of geode or a hollow concretion is called echites. The name is from the Greek root *aetites* and shares its translation, eagle-stone. These hollow forms contain loose crystal or pebbles that spill out when the stone is broken open. This characteristic symbolizes eggs, fertility and birth. The echites is said to bring fertility and ease childbirth.

Elopsides

Elopsides is a gem that cures headaches if suspended around the neck. It is cataloged in early writings but without description.

Epistides

A glittering red stone that exhibits many powers is called the epistides. The gem will keep one from all misfortunes if fastened over the heart with "magic straps." As the stone is placed on one's breast, a proper incantation must be repeated. The gem is said to take away all misfortunes and drive away locusts, harmful birds, winds, and storms.

Eumetis

The eumetis is a flint colored gem that allows one to speak prophecies. The stone must be placed under a pillow while the subject sleeps for its powers to have effect.

Exebonos

Powdered and mixed with milk, the white exebonos stone is a cure for insomnia.

Filaterius

The filaterius is bright yellow green, like peridot. It disperses terrors and melancholy. The gem also makes the wearer complaisant and comforted.

Fongites
If one wishes to have all ailments cured and live without the distraction of anger, he or she should carry the red fongites in hand.

Galactides
A powerful stone of magicians is called the galactides. It allows magical writings to be heard and brings ghosts to magicians to answer any question asked. The gem also stops quarrels, removes mischief and brings lovers together. If placed in the mouth, one is able to hear the unspoken opinions of others. To test the authenticity of the stone, it is placed in the mouth and honey is smeared on the body. If flies and bees do not attack, the stone is genuine.

Gargates
The gargates has many applications. If the stone is rubbed, smoke is generated which drives away all devils as it dissolves their spells. The stone also cures the stings of scorpions when powdered and mixed with the bone marrow from a stag. An additional attribute is its ability to tighten loose teeth.

Gasidana
Gasidana is a curious cream colored stone that has the ability to reproduce itself if it is shaken vigorously.

Glosopetra
The glosopetra is a gem shaped like a human tongue. It falls from the heavens as the moon wanes. Magicians claim it has the power to change the movement of the moon.

Hamonis
A gold colored stone from Ethiopia that is found in the shape of a ram's horn is called a hamonis. This is a sacred gem as it makes a man divinely inspired if he allows himself to go into deep contemplation.

Hyena Stone
The hyena stone is found within the eye of a hyena. The stone allows the owner to foretell the future if it is placed under the tongue. It also cures bodily pains, cramps and the gout.

Lippares or Liparia
The stone that is a protector of all animals is the lippares. It is found inside of a particular slug. If an animal is being chased, it will seek out this small white stone and become invisible to hunters. Worn around the neck, it cures fevers in people.

Peanita

The peanita is a geode stone, echites, or eagle-stone with foreign pebbles or crystals said to be born when the encapsuling mother stone is broken. The gem adds fertility and aides in child birth.

Quirinus or Quirus

The quirinus stone is harvested from the nest of the "hoopoe" bird. Its single power is to cause a person to confess their wrongs if the stone is placed on his or her chest while asleep.

Raven Stone

To make oneself invisible at will, a raven stone found in New Pomerania is carried. Accounts of how the stone is obtained are of interest. One should look for a raven's nest and determine that the adult birds are a least one hundred years old (no record is found as to how this is accomplished). Then kill one of the nestlings, it must be a male and no more than six weeks old. Upon descending the tree, it will become invisible, mark the location carefully. The adult raven will return and place a stone in the mouth of the dead chick. This is the raven stone which must be retrieved in a short period of time.

Serpent Stone

A gem is said to be found in the head of a serpent. This serpent stone appears in differing pieces of literature from various regions. The story of the ring of Gyges from Plato's *Republic* is supposed by some to be an opal (see the story of Gyges, page 158), but another author (Caeselii, 1516) states that the stone was a serpent-stone of India. In that land brilliant and beautiful stones were said to be found in the heads of serpents. Set in silver or gold at the time when the planetary or stellar control of the stone is in ascendence, the wearer of the ring is sure to have the fullest possible benefits of its powers. Indian legends say that these powerful reptiles are the guardians of diamonds, sapphires or glowing rubies. This gem was claimed to make Gyges invisible. Invisibility is the primary power of the stone. In Timberlake's *Discourse of the Travels of Two English Pilgrims to Jerusalem, Gaza, etc.* (1611), we find the account of another serpent stone.

> Among other stones, there is one in the possession of a conjurer, remarkable for its brilliancy and beauty, but more so for the extraordinary manner in which it was found. It grew, if we may credit the Indians, in the head of a monstrous serpent, whose retreat

was by its brilliancy discovered; but a great number of snakes attending him, he being, I suppose, by his diadem of a superior rank among the serpents, made it dangerous to attack him. Many were the attempts made by the Indians, but all frustrated, till a fellow more bold than the rest, casing himself in leather impenetrable to the bite of the serpent, or his guards, and watching a convenient opportunity, surprised and killed him, tearing the jewel from his head, which the conjurer had kept hid for many years, in some place unknown to all but two women, who have been offered large presents to destroy it, but steadily refused, lest some signal judgment or mischance should follow.[10]

A serpent stone is listed in the inventory of the Bishop of Ardfert's gifts to St. Alban's Abbey. It was mentioned as a spotted stone, square in form and set in silver. The stone was said to be effective against lunacy.

Swallow Stone

Two precious stones are gathered from a young swallow. These swallow stones must be obtained from the stomachs of young birds that are still in the nest. The quality of the stone increases if the bird has never touched the ground and if the gems are harvested in the month of August. The mothers of the birds should not be present when the stones are gathered. One of the stones is red and will work to restore sanity. The other is black and works to bring good luck to the wearer. Medicinally, either stone prevents fevers and jaundice if tied about the neck with a yellow linen thread. "According to some writers, the stones were to be wrapt in the skin of a calf or a hart, and bound to the left arm."[11]

Marbode's *De Lapidibus* is the basis for most of the lore regarding the swallow stone.

> The rapid swallow swifter than the airs
> Within her breast (Latin translates, belly) the Cheli-
> donian bears,
> A fatal gift, deep in her bowels pent,
> Which with her life is from the owner rent.
>
> The Chelidonian is of might supreme,
> Though not of those which shoot a brilliant gleam:

Yet many a gem that men for beauty praise,
Unshapen, small, and dull, its worth outweighs.

The feather'd victims in their bowels stored
Two different sorts — the white and red — afford:
The pining sickness feels their influence mild,
The moonstruck idiot, and the maniac wild.

With force persuasive orators they arm,
And grace the hearts of multitudes to charm:
Wrapped in a linen cloth this present rare,
Under thy left arm tied ne'er fail to wear;

The black, in woollen cloth thus too suspend,
And bring thy measures to the wished-for end.
It blunts the threats and cools the ire of kings,
And to the wearied sight refreshment brings.

This in a yellow cloth of linen laid
Will banish fevers that thy limbs invade,
or watery humours that with current slow
Obstruct the veins and stop their healthy flow.[12]

Toadstone or Chelonites or Crapaudina

The gem with the greatest lore and tradition is the toadstone. The composition and even the source of the gem is up for debate. As the name suggests, the stone is found in the head of a toad, however, some authors maintain that it is not from a toad, but is actually a fossilized tooth of a fish. The attributes of the gem are varied and numerous. It detects poisons by sweating and changing colors in their presence. It is also listed as an effective cure for cancers and growths, particularly warts. The idea that "like cures like" is an ancient tradition. The people of Surinam bind a dried toad to tumors to reduce their size. The stone also is supposed to prevent houses from burning and prevent the sinking of ships. A risk is taken by military commanders if they carry the stone. Tradition says the commander would either be assured of victory in battles or all his men would fall dead on the spot.

Gathering the stone is a ritual in itself. Baptista Porta gives the following method in his sixteenth century text.

> The stone Chelonites is found in the head of large old toads and is very curative if it can be taken from the living animal. The toad is placed on a red cloth

A toadstone is extracted from the head of a mature toad in this illustration from the Ortus Sanitatis by Johannis de Cuba, 1483. Cuba described the use of the stone, but called it a "bufotenine."

for the colour pleases it very much. When it jumps, the stone falls out of its head and must then drop through a hole in the cloth into a box that has been set beneath it, otherwise the toad will quickly swallow the stone. There is not the slightest doubt about the value and operation of the stone, although I have never found one in spite of the many toads I have dissected.[13]

The test of a stone's genuine quality follows a similar logic. If the gem is placed before a live toad, it will quickly be swallowed if it is true. Leonardus lists the stone as a principle ingredient in the incantations of "nocturnal hags." The spell is given as follows,

> Toad that under the cold stone
> Days and nights has thirty-on,
> Sweltered venom, sleeping got,
> Boil thou first i' the charméd pot.

From Gnostic writings found in Alexandria, called the *Kyranide*, comes the following prescription, "The earth-toad, called *saccos*, whose marrow is its head. If you take it when the moon is waning, put it in a linen cloth for forty days, and then cut it from the cloth and take the stone, you will have a powerful amulet."[14] The poet Shakespeare also mentions the stone in *As You Like It* Act I Scene 3

> Sweet are the uses of adversity;
> Which, like the toad, ugly and venomous,
> Wears yet a precious jewel in his head.

Tortoise Stone

Found in the brains of a tortoise is the tortoise stone. It is the enemy of fire and makes the wearer able to predict the future. The stone is to be placed in one's mouth at particular times to make its powers evident. The time when the stone must be in the mouth is during the first full day of a new moon and the succeeding fifteen days of lunar ascension. The stone should be held orally from sunrise to six o'clock in the morning of each of these days.

Vulture Stone

Found in the bird's brain, the vulture stone gives the owner good health and answers all his questions through visions in one's dreams.

The use of a toadstone to treat poisoning. From the fifteenth century treatise on drugs by de Cuba.

*"In Science, read, by preference, the newest works;
in literature, the oldest. The classic literature is always modern"*
Edward Bulwer Lytton

Gems in Literature

The Holy Bible 251

Stones of the New Jerusalem: Marbode 258
De Lapidibus: Marbode 259

Precious Stones in Shakespeare's Works

Plays 278

Poems 287

Gems in the Holy Qur'an 291

Biblical References

The Holy Bible contains numerous references to precious stones. As one of the oldest pieces of literature, it served as the foundation for many of the lapidaries of the ancient world and the Middle Ages. The following are all the references to gems found in the four commonly used translations. Differences in translation and tradition account for some variations. A discussion of translation variations may be found in the Birthstones section (pp. 294-319).

Material in standard type is from the Revised Standard Version. References to gems that differ in other translations are noted as follows: <u>Underlined, King James;</u> *Italic, New English;* <u>*Underlined and Italic, King James and New English*</u>; standard type in parenthesis, (New World Translation of the Holy Scriptures).

The Old Testament

Genesis 2:10-12
¹⁰A river flowed out of Eden to water the garden, and there it divide and became four rivers.
¹¹The name of the first is Pishon; it is the one which flows around the whole land a Hav´ilah, where there is gold;
¹²and the gold of that land is good; bdellium and onyx stone are there.

Exodus 24:9-10
⁹Then Moses and Aaron, Nadab, and Abi´hu, and seventy of the elders of Israel went up,
¹⁰and they saw the God of Israel; and there was under his feet as it were a pavement of sapphire stone, like the very heaven for clearness."

Exodus 25:3 and 7
³And this is the offering which you shall receive form them: gold, silver, and bronze, . . .
⁷onyx stones, and stones for setting, for the ephod and for the Breastplate.

Exodus 28:9-12
⁹And you shall take two onyx stones, and engrave on them the names of the sons of Israel,

¹⁰six of their names on the one stone, and the names of the remaining six on the other stone, in the order of their birth.

¹¹As a jeweler engraves signets, so shall you engrave the two stones with the names of the sons of Israel; you shall enclose them in settings of gold filigree.

¹²And you shall set the two stones upon the shoulder-pieces of the ephod, as stones of remembrance for the sons of Israel; and Aaron shall bear their names before the Lord upon his two shoulders for remembrance.

Exodus 28:15-21

¹⁵ And you shall make a breastplate of judgment, in skilled work; like the work of the ephod you shall make it; of gold, blue and purple and scarlet stuff, and fine twined linen shall you make it.

¹⁶It shall be square and double, a span its length and a span its breadth.

¹⁷And you shall set in it four rows of stones. A row of sardius (ruby), topaz, and carbuncle shall be the first row;

¹⁸and the second row an emerald, a sapphire, and a diamond (jasper);

¹⁹ and the third row a jacinth (hyacinth) (ligure), an agate, and an amethyst;

²⁰and the fourth row a beryl, an onyx, and a jasper (jade); they shall be set in gold filigree.

²¹There shall be twelve stones with their names according to the names of the sons of Israel; they shall be like signets, each engraved with its name, for the twelve tribes.

Exodus 35: 5, 9, and 27

⁵Take from among you an offering to the Lord; whoever is of a generous heart, let him bring the Lord's offering: gold, silver, and bronze; .

⁹and onyx stones and stones for setting, for the ephod and for the breastpiece.

²⁷And the leaders brought onyx stones and stones to be set, for the ephod and for the breastpiece.

Exodus 39: 6, 7, and 10-14

⁶The onyx stones were prepared, enclosed in settings of gold filigree and engraved like the engravings of a signet, according to the names of the sons of Israel.

⁷And he set them on the shoulder-pieces of the ephod, to be stones of remembrance for the sons of Israel; as the Lord commanded Moses. .

¹⁰And they set in it four rows of stones. A row of sardius(ruby), topaz, and carbuncle was the first row;

¹¹and the second row, an emerald, a sapphire, and a diamond;

Literature:

¹²and the third row, a jacinth (lesh´em) (<u>ligure</u>), an agate, and an amethyst;
¹³and the fourth row, a beryl, an onyx, and a jasper (jade); they were enclosed in settings of gold filigree.
¹⁴There were twelve stones with their names according to the names of the sons of Israel; they were like signets, each with its name, for the twelve tribes.

2 Samuel 12:29-30
²⁹So David gathered all the people together and went to Rabbah, and fought against it and took it.
³⁰And he took the crown of their king from his head; the weight of it was a talent of gold, and in it was a precious stone; and it was placed on David's head. And he brought forth the spoil of the city, a very great amount.

1 Kings 10:2
²She came to Jerusalem with a very great retinue, with camels bearing spices, and very much gold, and precious stones; and when she came to Solomon, she told him all that was on her mind.

1 Kings 10:10-11
¹⁰Then she gave the king a hundred and twenty talents of gold, and a very great quantity of spices, and precious stones; ...
¹¹Moreover the fleet of Hiram, which brought gold from Ophir, brought from Ophir a very great amount of almug wood and precious stones.

1 Chronicles 20:2
²And David took the crown of their king from his head; he found that it weighed a talent of gold, and in it was a precious stone; and it was placed on David's head. And he brought forth the spoil of the city, a very great amount.

1 Chronicles 29: 2 and 8
²So I have provided for the house of my God, so far as I was able, the gold for things of gold, the silver for the things of silver, and the bronze for the things of bronze, the iron for the things of iron, and wood for the things of wood, besides great quantities of onyx and stones for setting, antimony (<u>glistering stones</u>), colored stones, all sorts of precious stones, and marble. . . .
⁸And whoever had precious stones gave them to the treasury of the house of the Lord, in the care of Jehi´el the Gershonite.

2 Chronicles 3:6
⁶He adorned the house with settings of precious stones.

2 Chronicles 9:1
¹Now when the Queen of Sheba heard of the fame of Solomon she came to Jerusalem to test him with hard questions, having a very great retinue and camels bearing spices and very much gold and precious stones. When she came to Solomon, she told him all that was on her mind.

2 Chronicles 9:9-10
⁹Then she gave the king a hundred and twenty talents of gold, and a very great quantity of spices, and precious stones: . . .
¹⁰Moreover, the servants of Huram and the servants of Solomon, who brought gold from Opir, brought algum wood and precious stones.

2 Chronicles 32:27
²⁷And Hezeki´ah had very great riches and honor; and he made for himself treasuries for silver, for gold, for precious stones, for spices, for shields, and for all kinds of costly vessels; (and for all manner of pleasant jewels.)

Job 28:5-6
⁵As for the earth, out of it comes bread: but beneath it is turned up as by fire.
⁶Its stones are the place of sapphires, and it has dust of gold.

Job 28: 16-19
[referring to wisdom]¹⁶It cannot be valued in the gold of Ophir, in precious onyx or sapphire.
¹⁷Gold and glass (crystal) cannot equal it, not can it be exchanged for jewels of fine gold.
¹⁸No mention shall be made of coral or of crystal (pearls); the price of wisdom is above pearls (rubies).
¹⁹The topaz of Ethiopia cannot compare with it, nor can it be valued in pure gold.

Proverbs 3:15
¹⁵She is more precious than jewels, and nothing you desire can compare with her.

Proverbs 8:11
¹¹for wisdom is better than jewels (coral), and all that you may desire cannot compare with her.

Proverbs 20:15
¹⁵There is gold, and abundance of costly stones; but the lips of knowledge are a precious jewel (coral).

Proverbs 31:10
¹⁰A good wife who can find? She is far more precious than jewels (corals).

Literature:

Song of Solomon (Canticles) 5:14
¹⁴His arms are rounded gold, set with jewels (the beryl). His body is ivory work, encrusted with sapphires (Hebrew-lapis-lazuli).

Isaiah 54:11-12
¹¹O afflicted one, storm-tossed, and not comforted, behold, I will set your stones in antimony (<u>fair colours</u>), and lay your foundations with sapphires. (Hebrew-lapis-lazuli)
¹²I will make your pinnacles of agate and your gates of carbuncles(rubies), and all your wall of precious(<u>pleasant</u>) stones(gates of fiery glowing stones).

Jeremiah 17:1
¹The sin of Judah is written with a pen of iron; with a point of diamond it is engraved on the tablet of their heart, and on the horns of their altars,

Ezekiel 1:16
¹⁶As for the appearance of the wheels and their construction: their appearance was like the gleaming of a chrysolite (<u>colour of a beryl</u>); and the four had the same likeness, their construction being as it were a wheel within a wheel.

Ezekiel 3:8-9
⁸Behold, I have made your face hard against their faces, and your forehead hard against their foreheads.
⁹Like adamant harder than flint have I made your forehead; fear them not, nor be dismayed at their looks, for they are a rebellious house.

Ezekiel 10:9
⁹And I looked, and behold, there were four wheels beside the cherubim; and the appearance of the wheels was like sparkling chrysolite(as the colour of a beryl stone).

Ezekiel 27:16 and 22
¹⁶Edom trafficked with you because of your abundant goods; they exchanged for your wares emeralds, purple, embroidered work, fine linen, coral, and agate...
²²The traders of Sheba and Ra´amah traded with you; they exchanged for your wares the best of all kinds of spices, and all precious stones, and gold.

Ezekiel 28:13
¹³ You were in Eden, the garden of God; every precious stone was your covering, carnelian (ruby), topaz, and jasper, chrysolite, beryl, and onyx, sapphire, carbuncle, and emerald; (<u>the sardius, topaz, and the diamond, the beryl, the onyx, and the jasper, the sapphire, the emerald, and the</u>

carbuncle,) and wrought in gold were your settings and your engravings. On the day that you were created they were prepared.

Daniel 10:5-6
⁵"I lifted up my eyes and looked, and behold, a man clothed in linen, whose loins were girded with gold of Uphaz.
⁶His body was like beryl,...

Zechariah 7: 12
¹² They made their hearts like adamant lest they should hear the law and the words which the Lord of hosts had sent by his Spirit through the former prophets. Therefore, great wrath came from the Lord of hosts.

The New Testament

Matthew 7:6
⁶Do not give dogs what is holy; and do not throw your pearls before swine, lest they trample them under foot and turn to attack you.

Matthew 13:45-46
⁴⁵Again, the kingdom of heaven is like a merchant in search of fine pearls,
⁴⁶who, on finding one pearl of great value, went and sold all that he had and bought it.

1 Corinthians 3:12-13
¹²Now if anyone builds on the foundation with gold, silver, precious stones, wood, hay, straw–
¹³each man's work will become manifest; for the Day will disclose it, because it will be revealed with fire, and the fire will test what sort of work each one has done.

1 Timothy 2:9
⁹also that women should adorn themselves modestly and sensibly in seemly apparel not with braided hair or gold or pearls or costly attire.

Revelations 4:3
³And he who sat there appeared like jasper and carnelian (a sardine stone), and round the throne was a rainbow that looked like emerald.

Revelations 9:17
¹⁷And this is how I saw the horses in my vision: the riders wore breast-plates the color of fire and sapphire (jacinth) (*blue*) and of sulphur, and the heads of the horses were like lions' heads, and fire and smoke and sulphur issued from their mouths.

256
Literature:

Revelations 17:4-5

⁴The woman was arrayed in purple and scarlet, and bedecked with gold and jewels (precious stones) and pearls, holding in her hand a golden cup full of her abominations and the impurities of her fornication;
⁵and on her forehead was written a name of mastery: Babylon the great, mother of harlots and of earth's abominations.

Revelations 18:10-12

¹⁰they will stand far off, in fear of her torment, and say, 'Alas! alas! thou great city, thou mighty city, Babylon! In one hour has thy judgment come.'
¹¹ And the merchants of the earth weep and mourn for her, since no one buys their cargo any more,
¹²cargo of gold, silver, jewels (precious stones) and pearls, fine linen, purple, silk and scarlet, all kinds of scented wood, all articles of ivory, all articles of costly wood, bronze, iron and marble. . . .

Revelations 18:16

¹⁶Alas, alas, for the great city that was clothed in fine linen, in purple and scarlet, bedecked with gold, with jewels (precious stones), and pearls!

Revelations 21:10-11

¹⁰And in the Spirit he carried me away to a great, high mountain, and showed me the holy city Jerusalem coming down out of heaven from God,
¹¹having the glory of God, its radiance like a most rare (*priceless*) jewel (a stone most precious), like a jasper, clear as crystal.

Revelations 21:18-21

¹⁸The wall was built of jasper, while the city was pure gold, clear as glass.
¹⁹The foundations of the wall of the city were adorned with every jewel (precious stone); the first was jasper, the second sapphire (*lapis-lazuli*), the third agate (*chalcedony*), the fourth emerald,
²⁰the fifth onyx (*sardonyx*), the sixth carnelian (sardius) (*cornelian*), the seventh chrysolite, the eighth beryl, the ninth topaz, the tenth chrysoprase, the eleventh jacinth (*turquoise*), the twelfth amethyst.
²¹And the twelve gates were twelve pearls, each of the gates made of a single pearl, and the street of the city was pure gold, transparent as glass.

The Holy Bible

The Lapidary
Stones of the New Jerusalem in Verse

Marbode of Rennes' (1035-1123)
Translation from the original Latin by C. W. King (1870)

Citizens of the heavenly fatherland
Praise the king of kings
Who is the Supreme Builder
Of the heavenly city
In whose structure
Such a foundation appears:
    Jasper of green color
Reveals the greenness of faith
Which in all perfect men
Never grows weak deep within;
By its strong protection
The devil is resisted.
    Sapphire has an appearance
Similar to the heavenly throne;
It depicts the heart of simple men
Waiting with sure hope,
In whose life and ways
The Highest is pleased.
    Pale **Chalcedony**
Holds the image of fire;
It reddens in public
It gives a gleam in the cloud
It represents the virtue
Of the faithful serving in secret.
    Emerald, exceedingly green,
It is purest faith
Open to every good,
Which knows never to forsake
The work of piety.

    Sardonyx is always three-colored
Discloses the inner man,
Which baseness blackens
In which chastity becomes white.
Martyrdom also reddens
To the Crown of integrity.
    Sard is reddish purple
Whose color is the color of blood.
It displays the splendor
Of those suffering in the manner
of martyrs.
Sixth in the catalog, It holds fast to
the mystery of the cross.
    Gold-colored **chrysolite**
Gleams just like a furnace;
It represents the ways of men
Of perfect wisdom;
It gleams with the sacred
Light of seven-fold grace.
    Beryl is clear
Like the bright sun in water;
It represents the prayers of minds
Sagacious by nature.
What is better to please the mysterious
Leisure of the highest repose?
    Topaz, to whom more rare,
Is to him the more precious,
Blushing with golden splendor.

Both appearance and in ethereal brightness
It shows the steadfast duty
Of the contemplative life.
    Purple **chrysophras**
Portrays unity.
It is intervened with little golden drops
In a certain mixture.
This is perfect charity
Which no ferocity defeats.
    Jacinth is dark blue
With moderate verdure
[brightness used in oldest MS]
Its beautiful appearance
Is changed as the weather.
It signifies the angelic life
Endowed with capacity for discernment.
    Extraordinary **amethyst**,
Violet-colored in beauty,
Sends out golden flames
And little purple marks.

It signifies the heart of the lowly
Who die with Christ.
    These precious stones,
The diversity of colors and
The multiplicity of virtues,
Adorn carnal men.
Whosoever abounds in these
Will be able to be a fellow citizens.
    Peace-bringing Jerusalem
You have these foundations.
Happy and near to God
The soul which is given you,
The guard of your towers
Does not sleep forever.
Grant to us, O Holy One,
King of the heavenly city,
After the course of fleeting life,
Among the ranks of the saints,
Comrades in heaven,
Let us sing praises to you.
    Amen.

De Lapidibus

Marbode of Rennes' (1035-1123)
Translation from the original Latin by C. W. King (1870)

Foremost of all amongst the glittering race
Far India is the **Adamas'** (**diamond**'s) native place;
Produced and found within the crystal mines,
Its native source in its pure lustre shines:

Yet, though it flashes with the brilliant's rays
A steely tint the crystal still displays.
Hardness invincible which nought can tame,
Untouched by steel, unconquered by the flame;

But steeped in blood of goats it yields at length,
Yet tries the anvil's and the smiter's strength.
With these keen splinters armed, the artist's skill
Subdues all gems and graves them at his will.

Largest at best as the small kernel shut
Within th' inclosure of the hazel nut.
Another stone the swart Arabians find,
Broke without blood, of less obdurate kind;

Of duller luster and of lower price,
In weight and bulk it yet the first outvies.
A third gives Cyprus, girdled by the main;
The fourth Philippi's iron mines contain:

Yet all alike the obedient iron sway
As does the magnet, if this gem's away;
For in the presence of this sovereign stone
Robbed of its force an idle mass 'tis thrown.

In magic rites employed, a potent charm,
With force invincible it nerves the arm:
Its power will chase far from thy sleeping head
The dream illusive and the goblin dread;

Baffle the venom'd draught[1], fierce quarrels heal,
Madness appease and stay thy foreman's steel.
Its fitting setting, so have sages told
Is the pale silver or the glowing gold,

And let the jewel in the bracelet blaze
Which round the left arm clasped attracts the gaze.
Achates' stream, which through Sicilia's plains
Winds his soft course renowned in pastoral strains,

Named from himself the **Agate** first disclosed—
A jet black stone by milky zones enclosed:
With figured veins its varied surface strew'd,
Painted by nature in a sportive mood.

Now regal shapes, now gods its face adorn;
Such the fam'd Agate by King Pyrrhus worn,
Whose level surface the nine Muses graced,
Pound Phoebus with his lyre in order placed.

Strange to relate, 'twas to no artist due,
Nature herself the wondrous picture drew.
Another Agate yields the Cretan shore
As coral red, with gold-dust sprinkled o'er;

An antidote against the poisoned draught,
And for the treach'rous viper's venom'd shaft.
Whilst on the Agate which dark Indians praise
The woods arise, the sylvan monster strays:

Placed in the mouth 'twill raging thirst appease
And its mild radiance the tired eyeballs ease.
One fumes like myrrh if on the altar strewed;[2]
Another is besprent with drops of blood:

Whilst those which, like the comb, with yellow gleam,
Are most abundant, but in least esteem.
The Agate on the wearer strength bestows,
With ruddy health his fresh complexion glows;

Both eloquence and grace are by it given,
He gains the favour both of earth and heaven:
Anchises' son, by this attendant saved,
O'ercame all labours, every danger braved.

Not least the glory of the gem renowned
Within the belly of the capon found,
Which, made a eunuch when three years have flown,
Through twice two more in swelling bulk has grown;

Its utmost size no larger than a bean,
Like purest water or the crystal's sheen;
Hence **Alectorius** is the jewel hight,
For gifts of strength extolled, and matchless might.

De Lapidibus: Marbode

If parched with thirst place this within thy mouth,
'Twill in a moment quench thy burning drouth;
Aided by this on many a well-fought day
Crotonian Milo bore the palm away:

And many a prince, with laurel on his brow,
Returned victorious o'er a mightier foe.
The weary wretch who in far exile pines,
Restored to home, with pristine honours shines.

It gifts the pleader with persuasive art
To move the court and touch the hearer's heart:
Th' exhausted frame with youthful vigour filled
Exults once more the love's high rapture thrilled.

From this the bride full powerful aid may gain
To bind her spouse's heart with triple chain.
Bourne in the mouth the virtues of the stone
And all its mighty works are quickly shewn.

Of seventeen species can the **Jasper** boast
Of differing colours, in itself a host.
In various regions is this substance seen:
The best of all, the bright[3] translucent green;

The greatest virtue is to this assigned;
Fevers and dropsies fell its influence kind.
Hung round the neck it eases travail's throes,
And guards the wearer from approaching woes.

Power too it gives when blest by magic rite:
And drives away the phantoms of the night;[4]
But let the gem encased in silver shine,
And fortify thereby its force divine.

Fit only for the hands [fingers] of kings to wear,
With purest azure shines the **Sapphire** rare:
For worth and beauty chief of gems proclaimed,
And by the vulgar oft Syrtites named.

Oft in the Syrtes' midst their shifting sand
Cast by the boiling deep on Lybian strand;
The best the sort that Media's mines supply,
Opaque of colour which excludes the eye.

By nature with superior honours graced,
As gem of gems above all others placed;
Health to preserve, and treachery to disarm,
And guard the wearer from intended harm:

No envy bends him, and no terror shakes;
The captive's chains its mighty virtue breaks;
 The gates fly open, fetters fall away,
And send their prisoner to the light of day.

E'en Heaven is moved by its force divine,
To list the vows presented at its shrine.
Its soothing power contentions fierce controls,
And in sweet concord binds discordant souls;

Above all others this Magician's love,
Which draws responses from the realms above:
The body's ills its saving force allays
And cools the flame that on the entrails preys.

Can check the sweats that melt the waning force
And stay the ulcer in its festering course:
Dissolved in milk it clears the cloud away[5]
From the dimmed eye and pours the perfect day;[6]

Relieves the aching brow when racked with pain
And bids the tongue its wonted vigour gain,
But he who dares to wear this gem divine
Like snow in perfect chastity must shine.

Between the **Hyacinth** and **Beryl** placed,
With lustre fair is the **Calcedon** [Chalcedony] graced;
But pierced, and worn upon the neck or hand,
A sure success in lawsuits 'twill command.

Unlike the Jasper, of this precious stone
Three hues alone are unto merchants known.
Of all green things which bounteous earth supplies
Nothing in greenness with the **Emerald** vies;

Twelve kinds it gives, sent from the Scythian clime,
The Bactrian mountain, and old Nilus' slime;
And some from copper mines of viler race
Marked by the dross drawn from their matrix base:

The **Carchedonian** [chalcedony] from the Punic vale—
To name the others were a tedious tale.
From all the rest the Scythian bear the palm
Of higher value and of brighter charm,

From watchful gryphons in the desert isle
Stol'n by the vent'rous Arismaspian's guile.
Higher their value which admit the sight,
And tinge with green the circumambient light:

Unchanged by sun or shade their lustre glows,
The blazing lamp no dimness on it throws.
Such as a smooth or hollow surface spread
Like slumbering ocean in its tranquil bed,

These like a mirror the beholder's face
Exactly image with reflected rays:
And thus did Nero, if report say true,
The mimic warfare of the arena view.

But best the gem that shows an even sheen,
Lustrous with equal, never-varying green.
Of mighty use to seers who seek to pry
Into the future hid from the mortal eye.

Wear it with reverence due, 'twill wealth bestow
And word persuasive from thy lips shall flow
[As though the gift of eloquence inspired][7]
The stone itself or living spirit fired.

264
Literature:

Hung round the neck it cures the agues' chill,
Of falling sickness, dire mysterious ill;
Its hues so soft refresh the wearied eye,
And furious tempests banish from the sky:

So with chaste power it tames the furious mood
And cools the wanton thoughts that fire the blood.
If steeped in verdant oil or bathed in wine
Its deepened hues with perfect lustre shine.

The **Sard** and **Onyx** in one name unite,
And from their union spring three colours bright.
O'er jetty black the brilliant white is spread
And o'er the white diffused a fiery read:

If clear the colours, if distinct the line,
Where still unmixed the various layers join,
Such we for beauty and for value prize,
Rarest of all that teeming earth supplies,

Chief amongst signets it will best convey
The stamp impressed, nor tear the wax away.
The man of humble heart and modest face,
And purest soul the **Sardonyx** should grace;

A worthy gem, yet boasts no mystic powers:
'Tis sent from Indian and Arabian shores.
Called by the **Onyx** round the sleeper stand
Black dreams, and phantoms rise, a grisly band:

Who so on neck or hand this stone displays
Is plagued with lawsuits and with civil frays;
Round infants' necks if tied, so nurses show,
Their tender mouths with slaver overflow.

This is the Arabian, this the Indian sends,
And five the sorts to which its name it lends.
The red Sardian to its birthplaces owes
Its name, to **Sardis**, whence it first arose.

Cheapest of gems, it may no shore of fame
For any virtue save its beauty claim;
Except for power the onyx's spell to break:
Of this old sages five divisions make.

The golden **Chrysolite** a fiery blaze
Mixed with the hue of ocean's green displays;
Enchased in gold its strong protective might
Drives far away the terrors of the night:

Strung on the hairs plucked from an ass's tail,
The mightiest demons 'neath its influence quail.
This potent amulet, of old renowned,
Wear like a bracelet on thy left arm bound.

'Tis brought by merchants from those far off lands.
Where Ethiopia spreads her burning sands.
Cut with six facets shines the **Beryl** bright,
Else a pale dullness clouds its native light;

The most admired display of softened beam
Like tranquil seas or olives' oily gleam.
This potent gem, found in far India's mines,
With mutual love the wedded couple binds;

The wearer shall to wealth and honours rise
And from all rivals bear the wished-for prize;
Too tightly grasped, as if instinct with ire,
It burn th' incautious hand with sudden fire.

Lave this in water, it a wash supplies
For feeble sight and stops convulsive sighs.
Its species nine, for so the learned divide,
Avail the liver and the tortured side.

From the seas remote the yellow **Topaz** came,
Found in the island of the self-same name;
Great is the value for full rare the stone,
And but two kinds to eager merchants known.

One vies with purest gold, or orange bright;
The other glimmers with a fainter light:[8]
Its yielding nature to the file gives way[9]
Yet bids the bubbling caldron cease to play.

The land of gems, culled from its copious store,
Arabia sends this to the Latin shore;
One only virtue Nature grants the stone,
Those to relieve who under hemorrhoids groan.

Three various kinds the skilled as **Hyacinth's** name,
Varying in colour, and unlike in fame:
One, like pomegranate flowers a fiery blaze;
And one, the yellow citron's hue displays.

One charms with paley blue the gazer's eye
Like the mild tint that decks the northern sky:
A strength'ning power the several kinds convey
And grief and vain suspicions drive away.

Those skilled in jewels chief the **Granate** prize,
A rarer gem and flushed with ruby dyes.
The blue sort feels heaven's changes as they play
Bright on the sunny, dull when dark the day:

But best that gem which not too deep a hue
O'erloads, not yet degrades too light a blue;
But where the purple bloom unblemished shines
And in due measure both the tints combines.

No gem so cold upon the tongue can lie,
With greater hardness none the file defy;
The diamond splinter to th' engraver's use
Alone its hardened stubbournness subdues.

The citron-coloured, by their pallid dress,
Their baser nature openly confess;
With any kind borne on the neck or hand,
Secure from peril visit every land.

De Lapidibus: Marbode

On all the wand'rings honours shall attend
And noxious airs shall ne'er their health offend;
Whatever prince thy just petition hears
Fear no repulse, he'll listen to their prayers.

Midst other treasures to adorn the ring
This gem from Afric's burning sand they bring.
Parent of gems, rich India from her mines
The **Chrysoprase**, a precious gift, consigns,

As leaves of leeks in mingled shadows blent,
Or purple dark with golden stars besprent;
But what its virtue, rests concealed in night:
All things Fate grants not unto mortal sight.

The Tyrian purple the rich **Amethyst** dyes,
Or darker violet charms the gazer's eyes;
Bright as the ruby wine another glows,
Or fainter blush that decks the opening rose;

Like drops of wine with fountain streams allayed.
All these supplied by jewelled India's mart,
Easy to cut, yield to the graver's art:

The gem, if rarer, were a precious prize,
But now too common it neglected lies;
Famed for their power to check the fumes of wine,
Five different species yields the bounteous mine.

The rapid swallow swifter than the airs
Within her breast (Latin translates, belly) the Chelidonian bears,
A fatal gift, deep in her bowels pent,
Which with her life is from the owner rent.

The **Chelidonian** is of might supreme,
Though not of those which shoot a brilliant gleam:
Yet many a gem that men for beauty praise,
Unshapen, small, and dull, its worth outweighs.

The feather'd victims in their bowels stored
Two different sorts — the white and red — afford:
The pining sickness feels their influence mild,
The moonstruck idiot, and the maniac wild.

With force persuasive orators they arm,
And grace the hearts of multitudes to charm:
Wrapped in a linen cloth this present rare,
Under thy left arm tied ne'er fail to wear;

The black, in woolen cloth thus too suspend,
And bring thy measures to the wished-for end.
It blunts the threats and cools the ire of kings,
And to the wearied sight refreshment brings.

This in a yellow cloth of linen laid
Will banish fevers that thy limbs invade,
or watery humours that with current slow
Obstruct the veins and stop their healthy flow.

Lycia her **Jet** in medicine commends;
But dearest, that which distant Britain sends;
Black, light, and polished, to itself it draws
If warmed by friction near adjacent straws.

Though quenched by oil, its smouldering embers raise
Sprinkled with water a still fiercer blaze:
It cures the dropsy, shakey teeth are fixed
Washed with the powder'd stone in water mixed.

The female womb its piercing fumes relieve,
Nor epilepsy can this test deceive:
From its deep hole it lures the viper fell,
And chases far away the powers of hell;

It heals the swelling plagues that gnaw the heart
And baffles spells and magic's noxious art.
This by the wise the surest test is styled
Of virgin purity by lust defiled.

Three days in water steeped, the draught bestows
Ease to the pregnant womb in travail's throes.
Whilst rooted 'neath the waves[10] the **Coral** grows,
Like a green bush its waving foliage shows:

Torn off by nets, or by the iron mown;
Touched by the air it hardens into stone;
Now a bright red, before a grassy green,
And like a little branch its form is seen;

Of measure small, scarce half a foot in size,
A useful ornament the branch supplies.
Wondrous its power, so Zoroaster sings,
And to the wearer sure protection brings.

Its numerous virtues Metrodorus sage
Has told to mankind in his learned page:
How, lest they harm ship, land, or house, it binds
The scorching lightning and the furious winds.

Sprinkled 'mid climbing vines or olives' rows,
Or with the seed the patient rustic sows,
'Twill from the crops avert the arrowy hail
And with abundance bless the smiling vale.

Far from thy couch't will chase the shades of hell
Or monsters summoned by Thessalian spell;
Give happy opening, and successful end,
And calm the tortures that the entrails rend.

From Asia's climes rich Alabanda sends
The **Alabandine** and its name extends;
In fiery lustre with the Sard it vies
And leaves in doubt the skilled beholder's eyes.

Let not the Muse the dull **Carnelian** slight
Although it shines with but a feeble light;
Fate has with virtues great its nature graced,
Tied round the neck or on the finger placed.

Its friendly influence checks the rising fray,
And chases spites and quarrels far away:
That, where the colour of raw flesh is found,
Will stanch the blood fast issuing from the wound;

Whether from mangled limbs the torrents flow,
Or inward issues, source of deadly woe.
The **Carbuncle** eclipses by its blaze
All shining gems and casts its fiery rays.

Like to the burning coal; whence comes its name,
Among the Greeks as Anthrax known to fame.
Not e'en by darkness quenched its vigour tires;
Still at the gazer's eye it darts its fires;

A numerous race, within the Lybian ground
Twelve kinds by mining Troglodytes are found.
Voided by lynxes, to a precious stone
Congealed the liquid is Lyncurium (**amber**) grown;

This knows the lynx and strives with envious pride
'Neath scraped up sand and the precious drops to hide.
Surpassing amber in its golden hue[11]
It straws attracts if Theophrast says true:

The tortured chest it cures, their native bloom
Through its kind aid the jaundiced cheeks resume;
And let the patient wear the gem, its force
Will soon arrest the diarrhoea's course.

Chief amongst gems the **Aetites** stands
Borne by the bird of Jove from farthest lands:
As safeguard to his nest, and influence good
To ward off danger from the callow brood.

Shut in the pregnant stone another lies —
Hence pregnant women its protection prize;
With this gem duly round her left arm tied
Need no mischance affright the teeming bride.

Sober the wearer too shall ever prove,
Shall wealth amass, and reap his people's love:
Victory shall crown his brows; his offspring dear,
Shall healthy live nor fate untimely fear.

The epileptic wretch, saved by its worth,
No more shall fall and writhe upon the earth.
Should'st thou suspect they friend of treason foul,
The privy prisoner lurking in the bowl,

Thus prove his mind: him to thy banquet bid
And let this stone beneath the dish be hid,
When, if he harbour treachery in his thought,
Whilst there the stone lies he can swallow nought:

Remove the gem, delivered from its power
The tasted meats he'll greedily devour.
The stone they say is found, with scarlet dyed,
Hid on the margin of old ocean's tide.

In Persian lands, eagles' nests concealed,
And by the Twins its virtues first revealed.
Nor must we pass the Selenites (**moonstone**) by
Whose hues with grass or verdant jasper vie,

With the lov'd moon it sympathetic shines,
Grows with her increase —with her wane declines;
And since it thus for heav'nly changes cares
The fiting name of sacred stone it bears.

A powerful philtre to ensnare the heart,
It saves the fair from dire consumption's dart.
Long as the moon her wasted orb repairs
To pining mortals these effects it bears;

Yet ne'ertheless, when Luna's on the wane
Men from its use will diverse blessings gain.
This stone, a remedy for human ills,
Springs, as they tell, from famous Persia's hills.

Literature:

The **Heliotrope**, or "gem that turns the sun,"
From its strange power the name has justly won:
For set in water opposite his rays
As red as blood 'twill turn bright Phoebus' blaze.

And, far diffused the inauspicious light,
With strange eclipse the startled work affright.
Then boils the vase, urged by its magic power,
And casts far o'er the brim the sudden shower;

As when the gloomy air to rain gives way
It storms evokes, and clouds the fairest day;
It gifts the wearer with prophetic eye
Into the Future's darkest depths to spy.

A good report 'twill give and endless praise,
And crown they honour'd course with length of days.
It checks the flow of blood, the wearer's soul
Shall laugh at treason or the poison'd bowl.

Though with such potent virtues grac'd by heaven
One yet more wondrous to the gem is given.
This with the herb that bears its name unite
With incantation due and secret rite,

Then shalt thou mortal eyes in darkness shroud
And walk invisible amidst the crowd.
The stone for colour might an emerald seem,
But drops of blood diversify the green.

'T is sent sometimes from Ethiopia's land,
Sometimes from Afric or the Cyprian strand.
In Corinth's Isthmus springs the **Hephoestite**,
More precious than its brass, and ruddy bright.

The seething caldron bubbling o'er the blaze,
Cast in the stone, its fervent fury stays;
Tam'd by the virtue of the gem, as cool
It falls as water in a tranquil pool.

Nor flights of locusts, nor the scourging hail,
Nor whirlwinds fierce shall thy fair fields assail;
Nor falling rust the growing crops shall blight
That stand defended by its saying might.

Held to the sun it shoots out fiery rays
Dazzling the eye as with the furnace blaze:
This burning stone sedition's fury charms.
And 'gainst all danger its possessor arms.

But let this precept in thy mind be borne —
Right o'er the heart this mineral must be worn.
The Hoematite — named by the Greeks from blood —
Benignant nature formed for mortals' good:

Its styptic virtue many a proof will show
To heal the tumours that on th' eyelids grow.
And rubbed on darkening eyes it clears away
The gathering cloud and gives to see the day:

Rubbed in a mortar with tenacious glaire
And juice of pomegranates, and eye-salve rare.
Those who spit blood its healing power will own,
As those who under cankering ulcers groan.

It stays the flux that drains the female frame,
And, powdered fine, proud flesh in wounds can tame:
Dissolved in wine the oft repeated dose
Will stop all looseness that excessive flows;

Dissolved in water 'twill allay the smart
Of poisonous serpent's bite or aspic's dart.
If mixed with honey 't is an unction sure
All maladies that pain the eyes to cure.

This potent draught, as by experience shown,
Within the bladder melts the torturing stone.
Of red and rusty hue, in Afric found,
Or in Arabian, or in Lybian ground.

The mountains of the Macedonian bold
Within their mines the Poeanites hold,[12]
Unknown the cause, with imitative throes
It heaves, and all the pangs of childbirth knows.

From some mysterious seed the wondrous earth
Conceives, and in due time excludes the birth;
Hence teeming females its protection bless
In that last moment when their dangers press.

True to its name, the Hexacontalite (probably **opal**)
In one small orb doth sixty gems unite;
With numerous hues for scanty size atones
And singly shew the tints of many stones.

Mid Lybia's deserts parched by burning wind
The Troglodyte this rainbow jewel finds.
Midst precious stones a place the **Prase** may claim,
Of value small, content with beauty's fame.

No virtue has it; but it brightly gleams
With emerald green, and will the gold beseems;
Or blood-red spots diversify its green,
Or crossed with three white lines its face is seen.

Crystal is ice through countless ages grown
(So teach the wise) to hard transparent stone:
And still the gem retains its native force,
And holds the cold and colour of its source —

Yet some deny, and tell of crystal found
Where never icy winter froze the ground;
But true it is that held against the rays
Of Phoebus it conceives the sudden blaze,

And kindles tinder, which, from fungus dry
Beneath its beam, your skilful hands apply:
Dissolved in honey, let the lucious draught
By mothers suckling their lov'd charge be quaffed,

De Lapidibus: Marbode

Then from their breasts, as sage physicians show,
Shall milk abundant in rich torrents flow.
By the Red Sea the swarthy Arabs glean
Th' Iris (prism) resplendent with the Crystal's sheen;

Its form six-sided, full of heav'n's own light,
Has justly gained the name of rainbow bright;
For in a room held 'gainst the solar rays
It paints the wall with many-colour'd blaze,

And where the crystal its reflection throws
The heav'nly bow in all its splendour glows.
The' Androdamas (**pyrite**), in figure like a die,
In whiteness may with silver's lustre vie;

Hard as the Diamond, found in shifting sand
Tossed by the wind along the Red Sea's strand;
As Magians teach endued with might power,
To cool the soul with fury boiling o'er.

Though from thy eyes each ail th' **Opthalmius** chase
Yet 'tis the guardian of the thievish race:
It gifts the bearer with acutest sight
But clouds all other eyes with thickest night;

So that the plunderers bold in open day
Secure from harm can bear their spoil away.
The sea-born shell conceals the Union round,
Called by this name as always single found.

One in one shell, for ne'er a larger race,
Within their pearly walls the valves embrace.
Prized as an ornament its whiteness gleams,
And well the robe, and well the gold beseems.

At certain seasons do the oysters lie
With valves wide gaping towards the teeming sky,
And seize falling dews, and pregnant breed
The shining globules of th' ethereal seed.

Literature:

Brighter the offspring of the morning dew,
The evening yields a duskier birth to view;
The younger shells produce a whiter race,
We greater age in darker colours trace.

The more of dew the gaping shell receives,
Larger the **pearl** its fruitful womb conceives;
However favoring airs its growth may raise,
Its utmost bulk ne'er half an ounce outweighs,

If thunders rattle through the vaulted sky
The closing shells in sudden panic fly;
Killed by the shock the embryo pearls they breed,
Shapeless abortions in their place succeed.

These spoils of Neptune th' Indian ocean boasts;
But equal those from ancient Albion's coasts.
In the Pantheros (**opal**) varying colours meet,
Where black and red, and green and white compete:

Here rosy light, there brilliant purples play,
And blooms the gem with varying patterns gay.
At dawn of day its potent beauties view
So shall success they doings still pursue,

For all that day, defended by the charm,
No foe shall e'er prevail to work thee harm.
All travellers tell how 'midst far India's groves
Beauteous in spotted hide the panther roves.

How furious lions dread his piercing cry
And trembling at the sound in terror fly.
Marked like the beast that can the lion tame
The spotted gem obtains the self-same name.

The **Molachites'** virtue keeps from hurt
The infant's cradle, all mischance to avert,
Let spiteful witchcraft blast the tender frame.
Virtue with beauty joined exalt its fame.

De Lapidibus: Marbode

Opaque of hue, with th' Emerald's vivid green
It charms the sight, first in Arabia seen.
Named from the fire the yellow **Pyrite** spurns
The touch of man — and to be handled scorns:

Touch it with trembling hand and cautious arm,
For tightly grasped it burns the closed palm.
In Afric springs the **Chrysoprasion** bright,
Which day conceals but darkness brings to light:

By night a shining fire, it lifeless lies
Like golden ore when day illumes the skies.
Reversed is Nature's law where light reveals
Whate'er in darkness shrouding night conceals.

Footnotes are a compilation of translator's and editor's notes from various sources. Bold face type added to highlight gem names.

Precious Stones in Shakespeare's Works

The writings of the great poet Shakespeare attest to the practice of referencing precious stones in literature. The bard made over one hundred references to precious stones in his writings. Some gems are mentioned in the context of their lore and give insight into the popular beliefs of the day. Often gems serve as adjectives that convey more to the mind than corresponding adjectives of color or texture.

Each of the gems mentioned and corresponding passages are listed alphabetically by gem. The passages from plays are given first followed by an alphabetical list of all the citations found in Shakespeare's poems. If more than one gem is found in a particular passage, that passage is repeated under each of the gems mentioned.

Stones Mentioned in Plays
Agate
An agate very vilely cut.
Much Ado About Nothing, Act 3, Scene 1, line 25

His heart like an agate with your print impress'd
Love's Labour's Lost, Act 2, Scene 1, line 236[1]

I was never manned with an agate till now.
II Henry IV, Act 1, Scene 2, line 19

Agate-ring, pirke-stocking, caddis-garter, smooth-tongue.
I Henry IV, Act 2, Scene 4, line 78

In shape no bigger than an agate-stone
On the forefinger of an alderman.
Romeo and Juliet, Act 1, Scene 4, line 55

Amber
Her amber hair for foul hath amber quoted.
Love's Labour's Lost, Act 4, Scene 3, line 87

With amber bracelets, beads, and all this knavery.
Taming of the Shrew, Act 4, Scene 3, line 58

Their eyes purging thick amber and plum-tree gum.
Hamlet, Act 2, Scene 2, line 201[2]

Carbuncle
Her nose, all o'er embellished with rubies, carbuncles, sapphires.
Comedy of Errors, Act 3, Scene 2, line 138

A carbuncle entire, as big as thou art,
Were not so rich a jewel.
Coriolanus, Act 1, Scene 4, line 55

O'er sized with coagulate gore,
With eyes like carbuncles.
Hamlet, Act 2, Scene 2, line 485

Were it carbuncled
Like holy Phoebus' car.
Anthony and Cleopatra, Act 4, Scene 8, line 28

Had it been a carbuncle
Of Phoebus' wheel.
Cymbeline, Act 5, Scene 5, line 189

Chrysolite

If heaven would make me such another world
Of one entire and perfect chrysolite.
Othello, Act 5, Scene 2, line 145[3]

Coral

Of his bones are coral made.
The Tempest, Act 1, Scene 2, line 397

I saw her coral lips to move.
Taming of the Shrew, Act 1, Scene 1, line 179

Diamond

I see how thine eye would emulate the diamond.
Merry Wives of Windsor, Act 3 Scene 3, line 59

Give me the ring of mine you had at dinner,
Or, for my diamond, the chain you promised.
Comedy of Errors, Act 4 Scene 3, lines 61, 62

Sir, I must have that diamond from you.—
There, take it.
Comedy of Errors, Act 5, Scene 1, line 58

A lady walled about with diamonds!
Love's Labour's Lost, Act 5, Scene 2, line 6

A diamond gone, cost me two thousand ducats in Frankfort!
Merchant of Venice, Act 3, Scene 1, line 87

Literature:

Set this diamond safe
In golden palaces, as it becomes.
Henry VI, Part. 1, Act 5, Scene 3, line 169

A heart it was, bound in with diamonds.
Henry VI, Part. 3, Act 3, Scene 2, line 107

Not deck'd with diamonds and Indian stones,
Nor to be seen.
Henry VI, Part 3, Act 3, Scene 1, line 63

One day he gives us diamonds, next day stones.
Timon of Athens, Act 3, Scene 6, line 56

This diamond he greets your wife withal.
Macbeth, Act 2, Scene 1, line 15

Which parted thence,
As pearls from diamonds dropp'd.
King Lear, Act 4, Scene 3, line 24

This diamond was my mother's; take it, heart;
But keep it till you woo another wife.
Cymbeline, Act 1,Scene 1, line 112

She went before others I have seen,
as that diamond of yours outlustres many I have beheld.
Cymbeline, Act 1, Scene 4, line 81

I have not seen the most precious diamond that is, nor you the lady.
Cymbeline, Act 1, Scene 4, line 81
I shall but lend my diamond till your return.
Cymbeline, Act 1, Scene 4, line 153

My ten thousand ducats are yours; so is your diamond too.
Cymbeline, Act1, Scene 4, line 163

It must be married
To that your diamond.
Cymbeline, Act 2, Scene 4, line 98

That diamond upon your finger, say,
How came it yours?
Cymbeline, Act 5, Scene 5, line 137

To me he seems like diamond to glass.
Pericles, Act 2, Scene 3, line 36

You shall, like diamonds, sit about his crown.
Pericles, Act 3, Scene 2, line 102

The diamonds of a most praised water
Do appear, to make the world twice rich.
Pericles, Act 3, Scene 2, line 102

Emerald

In emerald tufts, flowers purple, blue, and white.
Merry Wives of Windsor, Act 5, Scene 5, line 74

Jet

There is more difference between thy flesh and hers than between jet and ivory.
Merchant of Venice, Act 3, Scene 1, line 42

What color is my gown of?—Black, forsooth: coal-black as jet.
II Henry VI, Act 2, Scene 1, line 112

Two proper palfreys, black as jet,
To hale thy vengeful waggon swift away
Titus Andronicus, Act 5, Scene 2, line 50

Opal

For thy mind is a very opal.
Twelfth Night, Act 2, Scene 4, line 77

Pearl

Full fathom five thy father lies;
Of his bones are coral made;
Those are pearls that were his eyes.
Tempest, Act 1, Scene 2, line 398

She is mine own,
And I as rich in having such a jewel
As twenty seas, if all their sand were pearl.
Two Gentlemen of Verona, Act 2, Scene 4, line 170

A sea of melting pearl, which some call tears.
Two Gentlemen of Verona, Act 3, Scene 1, line 224[4]

But pearls are fair; and the old saying is,
Black men are pearls in beauteous ladies' eyes
'Tis true; such pearls as put out ladies' eyes.
Two Gentlemen of Verona, Act 5, Scene 2, line 11

Like sapphire, pearl and rich embroidery
Buckled below fair knighthood's bending knee.
Merry Wives of Windsor, Act 5, Scene 5, line 75

Laced with silver, set with pearls.
Much Ado About Nothing, Act 3, Scene 4, line 20

Fire enough for a flint, pearl enough for a swine.
Love's Labour's Lost, Act 4, Scene 2, line 91

This and these pearls to me sent Longaville.
Love's Labour's Lost, Act 5, Scene 2, line 53

Will you have me, or your pearl again?
Neither of either.
Love's Labour's Lost, Act 5, Scene 2, line 458

Decking with liquid pearls the bladed grass.
Midsummer Night's Dream, Act 1, Scene 1, line 211

I must go seek some dewdrops here
And hang a pearl in every cowslip's ear.
Midsummer Night's Dream, Act 2, Scene 1, line 15[5]

And that same dew, which sometime on the buds
Was wont to swell, like round and orient pearls.
Midsummer Night's Dream, Act 4, Scene 1, line 57

Rich honesty dwells like a miser, sir, in a poor house; as your pearl in your foul oyster.
As You Like It, Act 5, Scene 4, line 63

Their harness studded all with gold and pearl.
Taming of the Shrew, Introduction, Scene 2, line 44

Fine linen, Turkey cushions boss'd pearls
Valance of Venice Gold.
Taming of the Shrew, Act 2, Scene 1, line 355[6]
Why, sir, what 'cerns it you if I wear pearl or gold?
Taming of the Shrew, Act 5, Scene 1, line 77

This pearl she gave me, I do feel't an see't.
Twelfth Night, Act 4, Scene 3, line 2

Draws those heaven-moving pearls from his poor eyes.
King John, Act 2, Scene 1, line 169

Our chains and our jewels.—
Your brooches, pearls and ouches.
II Henry IV, Act 2, Scene 4, line 53

The crown imperial,
The intertissued robe of gold and pearl.
Henry V, Act 4, Scene 1, line 279

Wedges of gold, great anchors, heaps of pearl,
Inestimable stones, unvalued jewels.
Richard III, Act 1, Scene 4, line 26

The liquid drops of tears that you have shed
Shall come again, transform'd to orient pearl.
Richard III, Act 4, Scene 4, line 322[7]

Her bed is India; there she lies, a pearl.
Troilus and Cressida, Act 1,Scene 1, line 103

She is as pearl
Whose price hath launch'd above a thousand ships.
Troilus and Cressida, Act 2, Scene 2, line 81

I will be bright, and shine in pearl and gold.
Titus Andronicus, Act 2, Scene 1, line 19[8]

This is the pearl that pleased your empress' eye.
Titus Andronicus, Act 5, Scene 1, line 42

I see thee compass'd with thy kingdom's pearl.
Macbeth, Act 5, Scene 8, line 56

Hamlet, this pearl is thine.
Hamlet, Act 5, Scene 2, line 293

What guests were in her eyes; which parted thence,
As pearls for diamonds dropp'd.
King Lear, Act 4, Scene 3, line 24

Like the base Iudean [Indian], threw a pearl away
Richer than all his tribe.
Othello, Act 5, Scene 2, line 347

He kiss'd,—the last of many doubled kisses,—
This orient pearl.
Anthony and Cleopatra, Act 1, Scene 5, line 41

I'll set thee in a shower of gold, and hail
Rich pearls upon thee.
Anthony and Cleopatra, Act 2, Scene 5, line 46

Rock Crystal

His mistress
Did hold his eyes lock'd in her crystal looks.
Two Gentlemen of Verona, Act 2, Scene 4, line 89

Me thought all his senses were lock'd in his eye
As jewels in crystal for some prince to buy.
Love's Labour's Lost, Act 2, Scene 1, line 243

To what, my love, shall I compare thine eye?
Crystal is muddy.
Midsummer Night's Dream, Act 3, Scene 2, line 139[9]

With these crystal beads heaven shall be bribed
To do him justice.
King John, Act 2, Scene 1, line 171[10]

Since the more fair and crystal is the sky,
The uglier seem the clouds that in it fly.
Richard II, Act 1, Scene 1, line 41

Go, clear thy crystals.
Henry V, Act 2, Scene 3, line 56

Comets, importing change of times and states,
Brandish your crystal tresses in the sky.
I Henry VI, Act 1, Scene 1, line 3

But in that crystal scales let there by weigh'd
Your lady's love against some other maid.
Romeo and Juliet, Act 1, Scene 2, line 101

Thy crystal window ope; look out.
Cymbeline, Act 5, Scene 4, line 81

Ruby

The impression of keen whips I'd wear as rubies.
Measure for Measure, Act 2, Scene 4, line 101

Her nose, all o'er embellished with rubies, carbuncles, sapphires.
Comedy of Errors, Act 3, Scene 2, line 138
Those be rubies, fairy favors.
Midsummer Night's Dream, Act 2, Scene 1, line 12

Over thy wounds now do I prophesy,—
Which, like dumb mouths, do ope their ruby lips.
Julius Caesar, Act 3, Scene 1, line 260[11]

And keep the natural ruby of your cheeks,
When mine is blanch'd with fear.
Macbeth, Act 3, Scene 4, line 115

But kiss; on kiss! Rubies unparagon'd,
How dearly they do't!
Cymbeline, Act 2, Scene 2, line 18

Sapphire

Like sapphire, pearl and rich embroidery
Buckled below fair knighthood's bending knee.
Merry Wives of Windsor, Act 5, Scene 5, line 75

Her nose, all o'er embellished with rubies, carbuncles, sapphires.
Comedy of Errors, Act 3, Scene 2, line 138

Turquois

It was my turquoise; I had it of Leah when I was a bachelor.
Merchant of Venice, Act 3, Scene 1, line 126

Stones Mentioned in Poems

Amber

With coral clasps and amber studs.
"Passionate Pilgrim", line 366

Favours from a maund[12] she drew
Of amber, crystal, and of beaded jet.
"Lover's Complaint", line 37

Coral

That sweet coral mouth
Whose precious taste her thirsty lips well knew.
"Venus and Adonis", line 542

Her alabaster skin,
Her coral lips, her snow white dimpled chin.
"Lucrece", line 420

Like ivory conduits coral cisterns filling.
"Idem", line 1234

Coral is far more red than her lips' red.
"Sonnet CXXX", line 2

A belt of straw and ivy buds.
With coral clasps and amber studs.
"Passionate Pilgrim", line 366

Diamond

The diamond—why 'twas beautiful and hard.
"Lover's Complaint", line 211

Emerald

The deep-green emerald, in whose fresh regard
Weak sights their sickly radiance do amend.
"Idem", line 213[13]

Jet

Favours from a maund[12] she drew
Of amber, crystal, and of beaded jet.
"Lover's Complaint", line 37

Pearls

Her tears began to turn their tide,
Being prison'd in her eye like pearls in glass.
"Venus and Adonis", line 980

Literature:

And wiped the brinish pearl from her bright eyes.
"Lucrece", line 1213

Those round clear pearls of his, that move thy pity,
Are balls of quenchless fire to burn thy city.
"Idem", line 1553

Of paled pearls and rubies red as blood.
"Lover's Complaint", line 198

Ah! but those tears are pearls which thy love sheds.
"Sonnet XXXIV", line 13

Bright orient pearl, alack, too timely shaded!
"Passionate Pilgrim", line 133

Rock Crystal

But hers through which the crystal tears gave light,
Shone like the moon in water seen by night.
"Venus and Adonis", line 491

Nor thy soft hand, sweet lips, and crystal eyne.
"Venus and Adonis", line 633

One, the hairs were fold, crystal the other's eyes.
"Idem", line 142

The crystal tide that from her two cheeks fair
In the sweet channel of her bosom dropt.
"Idem", line 957

Her eyes seen in the tears, tears in her eye;
Both crystals, where they view'd each other's sorrow.
"Idem", lines 962 and 963

Through, crystal walls each little mote will peep.
"Lucrece", line 1251

A closet never pierced with crystal eyes.
"Sonnet XLVI", line 6

Favours from a maund (basket or hamper) she drew
Of amber, crystal, and of beaded jet.
"Lover's Complaint", line 37

Who glazed with crystal gate the glowing roses.
"Lover's Complaint",' line 286

Ruby

Once more the ruby-colour'd portal open'd.
"Venus and Adonis", line 451

Of paled pearls and rubies red as blood.
"Lover's Complaint", line 198

Sapphire

The heaven-hued sapphire and the opal blend
With objects manifold.
"Idem", line 215

Gems in the Holy Qur'an

The Holy Qur'an (also known as the Koran) is the primary sacred text of Islam. These translations are those of Abdullah Yusuf Ali (1934), the most widely used in the Muslim world.

The Qur'an is arranged in chronological order by 114 Sûars, or chapters of varied length. Each Sûar is divided into Ayats, or verses. The Sûars are numbered in Roman numerals and the Ayats are noted by Arabic numbers. Using this method, VI. 10 refers to the tenth Ayat of the sixth Sûar.

Three gems are mentioned; pearls, rubies and coral.

XXII al-Hajj, The Pilgrimage

^{23}Godwill admit those
Who believe and work righteous deeds,
To Gardens beneath which
Rivers flow: they shall be
Adorned therein with bracelets
Of gold and pearls; and
Their garments there will be of silk.

XXXV al-Fatir, The Creation

^{33}Gardens of Eternity will they
Enter: therein will they
Be adorned with bracelets
Of gold and pearls;
And their garments there
Will be of silk.

LII at-Tur, The Mountain

^{24}Round about them will serve,
(Devoted) to them.
Youths (handsome) as Pearls
Well-guarded.

LV ar-Rahman, The Merciful

²¹Then which of the favours
Of your Lord will ye deny?

²²Out of them come
Pearls and Coral:

⁵⁷Then which of the favours
Of your Lord will ye deny?

⁵⁸Like unto Rubies and coral.

LVI al-Waqi`a, That Which is Coming

²²And (there will be) Companions
With beautiful, big,
And lustrous eyes,-

²³Like unto Pearls
Well-guarded.

LXXVI ad-Dahr, The Man

¹⁹And round about them
Will (serve) youths
Of perpetual (freshness):
If thou seest them,
Thou wouldst think them
Scattered Pearls.

Birthstones

The most generally accepted use of talismanic gems in the Twentieth Century is the wearing of birthstones. The tradition is followed by many who would not consider themselves practitioners of religious traditions or mystic sciences. It is common for a person to wear a natal stone, give one as a gift, or at least be aware of what gem is recommended according to their date of birth. Rare it is, however, to find someone who is familiar with the origins of the tradition or the significance of the stones they wear. The general public has been attuned to the practice through the efforts of the jewelry industry to market certain gems. Birth-pendants and charms are marketed to new parents. Mother's rings, bracelets and necklaces are sold to millions of customers who have no more knowledge of the subject than a chart in the retailer's showroom. Alternatives to the jeweler's recommendations are seldom considered. The reason a given stone is listed for a month or why such a list exists is rarely discussed. The majority of the birthstones sold are, in fact, synthetic or imitation gems. One would doubt that anything other than a genuine stone could offer metaphysical service. The history of natal stone use, and the traditions attached, offer insights to the applicability and appropriateness of these ubiquitous talismans.

The wearing of a particular stone according to the time of one's birth is a relatively recent practice in Western culture. George Frederick Kunz states the custom of wearing natal stones first appears in eighteenth century Poland. The Gemological Institute of America claims the origin of the custom in Germany about 1562. Which ever source is correct, both place the beginning of the Western tradition of natal stones in Europe only a few hundred years ago. As has been documented, the belief that stones possess mystic properties that influence the wearer's health, wealth, power, or prosperity long predates this time. For centuries the reason to wear a particular gem was influenced by its therapeutic value and ability to treat a particular disease or negative life condition. Only with a renewed interest and study of early Talmudic writings did European Jews begin to associate a single gem with one of the twelve months of the year.

The supposed connection was first proposed by the noted Jewish historian Flavius Josephus in the first century A.D. In his *Antiquities of the*

Jews, Josephus writes of the gems set in Aaron's Breastplate: "And for the twelve stones, whether we understand by them the months, or the twelve signs of what the Greeks call the zodiac, we shall not be mistaken in their meaning."[1] This reference to the months of the year and the signs of the zodiac was cited by St. Jerome* in a fourth century letter to Fabiola. In this letter he suggests the Foundation Stones of the New Jerusalem are the appropriate stones for Christian use. The writings of Josephus and St. Jerome are believed to be the basis for the custom of wearing one gem as a natal or birthstone.

The twelve stones of the Breastplate according to Josephus

Sardonyx	Topazos	Smaragdos
Anthrax	Jaspis	Sappheiros
Liguros	Amethystos	Achates
Chrysolithos	Onyx	Beryllus

There has long been a controversy as to which stones are appropriate. It would seem logical to simply follow the order given by Josephus or the two lists given in The Book of Exodus.

Exodus 28:15-21

> 15 And you shall make a breastplate of judgment, in skilled work; like the work of the ephod you shall make it; of gold, blue and purple and scarlet stuff, and fine twined linen shall you make it. 16It shall be square and double, a span its length and a span its breadth. 17And you shall set in it four rows of stones. A row of sardius, topaz, and carbuncle shall be the first row; 18and the second row an emerald, a sapphire, and a diamond; 19 and the third row a ligure, an agate, and an amethyst; 20and the fourth row a beryl, an onyx, and a jasper; they shall be set in gold filigree. 21There shall be twelve stones with their names according to the names of the sons of

---

*St. Jerome is considered one of the Fathers of the Latin Church. He is credited with the first translation and publication of the Latin version of the Bible, established as the official version of the Scripture for the Roman Catholic Church.

Israel; they shall be like signets, each engraved with its name, for the twelve tribes.
Exodus 39: 6-7, 10-13

"6The onyx stones were prepared, enclosed in settings of gold filigree and engraved like the engravings of a signet, according to the names of the sons of Israel. 7And he set them on the shoulder-pieces of the ephod, to be stones of remembrance for the sons of Israel; as the Lord commanded Moses. . . 10And they set in it four rows of stones. A row of sardius, topaz, and carbuncle was the first row; 11and the second row, an emerald, a sapphire, and a diamond; 12and the third row, a ligure, an agate, and an amethyst; 13and the fourth row, a beryl, an onyx, and a jasper; they were enclosed in settings of gold filigree.

The Holy Bible, King James Version

The question then arises, which interpretation? Throughout the centuries scholars have struggled with translations from Hebrew, to Greek, to Latin. An examination of Bibles used today gives differing results. *The King James Bible, The Revised Standard Version, The New English Bible* and *The New World Translation* all contain different lists. Each of these translations has relied upon interpreting the ancient names of stones. The names given in the earliest existing texts are not those used in modern languages. As has been shown in the section on individual gems, many terms used to refer to gems today had different meanings in ancient times. It is also speculated that some of the stones named in modern translations were probably not available in Egypt during the time of Moses and the crafting of the first breastplate. An examination of these ancient terms will shed further light on the subject.

The first stone listed of the twelve is the *odem*. Translated by Josephus as sardius or sardonyx, it is the red stone upon which Reuben's name was inscribed. Modern lists often translate this gem as carnelian, red jasper or red feldspar. The second stone is the *pitdah*, translated in modern texts as topaz, this stone is likely peridot. The Hebrew word is from the Sanskrit *pita*, meaning yellow, and may have referred to yellow serpentine, a stone readily available in ancient Egypt. The third stone is the *bareqeth*, translated to the Latin *smaragdus*. Generally thought to be emerald, this may be any number of green gems. The emerald was known in ancient

Egypt, but many green stones were given this same name by early writers. The second row of stones begins with the *nophek*, translated as carbuncle. The Hebrew word means red or bright-red stone. Translated by some as ruby, it is most likely almandine garnet. No evidence of rubies is found in ancient Egypt. *Sappir*, later called sapphire, is the center stone of the second row. As noted previously, Pliny and Theophrastus described this gem as a blue stone with golden spots. From this description it is assumed that the ancient sappir was actually lapis-lazuli. The last stone in the second row of the High-Priest's Breastplate is the *yahalom´*, translated in most Biblical versions as diamond. The New World Translation uses the gem jasper, reasoning that the diamond was probably unknown to the Israelites. It should also be noted that the diamond is too hard to have been engraved using any technology available at the time. The third row begins with a *leshem* stone, a name undefined by most Biblical scholars. It has been speculated that it was a jacinth, amber, opal or tourmaline. None of these can be confirmed, but most versions name jacinth. Second place in row three belongs to the *shevoh´*, meaning precious stone. No other definition of this stone exists. This generic gem has been called an agate in most modern translations. Pliny records the existence of red agates veined with white in the desserts of Thebes. These red agates may be the shevoh of the breastplate. Row three is completed with an *achlamah´*. translated in the Greek *Septuagint* to mean amethyst. All modern versions of the Bible list agate and amethyst as the second and third stones of the third row. The *tarshich* is the first stone of the fourth row. Translated as chrysolite by early writers, it was called beryl in later Protestant translations. All current versions of the Bible maintain the tenth stone was a beryl. Second in this row is the *shoham*, translated in all versions of the Bible as onyx. The final stone is the *yashepheh´*, translated from the Greek as jasper by most scholars, but called jade by Jehovah's Witness scholars in the *New World Translation*.

It may be seen that over the centuries the greatest Biblical scholars could not be certain what the stones of the High Priest's Breastplate were. The following lists show the variations found in translations of ancient and modern authorities:

Stones of the High-Priest's Breastplate

Hebrew c. 2000 B.C.	Septuagint (Greek) c. 250 B.C.	Josephus (Greek) c. 90 A.D.	Vugate (Latin) c. 400 A.D.	King James Version 1611 A.D.	Revised Version 1881 A.D.	New World Translation 1963 A.D.
Odem	Sardion	Sardonyx	Sardius	Sardius	Sardius	Ruby
Pitdah	Topazion	Topazos	Topazius	Topaz	Topaz	Topaz
Bareqeth	Smaragdos	Smaragdos	Smaragdus	Carbuncle	Carbuncle	Carbuncle
Nophek	Anthrax	Anthrax	Corbunculus	Emerald	Emerald	Emerald
Sappir	Sappheiros	Sappheiros	Sapphirius	Sapphire	Sapphire	Sapphire
Yahalom´	Iaspis	Iaspis	Jaspis	Diamond	Diamond	Jasper or Diamond
Leshem	Ligurion	Liguros	Ligurius	Ligure	Jacinth	Hyacinth or Leshem
Shevoh´	Achates	Achates	Achates	Agate	Agate	Agate
Ahlamah´	Amethystos	Amethystos	Amethystus	Amethyst	Amethyst	Amethyst
Tarshish	Chrysolithos	Chrysolithos	Chrysolithus	Beryl	Beryl	Beryl
Shoham	Beryllion	Onyx	Onychinus	Onyx	Onyx	Onyx
Yashapeh´	Onychion	Beryllos	Beryllus	Jasper	Jasper	Jade

Even if only one translation of the Bible is chosen as definitive, it would not correspond exactly to any modern commercial lists of birthstones.

The early twentieth century author, George Frederick Kunz wrote extensively about High Priest's Breastplate. In *The Curious Lore of Precious Stones*, Kunz reminds us that two breastplates have existed. The original, described in Exodus and known as the Mosaic Breastplate, was lost when the First Temple of Jerusalem was destroyed. The Breastplate of the Second Temple, dated to about the sixth century B.C., is the one witnessed and studied by Josephus. Kunz maintained that careful examination of the descriptions of both vestments, an understanding of ancient etymology, and knowledge of Egyptian geology allowed him to achieve the most accurate lists to date. His speculations rely heavily on the colors intended by the Hebrew terms, what gems would have been most revered by the Israelites, and stones that could be engraved using ancient technologies. His lists of the stones of the breastplates are as follows:[2]

The Mosaic Breastplate	The Breastplate of the Second Temple
Red Jasper	Carnelian
Light-green Serpentine	Peridot
Green Feldspar	Emerald
Almandine Garnet	Ruby
Lapis-lazuli	Lapis-lazuli
Onyx	Onyx
Brown Agate	Sapphire or Jacinth
Banded Agate	Banded Agate
Amethyst	Amethyst
Yellow Jasper	Topaz
Malachite	Beryl
Green Jasper or Jade	Green Jasper or Jade

Evidence exits that in early centuries all twelve stones chosen as appropriate were kept by the same person and worn in turn as the year progressed from month to month. Each respective stone was believed to demonstrate increased therapeutic power during the ascendancy of its corresponding zodiacal sign. As the practice grew it became popular for a person to wear one particular stone that held a more intimate association with the individual by virtue of his or her date of birth.

This fashion, once started, soon became a general custom. When adopted by European Christians, the order and assignment of stones to various months changed. The Foundation Stones of the New Jerusalem, listed by John in Revelations 21,19, replaced the stones of the breastplate. The gems included in the Foundation Stones were undoubtedly based on the earlier stones of Aaron's Breastplate, but the order and some material included differs. This may be attributed to an intended reordering by Saint John, errors in translation or the transcription of texts from memory.

The fourteenth verse of Revelations 21 states that each of the Foundation Stones will be inscribed with the names of the apostles. It was not until the eighth or ninth centuries that scholars tried to find analogies between the stones and the apostles. These early authors were concerned with the religious significance of the gems and cared little for the stones themselves. They knew little of their characteristics and may have never seen the stones about which they were writing. The pagan traditions of their inherent qualities and powers was ignored, and in many cases scorned, by early Christians. It was not until Christianity had established a firm hold in the Greek and Roman worlds that resistance to ancient traditions weakened.

Andreas, bishop of Cesarea, is credited as one of the first writers to associate the apostles with specific gems. His tenth century treatise gives a listing of the twelve stones and the reason each has a connection to a particular apostle.

> The jasper, which like the emerald is of a greenish hue, probably signifies St. Peter, chief of the apostles, as one who so bore Christ's death in his inmost nature that his lone for Him was always vigorous and fresh. By his fervent faith he has become our shepherd and leader.
>
> As the sapphire is likened to the heavens, I conceive it to mean St. Paul, since he was caught up to the third heaven, where his soul was firmly fixed. Thither he seeks to draw all those who may be obedient to him.
>
> The chalcedony was not inserted in the High-Priest's breastplate, but instead the carbuncle, of which no mention is made here. It may well be, however, that the author designated the carbuncle by the name

chalcedony. Andrew, then, can be likened to the carbuncle, since he was splendidly illumined by the fire of the Spirit.

The emerald, which is of a green color, is nourished with oil, that its transparency and beauty may not change; we conceive this stone to signify John the Evangelist. He, indeed, soothed the souls dejected by sin with a divine oil, and by the grace of his excellent doctrine lends constant strength to out faith.

By the sardonyx, showing with a certain transparency and purity the color of the human nail, we believe that James is denoted, seeing that he bore death for Christ before all others. This the nail by its color indicates, for it may be cut off without any sensible pain.

The sardius with its tawny and translucent coloring suggests fire, and it possesses the virtue of healing tumors and wounds inflicted by iron; hence I consider that it designates the beauty of virtue characterizing the apostle Philip, for his virtue, animated by the fire of the Holy Spirit, cured the soul of the wounds inflicted by the wiles of the devil, and revived it.

The chrysolite, gleaming with the splendor of gold, may symbolize Bartholomew, since he was illustrious for his divine preaching and his store of virtues.

The beryl, imitating the colors of the sea and of the air, and not unlike the jacinth, seems to suggest the admirable Thomas, especially as he make a long journey by sea, and even reached the Indies, sent by God to preach salvation to the peoples of that region.

The topaz, which is of a ruddy color, resembling somewhat the carbuncle, stops the discharge of the milky fluid with which those having eye-diseases suffer. This seems to denote Matthew, for he was animated by a divine zeal, and, his blood being fired because of Christ, he was found worthy to enlighten by his Gospel those whose heart was blinded, that they might like new-born children drink of the milk of faith.

Then chrysoprase, more brightly tinged with a golden hue than gold itself, symbolizes St. Thaddeus; the gold (chrysos) symbolizing the kingdom of Christ, and the prassius, Christ's death, both of which he preached to Agbar, King of Edessa.

The jacinth, which is of a celestial hue, signifies Simon Zelotes, zealous for the gifts and grace of Christ and endowed with a celestial prudence.

By the amethyst, which shows to the onlooker a fiery aspect, is signified Matthias, who in the gift of tongues was so filled with celestial fire and with fervent zeal to serve and please God, who had chosen him, that he was found worthy to take the place of the apostate Judas.[3]

Some in the Church took objection to the designation of apostles to each of the foundation stones. They argued that Christ alone should be held as the foundation of the church and that the color of each stone represents a facet of Christ's divine spirit. Green jasper, for satisfaction; the blue sapphire, the soul; bright red chalcedony, truth; green emerald, kindness and goodness; nail-colored sardonyx, strength of spiritual life; red sardius, the shedding of His blood; yellow chrysolite, His divine nature; sea-green beryl, moderation and control of passion; green topaz (chrysolite), uprightness; harsh colored chrysoprase, sternness toward sinners; violet or purple jacinth, royal dignity; and, finally, purple amethyst with a touch of red, perfection.[4]

Other works which list the foundation stones and associate each stone with Christian virtues are known collectively as *bestiaries*. A bestiary is a compilation of descriptions of animals, plants and minerals accompanied by explanations of, and correlations with, Christian dogma. The sources of these eleventh, twelfth, and thirteenth century collections include the third century *Liber Memorobilium* of Solinus, fourth century Greek *Physiologus*, the *Hexaemeron* of Ambrose, and the seventh century *Etymologies* of Isidore of Seville.

The *bestiary* housed in the Aberdeen University Library contains the following passages regarding the twelve foundation stones. The author first gives the reason for the foundation stones and what significance they should have to the practicing Christian.

'The foundations of the wall of the city were garnished with all manner of precious stones (Revelation, 21:19), that is, the prophets and Apostles on whose faith and doctrine the whole city of the church is founded. Of these it is said in the Psalms: it is founded on holy hills (see, eg, Psalms, 15:1); the wall was adorned with every precious stone (see Revelation, 21:18-19). They were furnished, that is, with every kind of virtue and good work. It is not only the prophets and apostles who are called "foundations," but lesser men also, who had or have a life and faith like theirs. They are called foundations not by virtue of their personalities, but the way in which they exercised their virtue; because it was through their virtue that they founded the church.

On account of this, John shows here in which virtues they were as a light in the church, reckoning their number as twelve, demonstrating that they shone in every virtue. For this number signifies the universe, because it is made from parts containing seven, that is by threes and by fours; and that faith is first among the virtues according to the statement of Prudentius.

Faith, the first of the virtues, ready to fight, takes to the field in battle with doubt. And because without faith it is impossible to please God (Hebrews, 11:6), faith is set in the first foundation. It should not trouble you that stones are called foundations, because by foundations are meant virtues. For this reason when the stone is said to be a foundation, it should be interpreted as a decoration of the foundation.

From this point forward the interpretation of each stone listed by John in Revelations is given. References to additional biblical passages, along with direct quotes, are included to assist the devout in understanding their faith. Along with moralized explanations, the descriptions and virtues of these foundation gems is also listed.

John says, therefore:
The first stone in the foundation of the wall is jasper. The first foundation, that is, the adornment of the first foundation, is jasper, that is, faith ever green, strengthening the sight, but whether it be faith in one's country or in the Church triumphant, which is in question here, does not primarily occur except to those coming to the Church; through it there will be entry to the aforesaid city, while he who does not have it will not be able to enter. Jasper is said to have seventeen species. It is also known to be of many colours. It is said to come from many regions of the world. The best is a translucent green in colour. It is shown to have more virtues than any other.

The second, sapphire
The second, that is, the second foundation, that is, the second decoration of the foundation, is sapphire. Its colour is similar to that of a clear sky; struck by the rays of the sun, it sends forth, burning, a flash of lightning, signifying the hope by which we are carried off to heaven; through it we are fired with a love of heavenly things, disdaining love of the present world, so that we can truly say with the apostle: "Our conversation is in heaven" (Philippians, 3:20); "I will lay thy foundations with sapphires" (Isaiah, 54:11). The image of the sapphire is most fitting for the fingers of kings. It shines in an outstanding way and resembles most a clear sky. The power of nature has endowed it with such honour that it is called sacred and deservedly the gemstone of gemstones.

The third, chalcedony
The third, that is, the third decoration of the foundation is

chalcedony; it is pale yellow, similar to lamp-light, and shines more under the open sky than indoors; warmed by the sun or by a rub of the fingers, it attracts particles to itself; it does not resist the subjects

of the engraver, and it signifies the charity which is within us, hidden in the heart. It is pale yellow like lamplight, but when it is forced into the open for the benefit of others, then what its virtue was inside is demonstrated outside. Touched by the sun, that is Christ, or the spirit, namely the finger, it attracts sinners to itself; that it cannot in any way be cut signifies that it is not wanting in times of adversity but is rather of advantage. In this context, it is said in the Song of Solomon: "Love is strong as death; jealousy is cruel as the grave: the coals thereof are coals of fire, which hath a most vehement flame. Many waters cannot quench love" (8:6-7). It cannot, therefore, be carved, because it is not shattered by adversity or even softened by fulsome praise. In this context, the psalms: "My head shall not be anointed with the oil of wicked men" (NEB, Psalms 141:5); 1 Corinthians, 13: "Charity is patient; it is kind; charity it suffers everything; it endures everything; it is not puffed up; it is not ambitious etc."
(see 13:4-5).

Chalcedony is a stone which shines with a faint paleness. It comes between the hyacinth and the beryl. Anyone who carries it will, it is said, be successful in lawsuits.

The fourth, smaragdus

The fourth foundation, smaragdus, outdoes in its greenness every kind of grass and the boughs of trees; it makes those who wear it appear attractive; it makes the air around grow green; it yields an image just as a mirror does; it signifies virginity, which wholly preserves the freshness of the flesh; and it surpasses all other virtues in a way. Because it preserves virginity it is more angelic than human; moreover, it is pleasing to angels and God and man and carries within itself the image of Christ because it follows the lamb wherever it goes; and for this reason this stone is called the fourth, because virginity is recommended in the four Gospels. The smaragdus surpasses every green thing in its greenness.

The fifth, sardonyx

The fifth foundation, sardonyx, gets its name from the association of two names, as Isidorus says; for it has the white of onyx and the red of sard; and it is three-coloured, as the Glossator says, black at the bottom, white in the middle and red at the top; and when used for sealing, it does not pull any of the wax away. From this, it signifies the suffering of the saints. At the bottom, that is in the world, they are considered worthless and despised; in this context, Job, 12: "The just upright man is laughed to scorn ... a lamp despised in the thought of him that is at ease" (12:4). In the middle, that is the righteous man in his heart or conscience, they are white, as a result of their innocence. At the top they are red, by reason of the zeal of their martyrdom for Christ. The stone does not pull any of the wax away, because the righteous man forgives his persecutors fully, from the heart, retaining no bitterness, according to Ecclesiasticus, 28: Forgive thy neighbour the hurt that he hath done unto thee, so shall thy sins also be forgiven when thou prayest (28:2). This virtue is said to be the fifth because it diminishes infirmity of the body, because it is ruled by the five senses. Two names, sard and onyx, make the sardonix. This single stone has taken from the two stones three colours. Alone of precious stones, it cannot pull away wax.

The sixth, sard

The sixth foundation, sard is so called because it was first found in Sardis; it is of the colour of blood only. For this reason it signifies the perfect constancy of the martyrs, who poured forth their blood for Christ, and for that reason it is placed in the sixth position, because Christ in the sixth age and on the sixth day consecrated his martyrdom with his blood. The sard gets its name from Sardis where it was first found. It gets its name from its reddish colour.

The seventh, chrysolite

The seventh foundation, chrysolite, is similar in colour to gold. For this reason its name comes from crisis [chrysos], which means "gold"; it seems to give out glittering sparks, as the Glossator says; and it signifies wisdom, which exceeds all other gifts, just as gold exceeds all other metals.

Wisdom, through the medium of preaching, gives out glittering sparks, that is, encouragement and doctrine, setting alight the hearts of those who hear them. In this context, Ezekiel 1, on the sacred animals: "They sparkled like the colour of burnished brass" (1:7); The Wisdom of Solomon, 3: "The righteous shall shine and run to and fro like sparks among the stubble" (see 3:1,7). This stone is placed in the seventh position, because it holds the seventh place in order of ascendancy among the gifts of the holy spirit. Chrysolite shines like gold and flashes like fire. It is similar to the sea, displaying something of its green colour. We read that the Ethiopians send us this stone.

The eighth, beryl

The eighth foundation is beryl. This stone is polished into a hexagonal shape; it shines like water struck by the sun; it is also said to be of such heat that it warms the hand of the holder; and it signifies the virtue of mercy. Mercy operates in six ways, warming the cold hearts of the infirm to a love of God and one's neighbour, according to Proverbs, 25: "If thine enemy be hungry, give him bread to eat; and if he thirsty give him water to drink: For thou shalt heap coals of fire upon his head" (25:21-23). Matthew, 5: "Let your light so shine before men, that they may see your good works and glorify your Father which is in heaven" (5:16). This virtue is placed in the eighth position, because not here but in the eighth age it expects its reward. In this context, the psalm: "Thou shall eat the labour of thine hands: happy shalt thou be, and it shall be well with thee" (Psalms, 128:2). Gregory: "It will be bad for those who eat their labours here, like hypocrites." Its hexagonal form causes beryl to shine brightly; otherwise it seems to have a faint pallor.

The ninth, topaz

The ninth foundation is topaz; this stone, although it is multi-coloured, has two colours especially, gold and a clear colour, as the Glossator of Exodus, 34, says: And it is touched by the splendour of the sun. It exceeds all other gemstones in clearness; its appearance is singularly pleasing to those who look at it; if it were polished, it would be dulled; left to its own nature, it is clearer; it is the largest of stones; and it is cherished by kings. Topaz signifies contemplation. The love with which contemplation burns, colours it gold; the understanding which illuminates contemplation, gives it its clear colour. In contemplation the Lord

is seen more clearly; and men are more especially drawn to his love when they are open to it. Their nature is such that, if they are embellished by the honours of this world, they see less clearly, because, like Martha, they have many distractions. This stone shines with every colour, because contemplation shines with the splendour of every virtue. It is the biggest of stones, because contemplation expands the heart greatly, and those who are truly kings think nothing of the flesh. It is placed in the ninth position, because contemplation aligns contemplative men with the nine orders of angels. Topaz comes from the island of the same name. It is all the more precious as it is rare. The land of the Arabs, rich in stones, produces it.

The tenth, chrysoprase

The tenth foundation is chrysoprase. This stone, according to Isidorus, comes from India, and is purple in colour with separate, small gold marks; for this reason it gets its name crisopassus, "scattered everywhere with gold". It signifies desire of the heavenly land, which burns the more brightly, the more it is affected by tribulation, because, as Gregory says: "What a bellows does to coal, tribulation does to love." Chrisoprase is placed in the tenth position, because holy men, in their desire for heaven, hasten to reach the tenth order of angels by observing the ten commandments. The tenth order is the one which will be renewed from men. In this context, man is called, in Luke, 15, the tenth piece of silver which the woman searched for and found (see 15:8-10). India, its home, sends us the stone called chrysoprase. It shines with the sap of the leek and is of mixed colour, tinted with purple and marked with gold.

The eleventh, hyacinth

The eleventh foundation is hyacinth. This stone changes in accordance with the weather: on clear days, it is transparent; when the sky is overcast, it is opaque. For this reason it signifies the judgment of holy men, who use it, as the Lord did, to adapt to all conditions of life, in order to win the hearts of all men; as the apostle says, 1 Corinthians, 9: "I am made all things to all men, that I might by all means save all" (see 9:22); Romans, 12: "Rejoice with them that do rejoice, and weep with them that weep" (12:15). This virtue enables holy teachers to know what, to whom, when and how to preach. This stone is placed in the eleventh position, because through it, especially, all manner of sin is avoided. The learned say that there are three kinds of hyacinth. The best is the kind whose colour is not so dense as to be obscure or so light as to be transparent but has a purple, myrtle-like bloom drawn from both parts of the spectrum. This stone, placed in the mouth, proves to be colder than others. It is very hard and resists cutting or engraving. But it can be marked by a fragment of diamond.

The twelfth, amethyst

The twelfth foundation is amethyst. Isidorus says of it: Among purple stones, the Indian amethyst holds first position; it is, indeed, purple but of mixed coloration, giving forth violet and rose-coloured lights; it is easy to engrave. For this reason the humility of the saints is signified by it; associated with humility is obedience, as Ambrose says: "Humility is small, like the violet, beautiful like the rose, easy to apply to all things"; or: "They are like burning flames, looking at love". For humility is acceptable to everyone, even to our enemies; as pride, in contrast, is viewed by everyone with detestation, as it says in Ecclesiasticus, 15: "Pride is hateful before God and men" (see 10: 7). For this reason the amethyst is placed in the final position, as if watching over all, and as if humility always reckons itself the least and always takes the last place.[5]

The veneration of gems and their metaphysical properties has been a long standing tradition in the Christian church. The most prominent authors on the subject have been persons of high status in the Holy Roman Church. With such convincing evidence of the Christian virtues expressed by each of these twelve stones, the wearing of birthstones has enjoyed general acceptance and popularity in the Christian world.

The connection of these twelve stones with the signs of the zodiac seems to be based on the ancient admonition of Josephus and the continued mixing of pagan and Christian tradition in the early church. Each of the twelve stones listed in Revelations was assigned to a month of the year. The early, pre-Julian, calendar began with the month of March. Starting in March and progressing through the year established this early arrangement of natal stones based on the gems given in the King James Version of the *Bible*.

Month	Gem
March	Jasper
April	Sapphire
May	Chalcedony
June	Emerald
July	Sardonyx
August	Sardius
September	Chrysolite
October	Beryl
November	Topaz
December	Chrysoprasus
January	Jacinth
February	Amethyst

It may be noted that this earliest list does not correspond to what may be called the "modern" order of birthstones presented by most retail jewelers. The modern list shows little relationship to the list given above. A table of various lists proposed throughout the centuries shows the latitude taken by numerous people and cultures in translating from the original texts. Confusion in translations of the Bible, cultural tastes, availability of gems, and regional custom have all contributed to the existence of many lists of natal stones.

The table on the following page represents the variety of natal stone lists found in various regions.

Studying these lists reveals the similarities to the order of the Foundation Stones. The exceptions are the preference of garnet for January, rather than jacinth, and ruby for December, instead of the chrysoprase. Despite the traditional pattern of these various natal stone lists, The National Association of Jewelers established a new "official" list in 1912.

Month	Gem
January	Garnet
February	Amethyst
March	Bloodstone or Aquamarine

Birthstones:

Month	Gem
April	Diamond
May	Emerald
June	Pearl or Moonstone
July	Ruby
August	Sardonyx or Peridot
September	Sapphire
October	Opal or Tourmaline
November	Topaz
December	Turquois or Lapis-lazuli

    The changes made from the traditional lists are quite evident. The ruby was moved from December to July and turquois moved to December. The rational given was that a red stone was more appropriate for the warm months and the cool blue of turquois better suited to December. This reasoning does not seem consistent with the rest of the list. One would probably not classify the pearl or moonstone as warm colored gems to represent the summer month of June, nor would garnet, amethyst or bloodstone seem compatible with the heart of winter. Diamond, the new stone of April, appears in few of the historical lists and was not a gem commonly known by the ancients. The pearl appears only three times in the lists and only once as the stone of June. Many argue that pearl is inappropriate as a natal stone since it is not a stone but organic matter. A stone that appears for the first time is tourmaline. The gem never appears in any early list and was unclassified until the eighteenth century. It is evident that public familiarity and potential profits from sales were two of the motives used to established the first list developed in the twentieth century.

    The 1912 National Association of Jewelers list did not become standard. Numerous other jewelry trade groups established their own lists. The National Association of Goldsmiths of Great Britain adopted a new birthstone list in 1937. The Jewelry Industry Council in America officially established this same list in 1952. This "modern" list was adopted by all the major trade organizations soon after. As may be seen, the modern list bears even less resemblance to the order of the Foundation Stones.

Worldwide Birthstone Lists

Months	Jewish	Roman	Ayurvedic	Old Spanish	Spanish
January	Garnet	Garnet	Garnet	Chalcedony	Hyacinth
February	Amethyst	Amethyst	Amethyst	Amethyst	Amethyst
March	Jasper	Bloodstone	Bloodstone	Unknown	Jasper
April	Sapphire	Sapphire	Diamond	Crystal	Sapphire
May	Chalcedony Carnelian Agate	Agate	Agate	Ruby	Agate
June	Emerald	Emerald	Pearl	Sapphire	Emerald
July	Onyx	Onyx	Ruby	Agate Beryl	Onyx
August	Carnelian	Carnelian	Sapphire	Topaz	Carnelian
September	Chrysolite	Sardonyx	Moonstone	Magnetite	Chrysolite
October	Aquamarine	Aquamarine	Opal	Jasper	Aquamarine
November	Topaz	Topaz	Topaz	Garnet	Topaz
December	Ruby	Ruby	Ruby	Emerald	Ruby

Birthstones:

Italian	Russian	Arabian	Polish	15-20 Century Europe
Jacinth Garnet	Garnet Hyacinth	Garnet	Garnet	Garnet
Amethyst	Amethyst	Amethyst	Amethyst	Amethyst Hyacinth Pearl
Jasper	Jasper	Bloodstone	Bloodstone	Bloodstone Jasper
Sapphire	Sapphire	Sapphire	Diamond	Sapphire Diamond
Agate	Emerald	Emerald	Emerald	Emerald Agate
Emerald	Agate Chalcedony	Agate Chalcedony Pearl	Agate Chalcedony	Agate Turquois Cat's-eye
Onyx	Ruby Sardonyx	Carnelian	Ruby	Onyx Turquois
Carnelian	Alexandrite	Sardonyx	Sardonyx	Carnelian Topaz Moonstone Sardonyx
Chrysolite	Chrysolite	Chrysolite	Sardonyx	Chrysolite
Beryl	Beryl	Aquamarine	Aquamarine	Opal Beryl
Topaz	Topaz	Topaz	Topaz	Topaz Pearl
Ruby	Turquoise Chrysoprase	Ruby	Turquoise	Ruby Bloodstone

Month	Gem
January	Garnet
February	Amethyst
March	Aquamarine or Bloodstone
April	Diamond (or Rock Crystal)
May	Emerald
June	Pearl, Moonstone or *Alexandrite
July	Ruby
August	Peridot or Sardonyx
September	Sapphire (or Lapis-lazuli)
October	Opal or *Pink Tourmaline
November	Topaz or *Citrine
December	Turquois or *Zircon

*Not included in the British Goldsmiths' list.
Those in parenthesis included in the British Goldsmiths' list but not retained in the Jewelry Industry Council's list.

This most recent listing contains many stones not named as original Foundation Stones or stones of Aaron's Breastplate. Only three gems from the Foundation Stones remain in their original positions; amethyst, sard(onyx) and topaz. Stones unknown to the ancients, alexanderite and tourmaline, or those never mentioned in the *Bible*; opal, citrine, and turquois have been added. It is difficult to recognize the source of our twelve natal stones by reviewing the lists used by modern retailers. If it is assumed that natal stones have some magical connection to those born in the assigned month, it is hard to understand how these alterations to the original "sacred" lists can be justified.

Along with the association of stones with each month, additional lists exist related to other divisions of the calendar or clock. One such division of the year is by season. G.F. Kunz proposed the four most highly regarded gems as representative of each of the seasons. The emerald, ruby, sapphire and diamond were chosen as primary stones, and each was assigned a season according to its sympathy of color.

The green of emerald is thought to represent the colors which appear in spring. The ruby represents the heat of summer. Kunz also reasons that the ruby "is born in the hot climates" of Burma, Ceylon and Siam (modern Thailand). The blue of autumn skies suggests the sapphire. Reasoning that sapphire represents calm, Kunz associates the gem with the decline in the sun's intensity. Finally, he assigns the diamond to winter as a symbol of the color and sparkle of ice.

Expanding upon these four primary stones of the seasons, additional gems are listed for each time of the year.[6]

The foliage and flowers of spring suggest:

Amethyst	Spinel
Green Diamond	Pink Topaz
Chrysoberyl	Peridot

The heat of summer is represented by:

Zircon	Spinel
Garnet	Pink Topaz
Alexandrite	Fire Opal

The colors of autumn are seen in the:

Hyacinth (reddish brown zircon)	
Topaz	Tourmaline
Smoky quartz	Chrysolite

The ice and cold of winter bring these stones to mind:

Colorless Quartz	Moonstone
White Sapphire	Pearl
Turquois	Labradorite

Natal Stones for the Days of the Week

Day	Gem	Phenomenal Gem
Sunday	Topaz	Sunstone
Monday	Pearl or Rock Crystal	Moonstone
Tuesday	Ruby or Emerald	Star Sapphire
Wednesday	Amethyst or Lodestone	Star Ruby
Thursday	Sapphire or Carnelian	Cat's-eye
Friday	Emerald or Cat's-eye	Alexandrite
Saturday	Turquois or Diamond	Labradorite

Astrological Signs for the Days of the Week.

Sunday
Day of the Sun.

Monday
Day of the Moon.

Tuesday
The day of Tyr, or the day of Mars in Latin.

Wednesday
The day of the Norse god Woden or Odin. In Latin countries the day of Mercury

Thursday
The day of Thor in Nordic lands and the day of Jupiter on the Latin calendar.

Friday
Named for the norse goddess Freja. Her counterpart in the Latin world in Venus.

Saturday
The day of Saturn.

A heptagram is used to represent the seven day week adopted in Hellenistic Greece

A curious use of the symbol is revealed when the signs of the planets are arrange as seen to the right.

If the planets known to the ancients are arranged around the heptagram in the order of the length of their astrological orbital times as viewed from earth, the following sequence results: Moon, Mercury, Venus, Sun, Mars, Jupiter, and Saturn.
Starting with the top of the diagram and proceeding through the heptagram, as if tracing with a pen in a continuous line, the seven planets that represent the days of the week are revealed in order.

316
Birthstones:

The use of a particular stone for a day of the week has a foundation in Oriental tradition. Fashion standards have developed over the centuries which determine a preference for certain gem accessories on each day. In Thailand rubies are considered Sunday's stone, moonstone is the gem of Monday, coral is chosen for Tuesday, cat's-eyes are Wednesday's preference, emeralds are favored on Thursday, diamonds are appropriate on Friday, and on Saturday dark blue sapphires are the gem of choice.

A list of natal stones related to each of the twenty-four hours of the day may be found in some references. This list appears in early twentieth century texts, but no origin is noted. No mention of stones of hours has been found in earlier works. The fact that Kunzite is included indicates the list was developed by G.F. Kunz. Kunzite was named for the gemologist and the same list occurs in his 1913 work *The Curious Lore of Precious Stones*.

Natal Stones for the Hours of the Day

Time AM		Time PM	
1	Smoky Quartz	1	Zircon
2	Hematite	2	Emerald
3	Malachite	3	Beryl
4	Lapis-lazuli	4	Topaz
5	Turquois	5	Ruby
6	Tourmaline	6	Opal
7	Chrysolite	7	Sardonyx
8	Amethyst	8	Chalcedony
9	Kunzite	9	Jade
10	Sapphire	10	Jasper
11	Garnet	11	Lodestone
12 Noon	Diamond	12 Midnight	Onyx

The second admonition of Josephus was to connect the twelve stones of Aaron's Breastplate with the "twelve signs of what the Greeks

Aries    *Taurus*    *Gemini*

call the zodiac." Archaeological evidence shows the Greeks long revered the times when the sun appeared in one of the twelve major constellations. Each of these twelve periods also held significance to those born during these times. Tradition has assigned a particular stone to each sign representing these constellations. The assignments have been based on the powers each stone represents or possesses, the color of the gem, and powers or colors associated with each sign. With the establishment of natal stones associated with each month, a problem arose. The obvious conflict is that the dates associated with the zodiac do not directly correspond to the calendar months. Each of the twelve periods covered by a sign begin and end on about the twenty-first or twenty-second of the month. As a result, a person born in any given month may be born under one sign if their birth is in the first part of the month, or under a different sign if born in the later part. The calendar of the solar or western zodiac is as follows:

Aries	March 21-April 20	Libra	Sept. 22 - Oct. 23
Taurus	April 20 - May 21	Scorpio	Oct. 23 - Nov. 21
Gemini	May 21 - June 21	Sagittarius	Nov. 21 - Dec. 21
Cancer	June 21- July 22	Capricorn	Dec. 21 - Jan.- 21
Leo	July 22 - Aug, 22	Aquarius	Jan. 21 - Feb. 21
Virgo	Aug. 22 - Sept. 22	Pisces	Feb. 21- March 21

Josephus stated a correspondence between the stones of the Breastplate and the zodiac in a list presented in his *Antiquities of the Jews*. He lists the stones believed to have been worn by Aaron in the original vestment and those he witnessed in the recreated Breastplate, known to have existed in the first century. Josephus includes the pairing of each stone in the first Breastplate with a tribe of Israel. His list is also of interest for its association of zodiacal signs with the Twelve Apostles.

Cancer   *Leo*   *Virgo*

Birthstones:

Signs of the Zodiac and their gem	First Breastplate and the Tribes of Israel	Second Breastplate and the Apostles
Aries-Bloodstone	Jasper-Benjamin	Jasper-Peter
Taurus-Sapphire	Sapphire-Issachar	Sapphire-Andrew
Gemini-Agate	Agate-Naphtali	Chalcedony-James
Cancer-Emerald	Emerald-Judah	Emerald-John
Leo-Onyx	Onyx-Josheph	Onyx-Philip
Virgo-Carnelian	Sardonyx-Rueben	Sardonyx-Bartholemew
Libra-Chrysolite	Chrysolite-Simeon	Chrysolite-Matthew
Scorpio-Beryl	Beryl-Asher	Beryl-Thomas
Sagittarius-Topaz	Topaz-Daniel	Topaz-James the Lesser
Capricorn-Ruby	Diamond-Zebulun	Chrysoprase-Jude
Aquarius-Garnet	Garnet-Levi	Hyacinth-Simon
Pisces-Amethyst	Amethyst-Gad	Amethyst-Judas

The Jews of the Middle Ages made adjustments to Josephus' assignment of gems to the Tribes of Israel. Various lists may be found in Rabbinical writings of the period. Originally included in *The Magus* by Barrett, the following list appears to be the most often cited as the gems commonly associated with the Tribes of Israel in Medieval and Renaissance Europe[7].

Aries-Daniel	Libra-Issachar
Taurus-Rueben	Scorpio-Benjamin
Gemini-Judah	Sagittarius-Nephtali
Cancer-Manasseh	Capricorn-Gad
Leo-Asher	Aquarius-Zebulun
Virgo-Simeon	Pisces-Ephraim

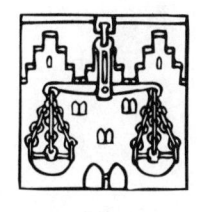

Libra    *Scorpio*    *Sagittarious*

The stones associated with signs of the zodiac have varied throughout the centuries. Lists of zodiacal stones from various sources show how many differences there have been. Some of the gems listed are not those commonly listed in tables of precious stones, others have non-traditional names. The most recent list, Crowley's of 1955, includes stones not known to the ancients and, in the case of alexandrite, others only recently distinguished.

Sign	Athanasius Kircher (1653)	Martianus Capella	CCAG XII*
Aries	Amethyst	Dendrites	Siberite
Taurus	Jacinth	Heliotrope	Yellow Jacinth
Gemini	Chrysoprase	Keraunos (red onyx)	Diamond Heliotrope
Cancer	Topaz	Lychnis (red stone)	Green Jasper Euchite
Leo	Beryl	Astrites (Cat's-eye)	Agate Selinite
Virgo	Chrysolite	Emerald	Corallite Dendrite
Libra	Sard	Scythian Emerald	Sardine Emerald
Scorpio	Sardonyx	Jasper	Hematite Pyrites
Sagittarius	Emerald	Rock Crystal	Amethyst
Capricorn	Chalcedony	water-colored gems	Ophite Chalcedony
Aquarius	Sapphire	Diamond	Magnet
Pisces	Jasper	Hyacinth	Beryl Jacinth

*CCAG, *Catalogus Codium Astrologorum Graecorum Vol 12*, Greek Astrology

Sign	Theosophical Society	G.F. Kunz (1913)	A. Crowley "777" (1955)
Aries	Sardonyx	Bloodstone	Ruby
Taurus	Carnelian	Sapphire	Topaz
Gemini	Topaz	Agate	Alexandrite Tourmaline
Cancer	Chalcedony	Emerald	Amber
Leo	Jasper	Onyx	Cat's-eye
Virgo	Emerald	Carnelian	Peridot
Libra	Beryl	Chrysolite	Emerald
Scorpio	Amethyst	Beryl	Snakestone
Sagittarius	Jacinth	Topaz	Jacinth
Capricorn	Chrysoprase	Ruby	Black Diamond
Aquarius	Rock Crystal	Garnet	Glass
Pisces	Sapphire	Amethyst	Pearl

The zodiacal authority Rupert Gleadow proposed a modern list in his *Origin of the Zodiac*. He states that in a "correct list" the color of gems is the critical attribute in discerning their value as stones of the zodiac. He further believed the Law of Correspondence, the *Qabalah*, and tradition were the only true way to know which stones hold power in relation to their sign. His list includes stones of different chemical and structural make-up primarily grouped by color. Gleadow's list also includes reasons for some of his choices.

Gems of the Zodiac, Rupert Gleadow[8]

Aries must obviously have a red stone, hence red jasper would be a good choice; and ruby, traditionally a solar stone, suits the exaltation of the sun here.

Capricorn *Aquarius* *Pisces*

Taurus, green jasper; emerald; malachite.

Gemini in the Middle Ages was thought to deserve a mixture of red, white, and dark stones, hence the best would be some variegated stone such as onyx (banded agate) or a striped chalcedony.

Cancer, in view of its association with the Moon and the sea, might well have moonstone, sea-green beryl, or turquoise.

Leo must have a yellow stone, presumably topaz, zircon, yellow jasper, or fire-opal.

Virgo, as sign of green corn, could have the apple-green chrysoprase, chrysoberyl, which is the same colour, or green feldspar, which was associated in ancient Egypt with fertility. If thought of as a sign of purity, diamond or an uncoloured chalcedony would be required. But since ripe corn is also a suitable colour for Virgo, one could use light brown agate or onyx.

For *Libra* the correct stone would be jade, from its use in treating kidney-disease in China, the kidneys being "ruled" by Libra; also chrysolite and peridot (olivine), which are transparent green stones, green being required because Libra is ruled by Venus.

Scorpio: bloodstone or haematite; carnelian (red chalcedony).

Sagittarius should for Qabalistic reasons have a blue stone, hence sapphire or star-sapphire; and this suits quite well in the sidereal zodiac, where the constellation is a mixture of air and earth, and associated with flight. Those who prefer the tropical zodiac, and think of this as a fiery sign, might choose rose-quartz.

For *Capricorn* tourmaline is appropriate by reason of its light-excluding qualities, in any colour including the dark blue indicolite; also the black opal.

Aquarius as the sky sign should have a sky-blue stone, but since it is a more watery sign than Sagittarius perhaps it should have the paler stone, aquamarine rather than sapphire. It might also have lapis-lazuli, which was much prized in antiquity and so ought not to be omitted.

Pisces- the amethyst, though attributed here for the wrong reason, suits very well on account of the Jupiterian colour and also because it was supposed to be a preventive of drunkenness, which is associated with the planet Neptune and the fishes. The opal also seems to belong here, and there is a purple fluorspar which much resembles amethyst.

The preceding lists demonstrate the variety of opinion as to the appropriate gem for any given sign of the zodiac. Further research and compilations would only serve to show more conflicting results.

Some scholars have chosen to assign gems to the planets. The stones are suppose to affect the reported influence of the planets over nature and human life. The 'planetary influences' have been associated with gems for centuries. In the eighth or ninth century a Syro-Arabic work, reported to have been originally written by Aristotle, lists rings, their material and appropriate settings which will bring the favor of the Sun, Moon, and the planets. The sun is favorably influenced by rock-crystal set in a gold ring, Mercury by magnetite in electrum and the Moon by one of the varieties of onyx set in silver.

Kunz includes an example of such a belief he gathered from an unnamed, "old work on the occult properties of gems."

The nature of magnet is in the iron, and the nature of the iron is in the magnet, and the nature of both polar stars is in both iron and magnet, and hence the nature of the iron and the magnet is also in both polar stars, and since they are Martian, that is to say, their region belongs to Mars, so do both iron and magnet belong to Mars.[9]

This example does not speak to materials we traditionally consider gems. Many occult writers have not drawn a distinction between gems and other mineral elements. The thirteenth century scholar Alfonso X was one of the first to give a complete listing of the seven planets (the Sun and Moon included) known in his day and the particular stones they influenced. The proposition was that the planet gave additional power to the stone and enhanced the gems inherent power. The Spanish King's *Lapidario* contains the following planet/gem connections:

The Sun influences	Diamond and Carnelian
Mercury	Emerald
Venus	Emerald, White Jargoon, Carnelian, Ruby, Coral, Lapis-lazuli and Chalcedony
The Moon	Coral
Mars	Red Jargoon
Jupiter	Emerald and Yellow Jargoon
Saturn	Diamond

The preceding list was said to be according to Chaldaic tradition, but another list is also credited to the Chaldeans:

The Sun	Diamond
Mercury	Agate
Venus	Emerald
The Moon	Selenite
Mars	Ruby
Jupiter	Jacinth
Saturn	Sapphire

A sixteenth century list contains different gems is association with heavenly bodies.

Birthstones:

The Sun	Jacinth and Yellow Chrysoberyl
Mercury	Agate and Magnetite
Venus	Sapphire, Carbuncle, Coral, Pearl and Emerald
The Moon	Beryl, Rock Crystal and Pearl
Mars	Diamond, Jacinth and Ruby
Jupiter	Emerald, Sapphire, Amethyst and Turquois
Saturn	All dark and black stones

In the Middle Ages, rings were a popular form of talismanic jewelry. The fact that they encircle a part of the body and were in direct contact to the skin added to their perceived powers. The following is a list of the "planets" influenced, the appropriate gems and the metals to be used as a setting.

Planet	Gem	Ring
The Sun	Diamond or Sapphire	Gold
Mercury	Magnetite	Quicksilver (mercury alloy)
Venus	Amethyst	Cyprium (copper, sacred metal of Cyprian goddess)
The Moon	Rock Crystal or Moonstone	Silver
Mars	Emerald	Iron
Jupiter	Carnelian	Tin
Saturn	Turquois	Lead

An entire text could be devoted to all the opinions included in the numerous works currently available on the subject of birthstones. This confusion is one of the reasons many have turned away from the Western tradition of the Greek zodiac and looked to the considerably more ancient writings of he East.

The Eastern zodiac and its planetary signs are estimated to have been established over six thousand years ago. The Western zodiac maintains that is the position of the sun in relation to various constellations that determines one's birth sign. The Eastern system relies on the position of the moon. Ancient Vedic authority contains the foundations of this sidereal science and the gems considered to give power to those influenced by a particular sign.

The basic premise is that the power of each gem emanates from the nine planets or *navra-graha*. Vedic literature recognizes various celestial bodies that Western science does not normally list as "planets." Besides those bodies which orbit the sun (Mercury, Venus, Mars, Jupiter and Saturn), the sun and moon are included. Vedic texts also distinguish shadowy planets, the descending node of the moon, *Rahu*, and the ascending node of the moon, *Ketu*.[10] Ancient beliefs state that these nine planets influence many aspects of out life by radiating cosmic energy expressed in colors. The colors of each celestial body correspond to the cosmic colors of different gems. Since gems are considered the most concentrated form of color, they provide a powerful source of cosmic color rays. These colors are not always the same as the visible sensations we receive by viewing stones. As may be expected, red is the cosmic color of rubies, but it is not readily perceived that orange is the cosmic color of pearls.

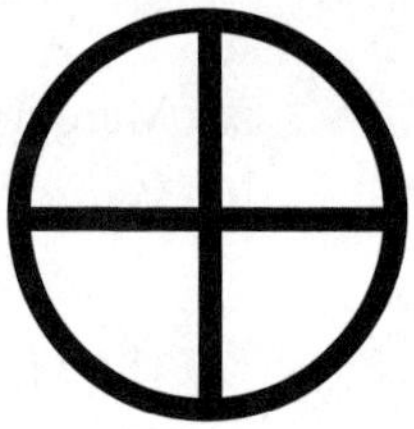

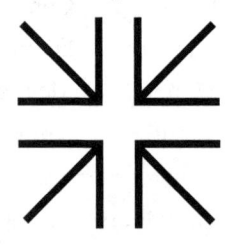

Gems of a particular cosmic color strengthen planetary powers and boost their influence in three ways. The gems add color to one's aura and strengthen its power. They astrologically enhance the power of each planet's influence over associated areas of one's being. For example, Venus rules art and sex, while Jupiter influences financial affairs and personal happiness. Gems also attract the attention of each planet's deity and bring its individual power to the wearer.

Tradition also speaks to the quality of gems and how this quality af-

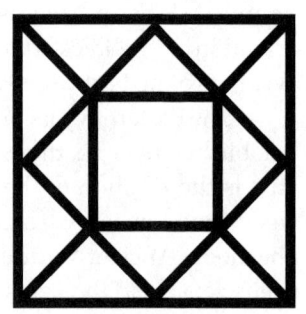

The ancient Egyptians and Greeks used the symbol shown above to plat natal charts. The division of a chart into twelve areas, as in the one shown below, was popular with astrologers of the middles ages.

This symbol was used in the seventeenth century to generate natal charts. The triangle between the outer square and inner square represented the twelve houses of the zodiac. The name and date of the subject was inscribed in the center.

326
Birthstones:

The Natal Stones, or *Navaratna*[12]

Zodiac Sign(s)	Planet	Cosmic Color	Primary Stone	Other or *Uparatna* Stones
Leo	Sun *Surya*	Red	Ruby	Red Spinal, Red Garnet, Red to Violet Tourmaline
Cancer	Moon *Chandra*	Orange	Pearl	Moonstone
Aries and Scorpio	Mars *Mangala*	Yellow	Coral	Carnelian
Gemini and Virgo	Mercury *Buddha*	Green	Emerald	Peridot, Green Jade, Tsavorite, Green Tourmaline, Green Diopside
Sagittarius and Pisces	Jupiter *Brihaspati*	Light Blue	Yellow Sapphire	Topaz, Citrine, Heliodor
Taurus and Libra	Venus *Shukra*	Indigo	Diamond	Colorless Quartz, Zircon, Topaz, White Sapphire, Goshenite
Capricorn and Aquarius	Saturn *Shani*	Violet	Blue Sapphire	Blue Spinel, Tanzanite, Iolite, Amethyst
+	*Rahu*	Ultraviolet	Hessonite	Hyacinth, Zircon, Spessartite
+	*Ketu*	Infrared	Precious Cat's-eye	Any stone that displays a strong chatoyant eye.

+ Rahu and Ketu, as shadowy planets, reflect the qualities of the sign they are in. For example, if Rahu is in Virgo or Gemini, it adopts the qualities of Mercury.

fects the gem's power. "Pure, flawless gems have auspicious powers which can protect one from demons, snakes, poisons, diseases, sinful reactions, and other dangers, while flawed stones have the opposite effect."[11] The following list of birthstones is common to many Asian countries.

The color intensity or hue of the gem also determines which is appropriate for one of the four social-economic-intellectual categories of life. The four classes distinguished in Vedic texts are as follows: educators, scientists and religious officials are in the highest caste; soldiers, bureaucrats and managers are in the second caste; farmers, bankers and merchants in the third; and servants, laborers and other workers are the fourth. All people must decide which hue best suits his or her class and individual needs. The classes and the cosmic color hues appropriate for the classes are listed below.

Planet	Cosmic Color	Gem Color	Cast	Hue
Sun	Red	Red	1	Pink
			2	Blood Red
			3	Orange Red
			4	Violet Red
Moon	Orange	White	1	White
			2	Orange White
			3	Greenish White
			4	Bluish or Black
Mars	Yellow	Red Orange	1	Light
			2	Blood Red
			3	Orange Red
			4	Brown Red
Mercury	Green	Green	1	Light Green
			2	Bluish Green
			3	Yellowish Green
			4	Dark Green
Jupiter	Light	Yellow	1	Light Yellow
			2	Yellow Orange
			3	Golden Yellow
			4	Greenish Yellow
Planet	**Cosmic**	**Gem**	**Cast**	**Hue**

Birthstones:

	Color	Color		
Venus	Indigo	Colorless	1	Colorless
			2	Pink Shade
			3	Yellowish Shade
			4	Bluish or Grayish
Saturn	Violet	Blue	1	Light Blue
			2	Violet Blue
			3	Greenish Blue
			4	Dark Blue or Gray
Rahu	Ultraviolet	Orange	1	Honey Orange
			2	Red Orange
			3	Golden Orange
			4	Brownish Orange
Ketu	Infrared	Chatoyant	1	Honey Yellow
			2	Honey Brown
			3	Honey Green
			4	Dark Green

Modern practitioners of crystal healing and gem therapy often adopt a list from one of the previously mentioned sources, However, consulting numerous texts on the subject, one will find nearly any gem or gem color may be connected to any zodiacal sign. Inconsistency of gem to zodiac correspondence is common. If one is in search of a natal stone to serve as a talisman, it would seem advisable to choose a stone of personal preference or on that "feels right." Certainly almost any stone can be found to align with one's birth if one or more of the many existing lists is consulted.

The Power of Shape

The shape and structure of a gem is thought to have influence over its metaphysical powers. Although uncut crystals are preferred by many contemporary practitioners, a wealth of lore refers to cut gems. Many of the gems favored over the centuries are, in fact, nearly unrecognizable as precious stones unless they are cut and polished. The natural crystal forms of gems are an interesting field of study on their own.

The habit, or most common form, of each of the gems covered by this text is listed with the gem. Each crystal structure is given attributes separate from those listed with the individual gems. The six major crystal systems are included here.

Solidity and Rigidity

The cube is considered a shape of structure and rigidity. Mystically it is considered a form of conflict. The solidity of the form and resulting hardness of many minerals of this structure make it an unyielding and egotistic form. The cube is difficult to move. It symbolizes materialism (building blocks) and a lust for possessions and power. It is the form and number (4) of Mars, planet and god of war. The isometric or cubic system is characterized by three axis of equal length set at right angle to one another. The external surfaces takes the form of a cube or, as illustrated, a regular form that has faces which allign with the faces of a cube.

Pyramid Power

The tetragonal system is a combination of three and four. This combination is found in architecture, particularly that with origins in Greece. Greek villages still feature homes with doors following this structure. The modern gable roofed house continues this form. The combining of three over four has been associated with spirit over matter. This pyramid,

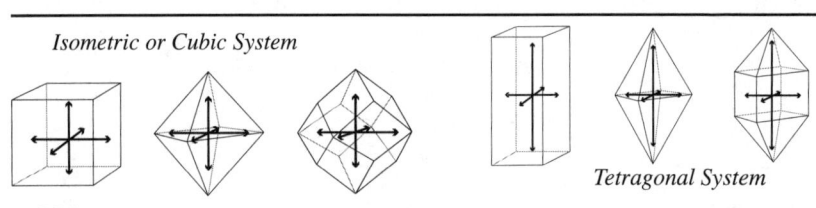

Isometric or Cubic System

Tetragonal System

330
Gems:

Hexagonal system

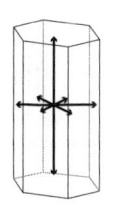

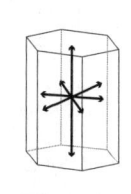

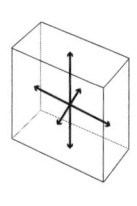

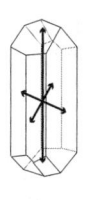

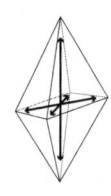

Orthorhombic system

attractor of ethereal power, is a perfect form to attract spiritual forces to the occupants of such a building. The same pyramidal attraction exists in many tetragonal crystals. Astrologically the crystal holds the power of Mars and that of Jupiter, the planet with the mystic number of three. Tetragonal system crystals have two axes of equal length and a third of unequal length all set at right angles. The resulting form resembles a column, a column terminated at each end with four-sided pyramids, or two four-sided pyramids joined at their bases.

The Number Six

The hexagonal systemfeatures six-sided columns or two six-sided pyramids joined at their bases results. The number six represent man's physical existence on earth. The triangles formed by the pyramids represent the spirit. The number six also represents harmony and peace. Bees live in six-sided structures and are examples of a productive, harmonious community. Six is the number of Venus and of love and peace. Long six-sided crystals are said to also contain the powers of Mercury. The crystal system has three axes on the same plane oriented at 120 degree angle with each other. A fourth axis of a different length crosses at their intersection at a 90 degree angle.

Radiant Energy

This orthorhombic system allows for many varied shapes. These radiating axes are connected with the sun. This is the crystal form of sulphur, "sun-bearer," and of peridot, another solar stone. The orthorhombic system generates remote forms with a heavenly connection. Crystals of this system have three axes at right angles to each other and of differing lengths.

Hidden Light

The monoclinic crystal system is associated with a hidden light. For

Monoclinic system

Triclinic system

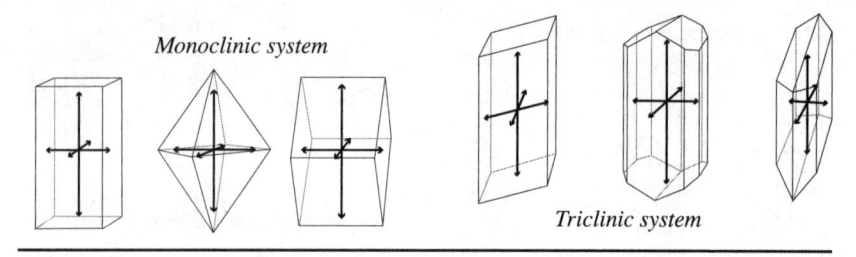

this reason, Saturn and Pluto have power over monoclinic gems. Crystals have three mutually unequal axes; two are set at an oblique angle to the remaining one. The third axis is perpendicular to the other two.

Impulsive Crystals

The most capricious crystal system is called triclinic. This system brings randomness, disorder, and upheaval. The gems of this system are connected to Uranus, the capricious planet. Triclinic crystals have three mutually unequal axes set at oblique angles to each other.

The Kindest Cut

The practice of gem cutting is centuries old. Beginning as a practice of polishing existing crystal faces, the art of fashioning gems has become very sophisticated. Many elaborate facet cuts have been developed, but they may all be classified by their basic geometric shape. Each of these shapes carries with it a unique set of attributes. These may be contrary to the powers of a gem and may tend to diminish a stone's usefulness. The greatest advantage to be found in a gem would be to add to a stone's internal strengths with a complimentary cut. The attributes of each of the basic geometric shapes is given here.

Shape of Magic

Round gems are the shape of magic. They symbolize the power of Mother Earth, the solar system, the galaxy, and the universe. They are considered to be the "shape of God." They are also claimed to be the key shape for unlocking spiritual or psychic awareness. Round gems are favored when casting love spells and are linked to the female reproductive system. Modern practice favors round gems as a symbol of promised love.

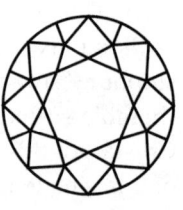

Round or Brilliant cut

Gems:

Unifying Sphere

Connected with round gems are those which are spherically cut. In a spherical configuration minerals transmit energy in all directions. Spheres are the most unified and compact of all forms. They represent totality or the whole person. They symbolize the Sun if set in gold and the Moon if set in white metal. As such, they represent light or enlightenment. Spheres represent the entirety of the universe and all the knowledge it contains. They act as a key to communication with spiritual forces and their vast knowledge. Spheres of clear crystal have long been favored as vehicles for contacting ethereal forces.

Oval cut

Mystery and Emotion

A symbol of fertility, both male and female, is the oval. This androgynous shape represents the mysteries of sex and the inner emotions of men and women. The oval shape acts to increase the talismanic power of a gem. Women have long carried oval stones to increase fertility and promote conception. The ancient Babylonians used oval or egg-shaped pearls as a male infant's amulet in the belief that it would hasten the desension of a baby boy's testicles. The amulet was usually tied to the child's neck with a thin leather cord. The enhancement fertility was not limited to humans. Oval gems would increase egg production if placed in a hen's nest and increase crop yields if buried in the corners of fields or orchards. The increased fertility also extends to the mind. Oval gems are said to enhance creativity.

Emerald cut

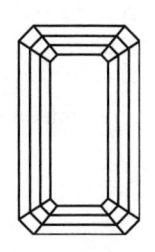

Control and Stability

Square or emerald cut gems demonstrate the ability of man to change nature. Nature favors curves and amorphic shapes while man tends to favor order, lines, and regularity. Both men and women can use the shape to improve their ability to modify or affect the proverbial "four corners" of the Earth. The square is a stable shape, used to anchor meditation and control magic. It is the second most powerful shape after the circle, the shape of God.

Square cut

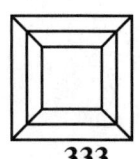

Symbol of Alchemy

The triangle is the shape that encloses the "Eye of God." In the United States this is seen on the back of the one dollar bill. It is the ancient symbol of alchemy. A triangle represents the outward flow of energy. Each of the three points stands for the mind, body, and soul. The three-dimensional form of the triangle is the pyramid, a form that represents stored energy. The upward point of a pyramid is said to release stored energy, or it may collect energy that is available. The bottom of a round-cut gem features a cone, a rounded pyramid. It is said the energy gathered by the top of a round gem is directed to the wearer by way of the tip of the cone, known as the culet.

Triangle cut

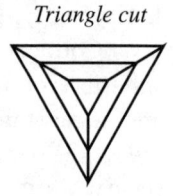

Adding a Point

The combination of a circle and triangle is a pear-shape. The shape is at once both projective and receptive of cosmic power. It also combines the masculine and feminine in one shape. This joining of the two primary shapes creates the natural shape of air and water. The pear shape has been used as the symbol of comets and shooting stars, and as the shape of tears and rain drops. The shape also makes a connection to the night and astral travel. The circle, or shape of God, is given a point to project spiritual power.

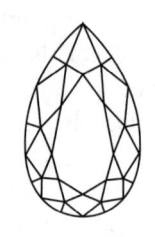

Pear-shape cut

Shape of Sex

The shape of sexual power, fertility and sexual mystery is the marquise or boat- shape. It is the rudimentary shape of the female sexual organ. As such, it represents being receptive of love and affection. As it is a female based shape, men rarely buy a marquise shape stone for their own wear. Marquise cut stones are second in popularity as the gem shape used to pledge romantic love.

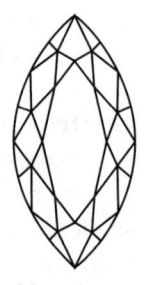

Marquise cut

334
Gems:

Appendices

The Foundation Stones

Assuming modern lists of birthstones are based on the Foundation Stones, an examination of various translations of Revelations reveals some differences.

King James Version 1611	Revised Version 1881	New English Translation 1961	New World Translation 1963
Jasper	Jasper	Jasper	Jasper
Sapphire	Sapphire	Lapis-lazuli	Sapphire
Chalcedony	Agate	Chalcedony	Agate
Emerald	Emerald	Emerald	Emerald
Sardonyx	Onyx	Sardonyx	Onyx
Sardius	Carnelian	Cornelian	Carnelian
Chrysolite	Chrysolite	Chrysolite	Chrysolite
Beryl	Beryl	Beryl	Beryl
Topaz	Topaz	Topaz	Topaz
Chrysoprase	Chrysoprase	Chrysoprase	Chrysoprase
Jacinth	Jacinth	Turquois	Jacinth
Amethyst	Amethyst	Amethyst	Amethyst

Patriotic Birthstones

G.F. Kunz suggested list of "patriotic" natal stones at the end of the nineteenth century. He proposed that "one born in the United States should wear a gem from among those which our country furnishes."[1]

Month	Stone	Source
January	Garnet, rhodolite	Montana, New Mexico, Arizona, North Carolina
February	Amethyst	North Carolina, Georgia, Virginia
March	Californite	California
April	Sapphire	Montana, Idaho
May	Green tourmaline	Lake Superior
June	Moss-agate	California, Montana, Wyoming, Arizona
July	Turquois	New Mexico, California, Arizona
August	Golden beryl	California, Connecticut, North Carolina
September	Kunzite*	California
October	Aquamarine	North Carolina, Maine, California
November	Topaz	Utah, California, Maine
December	Rubellite	Montana

*Named for G.F. Kunz

Hindu Gems of the Months[2]
from the *Mani-Málá*

Month	Gem	Month	Gem
April	Diamond	October	Coral
May	Emerald	November	Cat's-eye
June	Pearl	December	Topaz
July	Sapphire	January	Serpent-stone
August	Ruby	February	Chandrakanta*
September	Zircon	March	The Gold Siva-linga‡

*this name for an unknown stone does not translate to a modern gem name.
‡a Siva-linga is a phallus, a talismanic representation of the masculinity of the god Shiva.

Planetary Influences of Gems
A Seventeenth Century List

Planetary influences of stones as listed by Morales, 1604. The connections of some stone to planets differ from those most often listed. Morales listed other stones not of general interest. Those non-gemstones, such as asbestos, have been deleted.

Aetites	Sun	Jacinth	Mars, Jupiter
Agate	Venus, Mars	Jasper	Venus, Mercury
Alectoria	Sun	Jet	Saturn
Amber	Sun	Lapis-lazuli	Venus
Amethyst	Mars, Jupiter	Lyncurius	Sun
Beryl	Venus, Mars	Magnet	Mars
Bezoar	Jupiter	Opal	Sun, Mercury
Carbuncle	Mars, Venus	Pearl	Venus, Mercury
Carnelian	Jupiter, Mars, Venus	Pyrites in the colors of	
Chalcedony	Jupiter, Mercury, Saturn	-copper	Sun, Venus
		-gold	Sun
		-silver	Moon
Chrysolite	Mercury, Venus	-tin	Moon, Saturn
Chrysoprase	Mercury, Venus	-ash	Jupiter
Crystal	Moon, Mars	Sapphire	Jupiter, Mercury
Diamond	Jupiter	Sardonyx	Saturn, Mars
Emerald	Venus, Mercury	Topaz	Saturn, Mars
Garnet	Sun	Turquois	Venus, Mercury
Hematite	Mercury		

Planetary Influences of Gems
A List for the Twenty-First Century

The list on the following page is advocated by many contemporary gem therapists. The list of planets includes one not generally recognized by astronomers. The now destroyed planet Lucifer is included. Lucifer is the supposed planet that is now broken into fragments and exists as a belt of asteroids between Mars and Jupiter. Two other celestial bodies are also included, the Sun and Moon, as the "sources of light."

The correspondence of gems to planets in this modern list relies on the colors related to these bodies. The colors of the gems which enjoy a sympathy with these planets are not necessarily those perceived colors, but their cosmic colors.

The Planets and Their Sympathetic Colors

Planet	Color	Planet	Color
Mercury	yellow	Saturn	dark green & black
Venus	pink & blue	Uranus	blue-green
Earth	green & light green	Neptune	lilac
Mars	bright red	Pluto	dark red
Lucifer	mother-of-pearl	Sun	white
Jupiter	orange & gold	Moon	white

Gems and Their Corresponding Planets

Gem	Planet	Gem	Planet
Agate (dark red)	Pluto	Jade	
Amazonite	Uranus	blue	Venus
Amethyst	Neptune	green	Earth
Aventurine	Earth		
Bloodstone	Pluto	Jasper (bright red)	Mars
Carnelian (orange)	Jupiter	Jet	Saturn
Chalcedony (white)	Moon	Malachite	Uranus
Citrine	Mercury	Moonstone	Lucifer & Moon
Diamond	Sun		
Emerald	Venus	Onyx	Saturn
Garnet		Opal	
Almandine	Pluto	violet flashes	Neptune
bright red	Mars	white	Lucifer & Moon
Pyrope	Pluto		

Gem	Planet	Gem	Planet
Peridot	Earth	Spinel (red)	Saturn
Quartz		Topaz	
rose	Venus	orange	Jupiter
white or clear	Sun	yellow	Mercury
Ruby	Mars	Turquois	Uranus
Sapphire		Zircon (orange)	Jupiter
blue	Venus		
yellow	Mercury		

Gems and their Meaning in Dreams

Many "dream books" have been compiled over the centuries. The following is a collection of the significance of gems in dreams as listed in the majority of these texts.

Agate	a journey
Amber	a voyage
Amethyst	freedom from harm
Aquamarine	friends
Beryl	coming happiness
Bloodstone	bad news
Carbuncle	wisdom to come
Carnelian	impending doom
Cat's-eye	treachery
Chalcedony	friends rejoined
Chrysoberyl	a time of need
Chrysolite	caution should be taken
Coral	recovery from illness
Crystal	no enemies
Diamond	victory over enemies
Emerald	much to look forward to
Garnet	a mystery will be solved

Heliotrope	long life
Hyacinth	heavy storms
Jacinth	success
Jasper	love returned
Jet	sorrow and woe
Lapis-lazuli	faithfulness
Moonstone	danger is near
Moss-agate	an unsuccessful trip
Onyx	joy in marriage
Opal	wealth
Pearl	faithful friends
Ruby	unexpected guests
Sapphire	escape from harm
Sardonyx	faithful friends
Topaz	freedom from harm
Tourmaline	an accident comes
Turquois	prosperity

Gem Therapy Practices
Modern Considerations

The list on the following pages is a of the gems covered by this text and their attributes which are of interest to many who practice gem therapy. Three attributes are seen to have particular value: astrological sign, vibrational number and sympathy with chakras.

The astrological signs are those commonly assigned by contemporary practitioners and may not correspond to those listed in other parts of this text.

The Numerical Vibration of Each Gem

Numerologists believe that numbers hold a certain vibration or frequency. Each gem is believe to vibrate to a certain numerical frequency and, therefore, has numeric value which enhances one's life.

The chakras, Sanskrit for "wheel," are defined by Eastern mystics as those points on the body that form the intersections of energy flows. These points represent energy centers used in healing and in bringing the physical and emotional body into balance. Thousands of these chakras are described in Eastern literature. The seven major chakras will be addressed in this text. The placing of gems on these energy intersections is said to help clear and balance one's energy fields. Each chakra is given a number and is named.

Chakra Num.	Chakra Name	General Location	Controls or Energizes
first	base	at the base of the spine	vitality, physical energy, self preservation
second	sacral	1" to 2" below the navel	desire, emotion, creativity, sexuality
third	solar-plexus	below breast bone to navel	stomach, liver, gall bladder, pancreas, adrenal glands, sympathetic nervous system
fourth	heart	center of chest, over heart	compassion, love, spirituality associated with oneness, group consciousness
fifth	throat	throat just above collar bone	communication, sound, speech, writing, expression of creativity
sixth	third-eye	center of forehead	psychic power, the spirit, "the light," higher intuition or the "aha"
seventh	crown	top of the head	spirituality, enlightenment, dynamic thought

Gem	Astrological sign(s)	Chakra(s) Energized	Vibrational #
Agate	Gemini	all, according to color	7
Alexanderite	Scorpio	first, fourth & sixth	5
Amber	Leo & Aquarius	seventh	3
Amethyst	Pisces, Virgo, Aquarius & Capricorn	seventh	3
Aquamarine	Gemini, Pisces & Aries	all	1
Beryl, golden	Leo	third & seventh	1
Bloodstone	Aries, Pisces & Libra	first, second & fourth & fourth	4 & 6
Carnelian	Taurus, Cancer & Leo	first, second, third	5 & 6
Cat's-eye	Capricorn, Taurus & Aries	—	6
Chalcedony	Cancer & Sagittarius	sixth	9
Chrysoberyl	Leo	third & seventh	6
Chrysoprase	Libra	fourth	3
Citrine	Gemini, Aries, Libra & Leo	third	6
Coral	Pisces	sixth	22
Diamond	Aries, Leo & Taurus	seventh	33
Emerald	Taurus, Gemini & Aries	fourth	4
Garnet	Leo, Virgo, Capricorn & Aquarius	first & seventh	2
Hematite	Aries & Aquarius	third	9
Jade	Aries, Gemini, Taurus & Libra	third	11
Jasper	Leo	third	6
Jet	Capricorn	first	8
Lapis-lazuli	Sagittarius	fifth & sixth	3
Lodestone	Virgo & Gemini	aligns all	1
Malachite	Capricorn & Scorpio	all	9
Moonstone	Cancer, Libra & Scorpio	all	4
Obsidian	Sagittarius	sixth	1
Onyx	Leo	—	6

Gem	Astrological sign(s)	Chakra(s) Energized	Vibrational #
Opal	Cancer, Libra, Pisces & Scorpio	all	8
Pearl	Cancer & Gemini	—	7
Peridot	Virgo, Leo, Scorpio & Sagittarius	third & fourth	5, 6 & 7
Quartz (clear)	All	fourth & sixth	4
Ruby	Leo, Scorpio, Cancer & Sagittarius	fourth	3
Sapphire	Virgo, Libra & Sagittarius	all	1
Sardonyx	Aries	—	3
Serpentine	Gemini	seventh	8
Spinel	Sagittarius	all, according to color	3
Topaz	Sagittarius	all	6
Tourmaline	Libra	all	2
Turquois	Sagittarius	all with particular power	1
Zircon	Virgo, Leo & Sagittarius	first, third & fourth	4

Stones of a Theosophical Altar
The Altar of the Free Catholic Church

The following stones are arranged, as shown, on an altar in ceremonies of the Free Catholic Church. The temple is considered the church's body and the altar serves as its heart. From the energy of the stones, the heart of the church gains insight and wisdom. The same stones are set in the staff and pectoral cross of the Free Catholic Bishops.

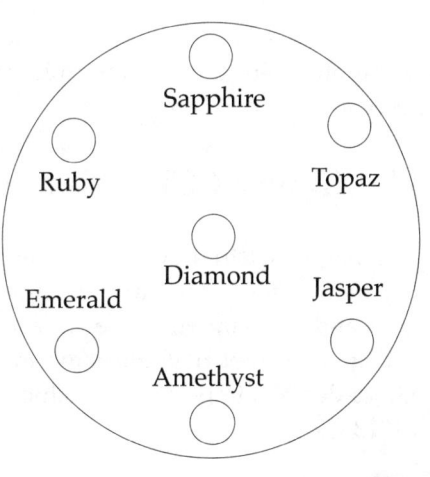

Sapphire
Ruby   Topaz
Emerald   Diamond   Jasper
Amethyst

The same gems are often set in the crozier and pectoral cross of Anglican and Roman Catholic Bishops. The seven gems are said to represent the seven "primary planets." These are the seven celestial bodies visible to the naked eye (Sun, Moon, Mercury, Venus, Mars, Saturn and Jupiter).

The "Nine-gem" Jewel The *Nararatna*

Ancient Hindu texts describe a sacred talisman known as the *nararatna* or *naoratna*, meaning the nine-gem jewel. The "gem" is most powerful when suspended from a mani málá, "chain of gems," or worn in a ring so the stones make contact with the skin. The nine gems represent the planets and correspond to the nine planets listed in vedic texts as the *navaratna*, or natal stones. The gems are arranged in a circle with their positions described as the points of a compass.

Position	Gem	Planet
Center	Ruby	Sun
East	Diamond	Venus
Southeast	Pearl	Moon
South	Coral	Mars
Southwest	Jacinth	Rahu*
West	Sapphire	Saturn
Northeast	Topaz	Jupiter
North	Cat's-eye	Ketu*
Northwest	Emerald	Mercury

*Rahu and Ketu are shadowy planets that exist at the ascending and descending nodes of the moon.

The Gems of Christian Names[3]

An old custom exists which entails assigning a gem to a person's Christian name. The tradition continues that the birthstone, patron saint's stone and the name stone are all worn together. It is considered particularly good fortune if all the gems are the same. It may be noted that the names deemed important to include are dated by this list quoted from a 1913 text.

Feminine Names

Adelaide	Andalusite	Gertrude	Garnet
Agnes	Agate	Gladys	Golden Beryl
Alice	Alexanderite	Grace	Grossularite Garnet
Anne	Amber	Hannah	Heliotrope
Beatrice	Basalt	Helen	Hyacinth
Belle	Bloodstone	Irene	Iolite
Bertha	Beryl	Jane	Jacinth
Caroline	Chalcedony	Jessie	Jasper
Catherine	Cat's-eye	Josephine	Jadeite
Charlotte	Carbuncle	Julia	Jade
Clara	Carnelian	Louise	Lapis-lazuli
Constance	Crystal	Lucy	Lepidolite
Dorea	Diamond	Margaret	Moss-agate
Dorothy	Diaspore	Martha	Malachite
Edith	Eye-agate	Marie	Moldavite
Eleanor	Elaeolite*	Mary	Moonstone
Elizabeth	Emerald	Olive	Olivine
Ellen	Essonite	Pauline	Pearl
Emily	Euclase	Rose	Ruby
Emma	Epidote	Sarah	Spodumene
Florence	Fluorite	Susan	Sapphire
Frances	Fire-opal	Theresa	Turquois

*Kunz lists a gem/mineral name not used in modern gemology. However, the prefix *elaeo* is a geological term meaning "oil."

Masculine Names

Abraham	Aragonite	Ambrose	Amber
Adolphus	Albite	Andrew	Aventurine
Adrian	Andalusite	Archibald	Axinite
Albert	Agate	Arnold	Aquamarine
Alexander	Alexanderite	Arthur	Amethyst
Alfred	Almandine Garnet	Augustus	Agalmatolite
		Benjamin	Bloodstone

Bernard	Beryl	John	Jacinth
Charles	Chalcedony	Joseph	Jargoon
Christian	Crystal	Julius	Jet
Claude	Cyanite	Lambert	Labradorite
Clement	Chrysolite	Lawrence	Lapis-lazuli
Conrad	Crocidolite	Leo	Lepidolite
Constantine	Chrysoberyl	Leonard	Loadstone
Cornelius	Cat's-eye	Mark	Malachite
Dennis	Demantoid Garnet	Matthew	Moonstone
		Maurice	Moss-agate
Dorian	Diamond	Michael	Microcline
Edmund	Emerald	Nathan	Natrolite
Edward	Epidote	Nicholas	Nephrite
Ernest	Euclase	Oliver	Onyx
Eugene	Essonite	Osborne	Orthoclase
Ferdinand	Feldspar	Oswald	Obsidian
Francis	Fire-opal	Patrick	Pyrope
Frederick	Flourite	Paul	Pearl
George	Garnet	Peter	Porphyry
Gilbert	Gadolinite	Phillip	Prase
Godfrey	Gagates	Ralph	Rubellite
Gregory	Grossularite Garnet	Raymond	Rose-quartz
		Richard	Rutile
Gustave	Galactite	Robert	Rock-crystal
Guy	Gold quartz	Roger	Rhodonite
Henry	Heliolite	Roland	Ruby
Herbert	Hyacinth	Stephen	Sapphire
Horace	Harlequin opal	Theodore	Tourmaline
Hubert	Heliotrope	Thomas	Topaz
Hugh	Heliodor	Valentine	Vesuvianite (Idocrase)
Humphrey	Hypersthene		
James	Jade	Vincent	Verd-antique
Jasper	Jasper	Walter	Wood-opal
Jerome	Jadeite	William	Willemite

346 Gems:

Endnotes

Introduction
[1] Geoffrey Chaucer, Walter W. Skeat ed., *Complete Works of Geoffrey Chaucer: The Hous of Fame* vol. 3 (London: Oxford at the Clarendon Press) 1894 p. 41

[2] George Frederick Kunz, *The Curious Lore of Precious Stones* (New York: Dover Publications Inc., 1913, 1938 ed.) p. 2

[3] Clinical studies by Carlos Vallbona M.D. and Carleton Hazelwood M.D. on the use of magnets in pain relief, published in the "Archives of Physical Medicine and Rehabilitaion," November, 1997.

[4] Ralph Waldo Emerson, *Poems by Ralph Waldo Emerson* (Cambridge, Massachusetts: Riverside Press, 1892) p. 106

A History of Writings
[1] *GIA Colored Stones Course* lesson 29 (Santa Monica, California: Gemological Institute of America 1980) p. 2

[2] John Sinkankas, *Emerald and Other Beryls* (Prescott, Arizona: Geoscience Press 1989) p. 24

[3] Sinkankas, p. 24

[4] Lynn Thorndike, *A History of the Magical Experimental Sciences* vol. I (New York: Columbia Press, 1923) p. 174

[5] Sinkankas, p. 26

[6] Dorothy Wyckoff, *Albertus Magnus Book of Minerals* (Oxford: Clarendon Press, 1967) p. 309

[7] F. D. Adams, *The Birth and Development of the Geological Sciences* (Baltimore: Williams & Wilkins, 1938) p. 309

[8] Joseph R. Strayer, editor, *Dictionary of the Middle Ages* vol. I (American Council of Learned Societies, 1982) p. 36

[9] Adams, p. 420

[10] Adams, p. 424

[11] Sinkankas, p. 46

Agate

[1] LeDande Quick, *The Book of Agates and Other Quartz Gems* (Philadelphia and New York: Chilton Books, 1963) p. 4 (modern names in parenthesis added by Quick to Pliny's text.)

[2] Berthold Laufer, *Agate, Archeology and Folklore* (Chicago: Field Museum of Natural History, publication #8, 1927) p. 3

[3] C.W. King, *The Natural History of Gems or Semi-Precious Stones* (London: Bell & Daldy, 1870) p. 6

[4] Damigeron, trans. by Patricia P. Tahil, *De Virtutibus Lapidum: The Virtues of Stones* (Seattle: Ars Obscura, 1989) p. 30

[5] *Encyclopedia of Superstitions, Folklore, and the Occult Sciences of the World*. Vol. 1 (Chicago: 1903) p. 13

[6] Quick, p. 7

[7] Laufer, p. 5

[8] Harriet Keith Fobes, *Mystic Gems* (The Gorham Press 1924) p. 58-59

[9] Kate Blanas, translation of Isaac del Sotto's, *Le Lapidaire Du Quatorziéme Siécle* (Geneva: 1862, Slatkine Reprints, 1974) p. 23

[10] George Frederick Kunz, *The Curious Lore of Precious Stones* (New York: Dover Publications Inc., 1913, 1938 ed.) p. 52

[11] Blanas, p. 23

[12] Camillus Leonardus, *Speculum Lapidum (The Mirror of Stones)* (Venice: 1502) p. 65

[13] Fobes, p. 55

[14] Fobes, p. 61

[15] Damigeron, p. 30

Alexandrite

[1] Melody, *Love is in the Earth- A Kaleidoscope of Crystals* (Wheat Ridge, Colorado: Earth-Love Publishing House, 1995) p.104

Amber

[1] George Frederick Kunz, *The Curious Lore of Precious Stones* (New York: Dover Publications Inc., 1913, 1938 ed.) p.56

[2] Patty C. Rice, Ph.D., *Amber, the Golden Gem of the Ages* (Van Nostrand Reinhold Co. 1980) p. 112

³ Rice, p. 112

⁴ Rosa Hunger, *The Magic of Amber* (Radnot PA: Chilton Books, 1979) p. 19

⁵ Sir Thomas Moore, *The Complete Poems of Sir Thomas Moore: Lalla Rookh: The Fire-Worshippers* 1817(London: A.L. Burt Company, n.d.imp.) p. 485

⁶ C.W. King, *The Natural History of Gems or Semi-Precious Stones* (London: Bell & Daldy, 1870) p. 303

⁷ *GIA Colored Stones Course* lesson 33 (Santa Monica, California: Gemological Institute of America 1980) p.1

⁸ William Jones F.S.A., *History and Mystery of Precious Stones* (London: Richard Bentley and Son, 1880) p. 31

⁹ Kate Blanas, translation of Isaac del Sotto's, *Le Lapidaire Du Quatorziéme Siécle* (Geneva: 1862, Slatkine Reprints, 1974) p. 67

¹⁰ Camillus Leonardus, *Speculum Lapidum (The Mirror of Stones)* (Venice: 1502) p.108

¹¹ Jones p. 31

¹² George Frederick Kunz, *The Magic of Jewels and Charms* (Philadelphia: J. B. Lippincott Company, 1915) p. 61

¹³ Leonardus, p. 227-228

¹⁴ Leonardus, p.108

¹⁵ Leonardus, p.125

¹⁶ King, p. 309-310

¹⁷ Melody, *Love is in the Earth— A Kaleidoscope of Crystals* (Wheat Ridge, Colorado: Earth-Love Publishing House, 1995) p. 107

Amethyst

¹ James Remington McCarthy, *Rings Through the Ages* (New York: Harper and Brothers, 1945) p. 19

² C.W. King, *The Natural History of Gems or Semi-Precious Stones* (London: Bell & Daldy, 1870) p. 27

³ Harriet Keith Fobes, *Mystic Gems* (The Gorham Press 1924) p. 26

⁴ King, p. 28

⁵ King, p. 36

⁶ William Jones F.S.A., *History and Mystery of Precious Stones* (London: Richard Bentley and Son, 1880) p. 36

⁷ George Frederick Kunz, *The Curious Lore of Precious Stones* (New York: Dover Publications Inc., 1913, 1938 ed.) p. 313

⁸ Fobes, p. 23

⁹ Camillus Leonardus, *Speculum Lapidum (The Mirror of Stones)* (Venice: 1502) p. 66

¹⁰ Kate Blanas, translation of Isaac del Sotto's, *Le Lapidaire Du Quatorziéme Siécle* (Geneva: 1862, Slatkine Reprints, 1974) p. 36

¹¹ Blanas, p. 67

¹² Melody, *Love is in the Earth— A Kaleidoscope of Crystals* (Wheat Ridge, Colorado: Earth-Love Publishing House, 1995) p. 109

¹³ Fobes p. 23

¹⁴*Encyclopedia of Superstitions, Folklore, and the Occult Sciences of the World* Vol. 1 (Chicago: 1903) p. 12

Aquamarine

¹ John Sinkankas, *Emerald and Other Beryls* (Prescott, Arizona: Geoscience Press, 1989) p. 614

Beryl

¹ Damigeron, trans by Patricia P. Tahil, *De Virtutibus Lapidum: The Virtues of Stones* (Seattle: Ars Obscura, 1989) p. 52

² Damigeron p. 52

³ John Sinkankas, *Emerald and Other Beryls* (Prescott, Arizona: Geoscience Press, 1989) p. 73

⁴ William Jones, F.S.A., *History and Mystery of Precious Stones* (London: Richard Bentley and Son, 1880) p. 33

⁵ George Frederick Kunz, *The Curious Lore of Precious Stones*, New York: Dover Publications Inc, 1913, 1938 ed.) p. 312

⁶ Sinkankas p. 71

⁷ George Frederick Kunz, *The Magic of Jewels and Charms* (Philadelphia: J. B. Lippincott Company, 1915) p. 130

⁸ Kunz, *The Curious Lore of Precious Stones*, p. 59

⁹ Damigeron, p. 52

[10] Kate Blanas, translation of Isaac del Sotto's, *Le Lapidaire Du Quatorziéme Siécle* (Geneva: 1862, Slatkine Reprints, 1974) p. 25

[11] Camillus Leonardus, *Speculum Lapidum (The Mirror of Stones)* (Venice. 1502) p. 77

Bloodstone

[1] Damigeron, trans by Patricia P. Tahil, *De Virtutibus Lapidum: The Virtues of Stones* (Seattle: Ars Obscura, 1989) p. 8

[2] Harriet Keith Fobes, *Mystic Gems* (The Gorham Press, 1924) p.32

[3] George Frederick Kunz, *The Curious Lore of Precious Stones* (New York: Dover Publications Inc, 1913, 1938 ed.) p. 61

[4] William Jones, F.S.A., *History and Mystery of Precious Stones* (London: Richard Bentley and Son, 1880) p. 28

[5] C.W. King, *The Natural History of Gems or Semi-Precious Stones* (London: Bell & Daldy, 1870) p. 134

[6] Jones, p. 29

[7] Fobes, pp. 33-34

[8] *Encyclopedia of Superstitions, Folklore, and the Occult Sciences of the World*. Vol. 1 (Chicago, 1903) p. 12

Carnelian

[1] Harriet Keith Fobes, *Mystic Gems* (The Gorham Press, 1924) p. 75

[2] George Frederick Kunz, *The Curious Lore of Precious Stones* (New York: Dover Publications Inc., 1913, 1938 ed.) p.63

[3] Kunz, p.65

[4] Fobes, p. 77

[5] Fobes, p. 75

[6] Kate Blanas, translation of Isaac del Sotto's, *Le Lapidaire Du Quatorziéme Siécle* (Geneva: 1862, Slatkine Reprints, 1974) p. 66

[7] Camillus Leonardus, *Speculum Lapidum (The Mirror of Stones)* (Venice:1502) p. 84

[8] C.W. King, *The Natural History of Gems or Semi-Precious Stones* (London: Bell & Daldy, 1870) p. 286

[9] James Remington McCarthy, *Rings Through the Ages* (New York: Harper and Brothers, 1945) p. 20

Chalcedony
[1] Damigeron, trans by Patricia P. Tahil, *De Virtutibus Lapidum: The Virtues of Stones* (Seattle: Ars Obscura, 1989) p. 48

[2] Aberdeen University Library Translation and Transcription, *Bestiary, The Aberdeen Manuscript* (http://www.clues.abdn.ac.uk:8080, 1995)

[3] George Frederick Kunz, *The Magic of Jewels and Charms* (Philadelphia: J. B. Lippincott Company, 1915) p. 131

[4] Melody, *Love is in the Earth— A Kaleidoscope of Crystals* (Wheat Ridge, Colorado: Earth-Love Publishing House, 1995) p. 193

[5] Kate Blanas, translation of Isaac del Sotto's, *Le Lapidaire Du Quatorziéme Siécle* (Geneva: 1862, Slatkine Reprints, 1974) p. 53

[6] Camillus Leonardus, *Speculum Lapidum (The Mirror of Stones)* (Venice: 1502) p. 81

[7] George Frederick Kunz, *The Curious Lore of Precious Stones* (New York: Dover Publications, 1913, 1938 ed.) p. 65

[8] William Jones, F.S.A., *History and Mystery of Precious Stones* (London: Richard Bentley and Son, 1880) p. 31

Chiastolite
[1] *Colored Stones Course*, Assignment 31, (Santa Monica, California: Gemological Institute of America, 1980) p. 2

Chrysoberyl
[1] Melody, *Love is in the Earth— A Kaleidoscope of Crystals* (Wheat Ridge, Colorado: Earth-Love Publishing House 1995) p. 203

Chrysoprase
[1] John Bartlett, *Familiar Quotations* (Boston: Little, Brown and Company, 1948) p. 432

[2] George Frederick Kunz, *The Curious Lore of Precious Stones* (New York: Dover Publications Inc., 1913, 1938 ed.) pp. 312-313

[3] Kunz, p. 68

[4] Kate Blanas, translation of Isaac del Sotto's, *Le Lapidaire Du Quatorziéme Siécle* (Geneva: 1862, Slatkine Reprints, 1974) p. 59

[5] William Jones, F.S.A., *History and Mystery of Precious Stones* (London: Richard Bentley and Son, 1880) p. 31

Citrine

[1] William Jones, F.S.A., *History and Mystery of Precious Stones* (London: Richard Bentley and Son, 1880) p. 31

[2] Melody, *Love is in the Earth— A Kaleidoscope of Crystals* (Wheat Ridge, Colorado: Earth-Love Publishing House 1995) p. 209

Coral

[1] C.W. King, *Natural History of Gems or Semi-Precious Stones* (London: Bell & Dalby, 1870) pp.101-102

[2] Damigeron, trans by Patricia P. Tahil, *De Virtutibus Lapidum: The Virtues of Stones* (Seattle: Ars Obscura, 1989) p. 16

[3] *Colored Stones Course*, assignment 33, (Santa Monica, California: Gemological Institute of American, 1980) p.6

[4] George Frederick Kunz, *The Magic of Jewels and Charms* (Philadelphia: J. B. Lippincott Company, 1915) p. 30

[5] Damigeron p. 16

[6] George Frederick Kunz, *The Curious Lore of Precious Stones* (New York: Dover Publications, 1913, 1938 ed.) p.68

[7] Camillus Leonardus, *Speculum Lapidum (The Mirror of Stones)* (Venice: 1502) p. 83

[8] Kunz, *The Curious Lore of Precious Stones*, p.70

[9] Kunz, *The Magic of Jewels and Charms*, p. 132-133

[10] Richard Cavendish, Ed. *Man Myth and Magic* Vol. 9 (New York: 1985) p. 2342

[11] Leonardus, p. 83

[12] Kunz, *The Magic of Jewels and Charms*, p. 131-132

[13] Edwin W. Streeter, *Precious Stones and Gems* (London: Chapman & Hall, 1877) p. 232

Diamond

[1] John Bartlett, *Familiar Quotations* (Boston: Little, Brown and Company, 1948) p. 430

[2] George Frederick Kunz, *The Magic of Jewels and Charms* (Philadelphia: J. B. Lippincott Company, 1915) p. 134

[3] *Encyclopedia of Superstitions, Folklore, and the Occult Sciences of the World*, Vol. 1, (Chicago: 1903) p. 13

[4] George Frederick Kunz, *The Curious Lore of Precious Stones* (New York: Dover Publications, 1913, 1938 ed.) p. 73

[5] Robert Maillard ed., *Diamonds: Myth, Magic, and Reality* (Crown Publishing, 1984) p. 20

[6] Kunz, *Curious Lore* p. 71

[7] Maillard, p. 20

[8] Maillard, p. 14

[9] Damigeron, trans by Patricia P. Tahil, *De Virtutibus Lapidum: The Virtues of Stones* (Seattle: Ars Obscura, 1989) p. 10

[10] Maillard, p. 14

[11] Thomas Moore, *The Complete Poems of Sir Thomas Moore* (New York, A. L. Burt Company, n.d.imp.) p. 524

[12] Damigeron, p. 10

[13] Aberdeen University Library Translation and Transcription, *Bestiary, The Aberdeen Manuscript* (http://www.clues.abdn.ac.uk:8080, 1995)

[14] Maillard p. 16

[15] John Bartlett, *Familiar Quotations* (Boston: Little, Brown and Company, 1948) p. 1198

[16] Kate Blanas, translation of Isaac del Sotto's, *Le Lapidaire Du Quatorziéme Siécle* (Geneva: 1862, Slatkine Reprints, 1974) p. 12

[17] Kunz, *Curious Lore...* p. 71

[18] Sir Walter Scott, *The Poetical Work of Sir Walter Scott: Bart. Bridal of Triermain: Canto Third: Chorus* 1813 (Boston: Phillips, Sampson, and Company, 1857) p. 404

[19] Kunz, *Curious Lore...* p. 157

[20] Blanas, p. 12

[21] Blanas, p. 12

[22] Kunz, *Curious Lore...* p.70

[23] Maillard, p. 41

[24] Kunz, *Curious Lore...* p. 378

[25] William Jones, F.S.A., *History and Mystery of Precious Stones* (London: Richard Bentley and Son, 1880) p. 39

[26] Kunz, *The Magic of Jewels...* p. 138
[27] Jones, p. 40
[28] Maillard, p. 16
[29] Kunz, *Curious Lore...* p. 75
[30] Maillard p. 18
[31] Kunz, *Curious Lore...* p. 74

Emerald

[1] Damigeron, trans by Patricia P. Tahil, *De Virtutibus Lapidum: The Virtues of Stones* (Seattle: Ars Obscura, 1989) p. 14

[2] Harriet Keith Fobes, *Mystic Gems* (The Gorham Press, 1924) p. 50

[3] I.A. Mumme, *The Emerald, Its Occurrence, Discrimination and Valuation* (Port Hacking, New South Wales, Australia: Mumme Publications, 1982) p. 1

[4] John Sinkankas, *Emerald and Other Beryls* (Prescott, Arizona: Geoscience Press, 1989) p. 62

[5] *GIA Colored Stone Course,* lesson 29 (Santa Monica, California: Gemological Institute of America, 1980) pp. 1-2

[6] *GIA*, p. 2

[7] Damigeron, p. 14

[8] Isaac del Sotto, *Le Lapidaire Du Quatorziéme Siécle* (Geneva: 1862, Slatkine Reprints, 1974) p. 28

[9] Sotto, p. 28

[10] Sotto, p. 76

[11] George Frederick Kunz, *The Curious Lore of Precious Stones* (New York: Dover Publications, 1913, 1938 ed.) p. 312

[12] *Encyclopedia of Superstitions, Folklore, and the Occult Sciences of the World*, Vol. 1 (Chicago: 1903) p. 13

[13] Fobes, p. 46

[14] William Jones, F.S.A., *History and Mystery of Precious Stones* (London: Richard Bentley and Son, 1880) p. 36

[15] Sotto, p. 28.

[16] Fobes, p. 49

[17] Kunz, p. 77

[18] Fobes, p. 49
[19] Fobes, p. 49
[20] Blanas, p. 28.
[21] Jones, p. 12
[22] Kunz, p. 158
[23] Kunz pp. 78-79
[24] George Frederick Kunz, *The Magic of Jewels and Charms* (Philadelphia: J. B. Lippincott Company, 1915) p. 136
[25] Edwin W. Streeter, *Precious Stones and Gems* (London: Chapman & Hall, 1877) p. 150
[26] Kunz, *Curious Lore...* p. 79
[27] Jones, p. 12-13
[28] Hartley Burr Alexander ed., *The Mythology of All Races* vol. VII (New York: Cooper Square Publishers Inc. 1964) p. 201.
[29] Sinkankas, p. 357

Garnet

[1] Harriet Keith Fobes, *Mystic Gems* (The Gorham Press, 1924) p. 15
[2] *Encyclopedia of Superstitions, Folklore, and the Occult Sciences of the World.* Vol. 1 (Chicago, 1903) p. 12
[3] C.W. King, *The Natural History of Gems or Semi-Precious Stones* (London: Bell & Daldy, 1870) p. 15
[4] George Frederick Kunz, *The Curious Lore of Precious Stones* (New York: Dover Publications Inc., 1913, 1938 ed.) p. 312
[5] Camillus Leonardus, *Speculum Lapidum (The Mirror of Stones)* (Venice: 1502) p. 80
[6] Fobes, p. 15-16
[7] William Jones F.S.A., *History and Mystery of Precious Stones* (London: Richard Bentley and Son, 1880) p. 32
[8] Fobes, p. 18
[9] George Frederick Kunz, *The Magic of Jewels and Charms* (Philadelphia: J. B. Lippincott Company, 1915) p. 130
[10] Fobes, p. 16
[11] Leonardus, p. 80

¹² Kate Blanas, translation of Isaac del Sotto's, *Le Lapidaire Du Quatorziéme Siécle* (Geneva: 1862, Slatkine Reprints, 1974) p. 3

¹³ Blanas, p. 9

¹⁴ Blanas, p. 37

¹⁵ Geoffrey Chaucer, Walter W. Skeat ed., *Complete Works of Geoffrey Chaucer: Romaunt of the Rose* vol. 1 (London: Oxford at the Clarendon Press, 1894) p. 139

¹⁶ Chaucer, vol. 1 p. 140

¹⁷ Fobes p. 19

¹⁸ Jones p. 82

Hematite

1 J. F. Borghouts, The Magical Texts of Papyrus Leiden I 348 (Leiden: E.J. Brill, Leiden, 1971) p.15

2 George Frederick Kunz, The Magic of Jewels and Charms (Philadelphia: J. B. Lippincott Company, 1915) p. 137

3 Kunz, p. 137

4 Melody, Love is in the Earth— A Kaleidoscope of Crystals (Wheat Ridge, Colorado: Earth-Love Publishing House, 1995) p. 316

Jade

¹ George Frederick Kunz, *The Magic of Jewels and Charms* (Philadelphia: J. B. Lippincott Company, 1915) p. 139

² Kate Blanas, translation of Isaac del Sotto's, *Le Lapidaire Du Quatorziéme Siécle* (Geneva: 1862, Slatkine Reprints, 1974) p. 69

³ Kunz, p. 383

⁴ Kunz, p. 384

⁵ George Frederick Kunz, *The Magic of Jewels and Charms* (Philadelphia: J. B. Lippincott Company, 1915) p. 140

⁶ John Goette, *Jade Lore* (New York: Reynal & Hitchcock, 1937) p. 297

⁷ John C. Ferguson, *The Mythology of All Races* Vol. 8 (Cooper Square Publishers Inc, 1964) pp. 46-47

⁸ Goette, p. 293

⁹ Kunz, *The Magic of Jewels and Charms*, p. 143

¹⁰ Ferguson, p.171

[11] Goette, p. 290

[12] Goette, p. 295

Jasper

[1] Harriet Keith Fobes, *Mystic Gems* (The Gorham Press, 1924) p. 55

[2] C.W. King, *Natural History of Gems or Semi-Precious Stones*, (London: Bell & Daldy, 1870) p. 139

[3] William Jones, F.S.A., *History and Mystery of Precious Stones* (London: Richard Bentley and Son, 1880) p. 35

[4] James Remington McCarthy, *Rings Through the Ages* (New York: Harper & Brothers, 1945) p. 19

[5] King, pp. 142-143

[6] George Frederick Kunz, *The Curious Lore of Precious Stones* (New York: Dover Publications, 1913) p. 90

[7] Kate Blanas, translation of Isaac del Sotto's, *Le Lapidaire Du Quatorziéme Siécle* (Geneva: 1862, Slatkine Reprints, 1974) p. 49

[8] Blanas, p. 78

[9] Jones, p. 35

[10] Kunz, pp. 311-312

[11] Camillus Leonardus, *Speculum Lapidum (Mirror of Stones)* (Venice: 1502) p. 113

[12] George Frederick Kunz, *The Magic of Jewels and Charms* (Phildelphia: J. B. Lippincott Company, 1915) p. 144

Lapis-lazuli

[1] Canon John Arnott MacCulloch, ed. *The Mythology of All Races* Vol. 5 (New York: Cooper Square Publishers Inc., 1964) p. 327

[2] George Frederick Kunz, *The Curious Lore of Precious Stones* (New York: Dover Publications, 1913, 1938 ed.) p. 92

[3] George Frederick Kunz, *The Magic of Jewels and Charms* (Philadelphia: J. B. Lippincott Company, 1915) p. 149

[4] Kunz,*The Curious Lore of Precious Stones*, p. 92

[5] Kate Blanas, translation of Isaac del Sotto's, *Le Lapidaire Du Quatorziéme Siécle* (Geneva: 1862, Slatkine Reprints, 1974) p. 111

[6] Melody, *Love is in the Earth— A Kaleidoscope of Crystals* (Wheat Ridge, Colorado: Earth-Love Publishing House, 1995) p. 371

Lodestone
[1] C.W. King, *The Natural History of Gems or Semi-Precious Stones* (London: Bell & Daldy, 1870) p. 173

[2] George Frederick Kunz, *The Curious Lore of Precious Stones* New York: Dover Publications, 1913, 1938 ed.) p. 94

[3] Kunz, p. 94

[4] Camillus Leonardus, *Speculum Lapidum (Mirror of Stones)* (Venice: 1502) p. 209

[5] Leonardus, p. 207

[6] King, p. 173

Malachite
[1] Hartley Burr Alexander ed., *The Mythology of All Races* Vol XII (New York: Cooper Square Publishers Inc. 1964) p. 367.

[2] Camillus Leonardus, *Speculum Lapidum (Mirror of Stones)* (Venice: 1502) p. 205

Moonstone
[1] William Jones, F.S.A., *History and Mystery of Precious Stones* (London: Richard Bentley and Son, 1880) p. 30

Obsidian
[1] Richard Cavendish, ed., *Man, Myth & Magic* Vol. 8 (New York: 1985) p. 2042

[2] Hartley Burr Alexander, ed., *The Mythology of All Races* Vol. XI (New York: Cooper Square Publishers Inc., 1964) pp. 179-180

[3] C.W. King, *The Natural History of Gems or Semi-Precious Stones* (London: Bell & Daldy, 1870) p. 209

Onyx
[1] Harriet Keith Fobes, *Mystic Gems* (The Gorham Press, 1924) p. 59

[2] Kate Blanas, translation of Isaac del Sotto's, *Le Lapidaire Du Quatorziéme Siécle* (Geneva: 1862, Slatkine Reprints, 1974) p. 57

[3] George Frederick Kunz, *The Curious Lore of Precious Stones* (New York: Dover Publications, 1913, 1938 ed.) p. 98

[4] Camillus Leonardus, *Speculum Lapidum (The Mirror of Stones)* (Venice: 1502) p. 213

⁵ George Frederick Kunz, *The Magic of Jewels and Charms* (Philadelphia: J. B. Lippincott Company, 1915) p. 152

⁶ Melody, *Love is in the Earth- A Kaleidoscope of Crystals* (Wheat Ridge, Colorado: Earth-Love Publishing House, 1995) p. 452

Opal

¹ Frank Leechman, *The Opal Book* (Sydney: Ure Smith, 1961) pp. 181-182

² George Frederick Kunz, *The Curious Lore of Precious Stones* (New York: Dover Publications, 1913, 1938 ed.) pp. 144-145

³ *GIA Colored Stones Course*, lesson 17 (Santa Monica, California: Gemological Institute of America, 1980) p. 2

⁴ Kunz, p. 147

⁵ William Jones, F.S.A., *History and Mystery of Precious Stones* (London: Richard Bentley and Son, 1880) p. 32

⁶ Kunz, p. 150

⁷ Leechman, p. 192

⁸ *Encyclopedia of Superstitions, Folklore, and the Occult Sciences of the World*, Vol. 1 (Chicago: 1903) p. 13

⁹ Fobes, p. 89

¹⁰ Harriet Keith Fobes, *Mystic Gems* (The Gorham Press 1924) p. 90-91

¹¹ Edwin W. Streeter, *Precious Stones and Gems* (London: Chapman and Hall, 1877) p. 161

¹² Nehemiah Bartley, *Opals and Agates* (Brisbane: Gordon & Gotch, 1892) p. 163

¹³ Kunz, p. 143

Pearl

¹William Jones, F.S.A., *History and Mystery of Precious Stones* (London: Richard Bentley and Son, 1880) p. 114

²Richard Cavendish ed., *Man, Myth & Magic* vol. 8 (New York: 1985) p. 2156

³ Jones, pp. 116-117

⁴ Jones, p. 116

⁵ Thomas Moore, *The Complete Poems of Sir Thomas Moore* (New York, A. L. Burt Company, n.d.imp.) p. 524

⁶ Jones, p. 118

⁷ Kate Blanas, translation of Isaac del Sotto's, *Le Lapidaire Du Quatorziéme Siécle* (Geneva: 1862, Slatkine Reprints, 1974) p. 45

⁸ Camillus Leonardus, *Speculum Lapidum (The Mirror of Stones)* (Venice: 1502) p. 200

⁹ Sir Walter Scott, *The Poetical Work of Sir Walter Scott: Bart. Bridal of Triermain: Canto Third: Chorus* (Boston: Phillips, Sampson, and Company, 1857) p. 404

¹⁰ Leonardus, pp. 201-202

¹¹ *GIA Colored Stone Course*, lesson 13 (Santa Monica, California: Gemological Institute of America, 1980) p. 1

¹² Cavendish, p. 2157

¹³ Cavendish, p. 1514

¹⁴ John Milton, F. A. Patterson ed., *The Work of John Milton: An Epitaph on the Marchioness of Winchester* (New York: Columbia University Press, 1931) vol. 1 part 1 p. 30

¹⁵ Jones, p. 123

¹⁶ Blanas, p. 45

¹⁷ Jones p. 266

¹⁸ Hartley Burr Alexander, ed., *The Mythology of All Races* vol. VIII (New York: Cooper Square Publishers Inc., 1964) p. 273

Peridot

¹ George Frederick Kunz, *The Curious Lore of Precious Stones* (New York: Dover Publications, 1913, 1938 ed.,) p. 66

² Kunz, p. 312

³ Damigeron, trans by Patricia P. Tahil, *De Virtutibus Lapidum: The Virtues of Stones* (Seattle: Ars Obscura, 1989) p. 60

⁴ Kate Blanas, translation of Isaac del Sotto's, *Le Lapidaire Du Quatorziéme Siécle* (Geneva: 1862, Slatkine Reprints, 1974) p. 56

⁵ Blanas, p. 79

⁶ Kunz, p. 67

⁷ William Jones, F.S.A., *History and Mystery of Precious Stones* (London: Richard Bentley and Son, 1880) p. 31

Quartz:Rock Crystal

[1] Edwin W. Streeter, *Precious Stone and Gems* (London: Chapman & Hill, 1877) p. 17

[2] Frances Rogers and Alice Beard, *5000 Years of Gems and Jewelry* (New York: Frederick Stokes Company, 1940) p. 245

[3] Richard Cavendish, ed., *Man Myth and Magic*, Vol. 6 (New York, 1985) p. 1510

[4] Camillus Leonardus, *Speculum Lapidum (The Mirror of Stones)* (Venice: 1502) p. 85

[5] Janet and Colin Bord, *The Secret Country* (New York: Walker and Co.,1977) p.160

[6] Cheri Lesh, "Quartz: Myth and Magic, Science and Sales" *Gems and Gemology* Vol XVI #6, Summer 1979: p. 176

[7] Melody, *Love is in the Earth- A Kaleidoscope of Crystals* (Wheat Ridge, Colorado: Earth-Love Publishing House, 1995) p. 509

[8] Cavendish, p. 2316

[9] C.W. King, *The Natural History of Gems or Semi-Precious Stones* (London: Bell & Daldy, 1870) pp. 119-120

[10] George Frederick Kunz, *The Magic of Jewels and Charms* (Philadelphia: J. B. Lippincott Company, 1915) pp. 154-155

[11] Kate Blanas, translation of Isaac del Sotto's, *Le Lapidaire Du Quatorziéme Siécle* (Geneva: 1862, Slatkine Reprints, 1974) p. 89

[12] Leonardus, p. 85

[13] Kunz, p. 154

[14] Rogers and Beard, p. 72

[15] King, p. 122

Ruby

[1] Rev. Walter Skeat, ed., The Complete Works of Geoffrey Chaucer vol. I-VII (Oxford: Clarendon Press, 1894) (Chaucer used rubyes once in *Romaunt of the Rose* and the singular form, ruby, three times in *Troilus and Criseyde*.)

[2] George Frederick Kunz, *The Curious Lore of Precious Stones* (New York: Dover Publications, 1913, 1938 ed.) p. 101

[3] *GIA Colored Stones Course,* lesson 30 (Santa Monica, California: Geomological Institute of America, 1980) p. 1

[4] *Encyclopedia of Superstitions, Folklore, and the Occult Sciences of the World,* vol. 1 (Chicago, 1903) p. 13

[5] GIA, p. 2

[6] Harriet Keith Fobes, *Mystic Gems* (The Gorham Press, 1924) p. 65

[7] Kunz, p. 102

[8] Bharata, a region of Asia, or ancient India, included Burma, Thailand, Afghanistan, Pakistan, Nepal, Tibet and Sri Lanka.

[9] Kunz, p. 386

[10] John M. Riddle, *Marbode of Rennes' (1035-1123) De Lapidibus Considered as a Medical Treatise with Text, Commentary and C. W. King's Translation* (Wiesbaden: Franz Steiner Verlag GMBH, 1977) pp.68-69

[11] GIA, p. 3

[12] Kunz, p. 102

[13] Ralph Waldo Emerson, *Poems by Ralph Waldo Emerson* (Cambridge, Massachusetts: Riverside Press, 1892) p. 287

[14] Kate Blanas, translation of Isaac del Sotto's, *Le Lapidaire Du Quatorziéme Siécle* (Geneva: 1862, Slatkine Reprints, 1974) p. 5

[15] Fobes, p. 64

[16] William Jones, F.S.A., *History and Mystery of Precious Stones* (London: Richard Bentley and Son, 1880) p. 35

[17] Fobes, p. 66

[18] Kunz, p. 158-159

[19] Ralph Waldo Emerson, p. 217

[20] Jones, p. 68

Sapphire

[1] Damigeron, trans. by Patricia P. Tahil, *De Virtutibus Lapidum: The Virtues of Stones,* (Seattle, Ars Obscura, 1989) p. 24-25

[2] Harriet Keith Fobes, *Mystic Gems* (The Gorham Press, 1924) p. 82-83

[3] Fobes, p. 83

[4] John M. Riddle, *Marbode of Rennes' (1035-1123) De Lapidibus, Considered as a Medical Treatise with Text, Commentary and C. W. King's Translation* (Wiesbaden, Franz Steiner Verlag GMBH, 1977) p. 42

⁵ George Frederick Kunz, *The Magic of Jewels and Charms* (Philadelphia: J. B. Lippincott Company, 1915) p. 158.

⁶ George Frederick Kunz, *The Curious Lore of Precious Stones* (New York: Dover Publications, 1913, 1938 ed) p. 105

⁷ *Encyclopedia of Superstitions, Folklore, and the Occult Sciences of the World.* Vol. 1 (Chicago: 1903) p. 13

⁸ Kate Blanas, translation of Isaac del Sotto's, *Le Lapidaire Du Quatorziéme Siécle* (Geneva: 1862, Slatkine Reprints, 1974) p. 73

⁹ Camillus Leonardus, *Speculum Lapidum (The Mirror of Stones)* (Venice: 1502) p. 224-225

¹⁰ Kunz, *The Curious Lore of Precious Stones* p. 389

¹¹ Fobes, p. 87

¹² Riddle, p. 42

¹³ Fobes, p. 80-81

¹⁴ Kunz, *The Curious Lore of Precious Stones* p. 312

¹⁵ Kunz, *The Curious Lore of Precious Stones* p.106-107

¹⁶ Fobes, p. 82

¹⁷ Melody, *Love is in the Earth — A Kaleidoscope of Crystals* (Wheat Ridge, Colorado: Earth-Love Publishing House, 1995) p. 586

Sard

¹ Damigeron, *De Virtutibus Lapidum*, located in Vol. III of *Spicilegium Solesmense* (JB Pitra, 1855) p. 62

² John M. Riddle, *Marbode of Rennes' (1035-1123) De Lapidibus, Considered as a Medical Treatise with Text, Commentary and C. W. King's Translation* (Wiesbaden, Franz Steiner Verlag GMBH, 1977) p. 48

³ Kate Blanas, translation of Isaac del Sotto's, *Le Lapidaire Du Quatorziéme Siécle* (Geneva: 1862, Slatkine Reprints, 1974) p. 55

⁴ Camillus Leonardus, *Speculum Lapidum (The Mirror of Stones)* (Venice: 1502) p. 228

⁵ George Frederick Kunz, *The Curious Lore of Precious Stones* (New York: Dover Publications, 1913, 1938 ed) p. 312

Sardonyx

¹ John M. Riddle, *Marbode of Rennes' (1035-1123) De Lapidibus, Considered as a Medical Treatise with Text, Commentary and C. W. King's Translation* (Wiesbaden, Franz Steiner Verlag GMBH, 1977) p. 46

² Camillus Leonardus, *Speculum Lapidum (The Mirror of Stones)* (Venice: 1502) p. 229

³ George Frederick Kunz, *The Curious Lore of Precious Stones* (New York: Dover Publications, 1913, 1938 ed) p. 312

⁴ *Encyclopedia of Superstitions, Folklore, and the Occult Sciences of the World* Vol. 1 (Chicago, 1903) p. 13

Serpentine

¹ George Frederick Kunz, *The Curious Lore of Precious Stones* (New York: Dover Publications, 1913, 1938 ed.) p. 108

Spinel

¹ Kenneth Mears, *The Crown Jewels* (London: Historic Royal Palace Agencies, 1994) pp. 29-30

Staurolite

¹ Jim and Jenny Monaco, "The Mythical Magic of Fairy Stones", *Rock and Gem* (Ventura, California: Miller Publishing, Vol. 25 Number 8, August 1995) p. 8

² Anonymous passage, 1968

Topaz

¹ The literal translation would be "it is thought to feel the moon." King's translation was based on a corrupted edition of 1799.

² John M. Riddle, *Marbode of Rennes' (1035-1123) De Lapidibus, Considered as a Medical Treatise with Text, Commentary and C. W. King's Translation* (Wiesbaden: Franz Steiner Verlag GMBH, 1977) p. 51

³ George Frederick Kunz, *The Curious Lore of Precious Stones* (New York: Dover Publications, 1913, 1938 ed.) p. 312

⁴ Harriet Keith Fobes, *Mystic Gems* (The Gorham Press 1924) p. 102

⁵ Kunz, p. 389

⁶ Kate Blanas, translation of Isaac del Sotto's, *Le Lapidaire Du Quatorziéme Siécle* (Geneva: 1862, Slatkine Reprints, 1974) p. 33

⁷ Fobes, p. 99

⁸ George Frederick Kunz, *The Magic of Jewels and Charms* (Philadelphia: J. B. Lippincott Company, 1915) p. 158

⁹ *Encyclopedia of Superstitions, Folklore, and the Occult Sciences of the World* Vol. 1 (Chicago: 1903) p. 13

[10] Melody, *Love is in the Earth — A Kaleidoscope of Crystals* (Wheat Ridge, Colorado: Earth-Love Publishing House, 1995) p. 648

Tourmaline

[1] George Frederick Kunz, *The Magic of Jewels and Charms* (Philadelphia: J. B. Lippincott Company, 1915) pp. 51-52

[2] G. F. Kunz, *Rings for the Finger* (Philadelphia: J. B. Lippincott Company, 1917) p. 304

[3] Melody, *Love is in the Earth — A Kaleidoscope of Crystals* (Wheat Ridge, Colorado: Earth-Love Publishing House, 1995) p. 653

[4] George Frederick Kunz, *The Magic of Jewels and Charms* (Philadelphia: J. B. Lippincott Company, 1915) p. 54

Turquois

[1] Valentin Rose, *Aristoteles de Lapidibus und Arnoldus Saxo* vol. 18 (Berlin: Alterthum, 1875) p. 446

[2] Pliny, trans. by Bostock and Riley, *Natural History* book 37, chap. 56, vol. 6 (London, 1857) pp. 444-445

[3] Harriet Keith Fobes, *Mystic Gems* (The Gorham Press 1924) p.107

[4] Joseph E. Pogue, *Turquois* Memoirs of the National Academy of Sciences Vol. XII Part II (1974) p. 111

[5] Pogue, p. 12

[6] Pogue, p. 13

[7] *Encyclopedia of Superstitions, Folklore, and the Occult Sciences of the World* Vol. 1 (Chicago: 1903) p. 13

[8] Pogue, p. 14

[9] Pogue, p. 72

[10] Pogue, p. 13

[11] Kate Blanas, translation of Isaac del Sotto's, *Le Lapidaire Du Quatorziéme Siécle* (Geneva: 1862, Slatkine Reprints, 1974) p. 109

[12] Bernardus Caesius *de Mineralibus* (Lugduni, 1636) p. 601

[13] George Frederick Kunz, *The Curious Lore of Precious Stone*s (New York: Dover Publications, 1913, 1938 ed.) p. 110

[14] John Bartlett, *Familiar Quotations* (Boston: Little, Brown and Company, 1948) p. 118

[15] William Jones, F.S.A., *History and Mystery of Precious Stones* (London: Richard Bentley and Son, 1880) p. 37

[16] Fobes, p.106

[17] Pogue, p. 115

[18] Pogue, p. 115

[19] Pogue, p. 122

[20] Edna Mae Bennett, *Turquoise and the Indian* (Chicago: The Swallow Press Inc. 1970) p. 116

Zircon

[1] C.W. King, *Natural History of Gems or Semi-Precious Stones* (London: Bell & Dalby, 1870) p. 168

[2] George Frederick Kunz, *The Curious Lore of Precious Stones* (New York: Dover Publications, 1913, 1938 ed.) p. 82

[3] Kunz, p. 313

[4] Camillus Leonardus, *Speculum Lapidum (Mirror of Stones)* (Venice: 1502) p. 112

[5] Kate Blanas, translation of Isaac del Sotto's, *Le Lapidaire Du Quatorziéme Siécle* (Geneva: 1862, Slatkine Reprints, 1974) p. 42

[6] King, p. 168

[7] William Jones, F.S.A., *History and Mystery of Precious Stones* (London: Richard Bentley and Son, 1880) p. 29

Mythical Gems

[1] John M. Riddle, *Marbode of Rennes' (1035-1123) De Lapidibus Considered as a Medical Treatise with Text, Commentary and C. W. King's Translation* (Wiesbaden: Franz Steiner Verlag GMBH, 1977) pp.63-64

[2] Riddle, 39

[3] Camillus Leonardus,*Speculum Lapidum (The Mirror of Stones)* (Venice: 1502) p. 78

[4] Patty C. Rice, Ph,D., *Amber the Golden Gem of the Ages* (???: Van Nostrand Reinhold Co. 1980) p. 124

[5] George Frederick Kunz,*The Curious Lore of Precious Stones* (New York: Dover Publications Inc, 1913, 1938 ed.) p. 340

[6] Leonardus, p. 89

[7] Leonardus, p. 84

[8] Mellie Uyldert, *The Magic of Precious Stones* (Wellingborough, England: Turnstone Press Limited, 1981) p. 156-157

[9] Leonardus, p. 91

[10] Jones p. 9-10

[11] William Jones, F.S.A., *History and Mystery of Precious Stones* (London: Richard Bentley and Son, 1880) p. 22

[12] Riddle, pp. 54-55

[13] Uyldert, p. 156

[14] Frances Rogers and Alice Beard, *500 Years of Gems and Jewelry* (New York: Frederick Stokes Co., 1940) p. 114

Literature
De Lapidibus

[1] better translated as, "it chases away black poisons."

[2] a possible reference to jet

[3] the Latin has only *veridi* meaning "green," and not necessarily "bright green."

[4] the Latin reads *noxia*, "harmful," King used the form *nox*, "of the night."

[5] original translates literally to "Ground and rubbed on with milk, it makes well sores."

[6] original translates literally to "It lifts dirt from the eyes and it cures headaches."

[7] King added a line to complete the rhyme.

[8] first stone probably citrine, second stone chrysolite or peridot

[9] literal translation would be "it is thought to feel the moon."

[10] literally, "The coral stone is a twig whilst it is living in the seas."

[11] original says color is "similar to amber."

[12] *peanita* is a geode, like eaglestone, with foreign pebbles or crystals said to be born when the stone is broken.

Precious Stones in Shakespeare's Works

[1] A reference to the practice of carving agate cameos.

[2] The belief in amber as the "tears of trees" is demonstrated.

[3] Chrysolite was regarded as exceptionally rare.

⁴ Comparison of tears and pearls lead to German proverb, "Perlen bedeuten Tränen" (pearls mean tears) taken to signify that pearls portend sadness and tears.

⁵ Indirect reference to Pliny's tales of pearl's origins in dew.

⁶ Alludes to the use of pearls in the embroidery of garments and upholstery.

⁷ Another comparison of tears to pearls and their symbolism of sorrow.

⁸ Reflects the belief in the natural luminous quality of gems.

⁹ Rock crystal stood for clarity and purity, as in "crystal clear."

¹⁰ Rock crystal was highly valued in the England of Elizabeth and James I.

¹¹ Shakespeare seems to know little of ruby except as a rich color.

¹² a basket or hamper

¹³ Demonstrates a familiarity with ancient accounts of the gem's ability to soothe eyes.

Birthstones

¹ Rupert Gleadow, *The Origin of The Zodiac* (New York: Athenum, 1969) p. 131

² George Frederick Kunz, *The Curious Lore of Precious Stones* (New York: Dover Publications, 1913, 1938 ed.) p. 301

³ Kunz, pp. 311-313

⁴ Kunz, p. 313

⁵ *Bestiary, The Aberdeen Manuscript*, Aberdeen University Library Translation and Transcription, 1995, http://www.clues.abdn.ac.uk:8080

⁶ Kunz, p. 326

⁷ George Frederick Kunz, *The Magic of Jewels and Charms* (Philadelphia: J. B. Lippincott Company, 1915) p. 248

⁸ Gleadow, pp. 133-134

⁹ Kunz, *Curious Lore of Precious Stones*, p. 338

¹⁰ Prof. B.V. Ramn, *Astrology For Beginners* (Bangalore, India: IBH Praksshana. 1983) p. 5

¹¹ M.N. Dutt, trans., *Garuda-Puranam* (Varanasi, India: The Chowkhamba Sanskrit Series Office, 1968 Chapter 68, verse 17

[12] Surindro Mohun Tagore, *Mani-Mala* (Calcutta, India: I. C. Bose & Co., 1879) p. 575

Appendices

[1] George Frederick Kunz, *The Curious Lore of Precious Stones* (New York: Dover Publications, 1913, 1938 ed.) p. 323

[2] Surindro Mohun Tagore, *Mani-Mala* (Calcutta, India: I. C. Bose & Co., 1879) pp. 616, 621

[3] Kunz, pp. 47-50

Index

Line art and woodcuts, page number underlined

A

Aaron's Breastplate 44, 294, 315, 317 see also, Breastplate of the High Priest
Abbey of St. Alban 152
Abenzoar, a.k.a. Abû Meruân (on tincture of emerald) 103
"Achates" (agate) 294, 297
Achates River (fabled source of agate) 24
"Achlamah'" (Greek, amethyst) 296
"Achlamath'" (Hebrew, amethyst) 43
Achroite 211
"Açmagarbhaja" (emerald) 93
Adamas (diamond) 76, 80, 255, 259
Adams, F. D. (on Marbode and de Boodt) 15, 18
"Aeroldes" (emerald, Pliny) 51
"Aethiopian" (emerald, Pliny) 95
Aetites, eagle-stone 235, 236, 271
Against the Grain by Joris Karl Huysmans on diamond 81
Agastimata, 78
Agate 23-30, as chalcedony 62, dyed 151, in the *Bible* 253, 255,257,
    in literature 260,279, as birthstone 294-298, 312-320, 324
Agatharcides (author) on peridot 169
Agricola, Georgius (mineralogist) 10,17
Agn'es Sorel, see Sorel, Agn'es
Agni or Agnideva (India, fire-god) 56, 79
"Ahlamah" or "Allamah" (Hebrew, amethyst) 297
Aistian (ascestors of Lithuanians) 39
Alabandine (garnet?) 270
Alabaster 95, 288
Albertus Magnus, see Magnus, Albertus
Aldrovandi, Ulyssis author of *Musaeum Metallicum in Libros IV Distributum* 18
Alectorius, also Allectory 237, 261
"Aleppo" (eye-agate) 26
Alexander the Great 89, 104, 143
Alexandrite 31-32, as chrysoberyl 66, as birthstone 313, 315, 321
Alfonso X 16, 162, 324
Allectory 237
Almandine (garnet) 108, 118, as birthstone 296, 298

Almandite (garnet) 108
Alphonso XII (on opal) 158
Amber 38, 32-42, compared to jet 137, ingredient in bezoar 239,
    in literature 271, 279, 287, 288, 290, as birthstone 296, 320
Amberella, Myth of (on amber) 41
Ambrose (author of *Hexaemeron*) 301
"American ruby" 301
"American turquois" 215
Amethyst 42-46, treated to citrine 70, as quartz 172, compared to zircon 230,
    in the *Bible* 253, 257, in literature 259, 268,
    as birthstone 294-298, 301, 309, 310, 312-325, 327
Amethyst, The Myth of (on amethyst) 46
Amun (Egyptian god) 217
Anatomy of Melancholy by Burton 114, 196
Anatomy of the World, An by John Donne 220
Andradite (garnet) 109, 220
Andreas, Bishop of Caesarea, regarding individual gems 44, 49, 68, 99, 112,
    135, 170, 197, 200, 208, 231,
    treatise on gems and the Apostles 299-301
Andrew, Saint 112, 300, 319
"Androdamus hematite" 121, 276
Anglicus, Bartolomæus (poet) 135, 193
Anne of Geierstein, Legend of (opal) 157
Anselmus Boëtius de Boodt, see de Boodt
Anthony and Cleopatra by Shakespeare 280, 285
Anthrax (birthstone) 294, 296
Antiquities of the Jews by Josephus 293-294, 318
Apache tears, legend of (obsidian) 150
Apollinus of Tyana (on quartz) 173
Apostles, the (with associated birthstones) 319
    St. Andrew 112, 205, 300,
    St. Bartholomew 170, 300,
    St. James 300,
    St. John 68, 99, 300,
    St. Matthew 208, 300,
    St. Peter 135, 299,
    St. Philip 300,
    St. Simon Zelotes 300,
    St. Thaddeus 68, 300,
    St. Thomas 49, 300,
Aquamarine 46-51, as birthstone 312-314, 323,
    aquamarine-chrysolite (peridot) 169

Aquinas, Thomas (student of Albertus Magnus) 15
Arcula Gemmea by Thomas Nicols (names of turquois) 216
Aristotle, regarding the physical world and the elements 10, 11,
    regarding gems 43, 93, 97, astrological rings 323
Aristatalis, Hakim (Aristotle?) on turquois 223
"Arizona rubies" 115
Aryama (India, lord of ancestors) 79
As You Like It by Shakespeare 248, 284
"aschentrekker" (tourmaline) 212
Asclepiades (author) on amethyst 43
Asher (tribe of Israel) connection to gems 317
Aspilates (Arabian version of aetites) 238
Astrites (cat's-eye) as birthstone 320
Astronomica by Roman poet Manilus, 80
Attican (emerald, Pliny) 95
"Autachates" (agate) 24
Author of *Lithica* (on jasper) 133
Autumn, gems of 315
Azchalias (author) on garnet 120
Azurite (confused with emerald, Pliny) 95

B

"Babylonian gem" (bloodstone) 52
Bacchus (in myth of Amethyst) 46
Baccio, Andrea, author of *De Gemmis* 188
"Bactrian emerald" 95, 264
Badakhshan (India, source of spinel) 185, 203
"Balas ruby" 185, 203
Balascia (source of term balas) 185,203
"Banded obsidian" 148
"Bareqeth" (Hebrew, green gem) 295, 297
Barrett, Eaton, author of *The Magnus* 319
Bartholemew, Saint 170, 300, 319
"Baseler taufstein" (staurolite) 205
Batman, Dr. Stephen (author, on opal) 156
Baufra's Tale (Egyptian legend, turquois) 217
Bela King of Hungary, legend of 99
Benjamin (tribe of Israel) connection to gems 319
Benjamin of Tudela (author, on pearls) 162
Benoni, Rabbi (mystic, on diamonds) 84

"Bernstein" (amber) 36
Beryl 47-51, in Pliny's work 13, as aquamarine 46, as emerald 92-93, 95-96, peridot pseudonym 169, in the *Bible* 253,255,257, in literature 258,263,266, as birthstone 293-295, 296-298, 300-301, 306, 310, 313, 319-321
"Beryllus" (beryl) 48, 294
Bestiary of Aberdeen University Library 301-309
"Bezoar" 238-239
Black Prince's Ruby 203
Bloodstone 52-56
Body color (pearl) 160-161
Book of Marvels, The by Marco Polo (on diamond) 81, 90
Book of Seals, The 45, 113-114
Book of Stones by Dioscorides 218
Book of Various Acts, The by Theophilus (on quartz) 173, 178
Book of Wings by Rabbi Ragiel 59-60, 196, 209
Bord, Colin (quartz in Britain) 175
Bord, Janet (quartz in Britain) 175
Boreas (turquois, Pliny) 216
Boyle, Robert author of *An Essay about the Origine and Virtues of Gems* 19
"Brahmin ruby" 185
"Brazilian Agate" 30
"Brazilian chrysoberyl" (peridot) 169
Breastpiece of the High Priest 13, 44, 58, 76, 186, 251-252, 294, 297, 314, 317
Breastplate of the Second Temple 296
"Breciated jasper" 136
Bridal of Triermain by Sir Walter Scott 84, 163
Brihaspati (India, Jupiter) 327
Brihatsamhita (India, gem texts) 78
"Brown spinel" 204
Brunswick Blue Diamond 88
Buddha (India, Mercury) 327
Buddhabhatta (Hindu author) on diamond 80
Buffalo Horns of Turquois, Legend of 228-229
Burton, Elver, author of *Anatomy of Melancholy* 114, 196
"Butterfly jade" (Japanese burial stone) 127

C

CCAG, see *Catalogus Codium Astrologorum Graecorum*

Gems:

"Cabot stone" also cimedia 239
"Cairngorm quartz" 70, 172
Calcedon (chalcedony, Marbode) 214
Calcite 139
Calkedon (Greek, chalcedony) 63
Callais (turquois, Pliny) 216
Callistratus (author) on amber 34
Cameo (carved of onyx) 152
Camillus Leonardus see Leonardus, Camillus
"Cape rubies" (garnet) 115
Capella, Martianus (on stones of zodiac) 318
Carbuncle (garnet) 56, 108, 112, 115, 117, color of amber 35,
    like spinel 203, in the *Bible* 252, 255, 256,
    in literature 271, 279-280, 287,
    as birthstone 294, 295-297, 299, 325,
Cardano, Girolamo 85, 86, 102, 135, 152, 231
Carnelian 57-60, banded with onyx 57, similar to sard 162, 198,
    in the *Bible* 255, 256, in literature 270,
    as birthstone 296, 298, 312-313, 314, 319, 321, 325, 327
"Carnelian onyx" 62, 151
Casesius, Bernardus (author) on turquois 219, 220
Caste, proper owner of gems 328-329
Catalogus Codium Astrologorum Graecorum (astrology) 320
"Cat's-eye apatite" 61
Cat's-eye Chrysoberyl 60-62, as birthstone 315, 320-321, 327
"Cat's-eye moonstones" 147
"Cat's-eye quartz" 61
"Cat's-eye tourmaline" 61
Caxton, William author of *Roman de Renard* (tourmaline) 212
Cellini, Benvenuto on diamond as poison 82, tale of glowing garnets 117
"Cerachates" (agate, Pliny) 24
Chalcedonian (emerald, Pliny) 95
Chalcedony 62-64, as agate 23, 25, as other gems 52, 57, 68, 112, 132, 151,
    198, in the *Bible* 257, in literature 258, 263, 264,
    as birthstone 299, 300, 303-304, 310, 312, 317, 319, 321, 324
"Chalchihuitl" (Aztec, jade) 124, 129
Chalcosmaragdus (emerald, Pliny) 96
Chandra (India, the Moon) 327
Chang (Chinese, red jade tablet) 128
Chaucer, Geoffrey 1, 116, 184, 188
Chelonites (toadstone) 246-247
Chiastolite 64-65, misnomer for staurolite 205

Chief Morning Green, Legend of (turquois) 227-228
Child of the Sun, by Fray Simon *(emerald)* 106-107
Chivers, Thomas Holley, author of *Lily Adair* 75, and *Rosalie Lee* 67
"Chlorite quartz" 178
Chow Li, Chinese writings 128
"Chryselectrum" (amber) 34
Chrysoberyl 66, alexandrite form 31, cat's-eye form 60-62,
    misnomer peridot 169, as birthstone 315, 325
"Chrysoberyllus" (emerald, Pliny) 51
Chrysocolla (chacedony) 62, confused as emerald 95
Chrysolite 67, as peridot 169-170, in the Bible 255, 257,
    in literature 258, 266, 280,
    as birthstone 296, 300, 301, 306, 310, 312, 315, 317, 319-321
Chrysolithos (Greek, chrysolite) 294, 297
Chrysolithus (Roman, chrysolite) 297
Chrysoprase 67-69, misidentified (emerald, Pliny) 51, as chalcedony 62,
    in the Bible 301, in literature 258, 268, 278,
    as birthstone 301, 308, 310, 313, 320, 321
Cimedia, see also cabot stone 239
"Cinnamon stone" (garnet) 110, 118
Citrine 69-70, type of quartz 172, misnomer for topaz 207,
    as birthstone 314, 327
"Citrine topaz" (yellow topaz) 207
"Citrini" (yellow corundum) 70
Claudian (Roman poet) 143, 181-182
Clear or colorless quartz (rock-crystal) 172, 178, 313
Cleopatra (pearl toast) 164
"Cloras" (emerald, Pliny) 95
Cock-stone 239
Comedy of Errors by Shakespeare 279-280, 287
Coral 71-74, attracts robbers 222, ingrediant in bezoar 239,
    in the Bible 254, 255, in literature 270, 280, 283, 287, 288, 292
    as birthstone 324, 325, 327
Corallite (birthstone in *CCAG*) 320
"Coralloachates" (lapis-lazuli) 24
"Coriolanus" (coral? Shakespeare) 279
Corundum, as ruby 183-189, as sapphire 189-197
Corvia 239
Cosmic colors (India, in the *Navaratna*) 328-329
Countess of Blessington, author of *Gems of Beauty* 159
Crab's eye 240
Craft and Frauds of Physick Expos'd by Dr. Robert Pitt 19-20

Crapaudina (toadstone) 246
Cross-stone (chiastolite) 65, (staurolite) 205
Crowley, Alexander, author of *"777"* 320-321
Crystal (colorless quartz) 177, in the *Bible* 254, 257,
    in literature 259, 261, 275, 286, 289-290
Crystal ball, used for scrying 178
Curious Lore of Precious Stones, The by George Frederick Kunz 20-21
Cymbeline by Shakespeare 189, 280-282, 286
Cyprian emerald (emerald, Pliny) 95, (tourmaline, Theophrastus) 212
Cyprium (copper) as planetary ring 325
Czar Alexander II (namesake, alexandrite) 31

D

Damigeron, author of *De Virtutibus Lapidum* 15, 24, 48, 50, 52, 63, 72, 76,
    80, 93, 97, 170, 193, 198,
Damour, Professor A. (mineralogist) on jade 125
Daniel (tribe of Israel) connection to gems 319
de Boniface, Pierre (on diamond) 84
de Boodt, Anselmus Boëtius author of *Gemmarum et Lapidum Historia* 18,19,
    name aquamarine 47, 51, regarding gems 135, 197, 205, 220
de Cantimpré, Thomas (author of *De Rerum Natura*) 50
de Cuba, Johannis, author of *Ortus Sanitatis* 4, 38, 55, 113, 167, 195, 233,
    235, 237, 241, 247, 249
de Valois, Phillipe (author of *Lapidaire*) on ruby 186
De Gemmis by Andrea Boccio 188
De Gemmis Aliquot by Franciscus Rues 17
De Laet, Johann (author) on carnelian 60, on jade 126, on emerald 212
De Mineralibus by Albertus Magnus 15
De Natura Fossilium by Georgius Agricola 17
De Omne Rerum Fossilium by Conrad Gesner 14, 17
De Proprietatibus Rerum by Bartholomæus Glanvilla 36
De Re Mettallica by Georgius Agricola 10
De Rerum Natura by Thomas de Cantimpré 50
De Virtutibus Lapidum by Damigeron 48, 63, 170
Demantoid (garnet) 109
Demoniu 240
Den Sina (author) on zircon 232
Dendrachate (moss agate, Pliny) 24
Dendrites (birthstone in *CCAG*) 320

"Dendritic Agate" 30
Diadochus 240
Diamond 75-91, associated with lodestone 143, loss of special powers 222,
  zircon confused with 230, in the *Bible* 252, 255,
  in literature 259, 267, 276, 280-282, 285, 288,
  as birthstone 293-297, 310-315, 317, 319, 320, 324-325, 327
Diana (in myth of Amethyst) 46
Diopside 327
Dioscorides, author of *Book of Stones*. on turquois 218, on aetites 235
Discourse of the Travels of Two English Pilgrims to Jerusalem, Gaza, etc. by
  Timberlake 244-245
Discoverie of Witchcraft by Sir Reginald Scot 49, 73
Don Pedro the Cruel (source of Black Prince's Ruby) 203
Donne, John author of *An Anatomy of the World,* on turquois 220
Doors of the Temple of Bacchus, Legend of *The* by Francois Rabelais 91-92
Doriatides 240
Dorland, Frank (curator) on Mitchell-Hedges Skull 179-180
Draconius 241-242
Dravite 211
Du Ble (artist/poet) on opal 153

E

Eagle-stone, aetites 235, 242, 244
Echites, eagle-stone 242, 244
Egyptian Tales by Petrie 217
"Egyptian turquois" 216
Elatite (hematite, Pliny) 121
Elbite 211
Electrum 323
Elektron (amber) 34
"Elestials quartz" 178
Elopsides 242
Emerald 92-107, Pliny names 12, 95-96, classified by place (Brazilian, Co-
  lumbian, Occidental, Oriental 17, as beryl 48,
  value related to diamond 81, demantoid garnet called 115,
  peridot confused for 170, tourmaline confused for 211-212,
  turquois displayed with 224, in the *Bible* 252, 255, 257,
  in literature 258, 264, 278, 282, 288,
  as birthstone 294-301, 304, 310-315, 317, 319-322, 324-325, 327
Emerald Buddha, Legend of 106

378
Gems:

Emerald Tablet, Legend of 104
Emerson, Ralph Waldo (on amulets) 8, (on rubies) 186, 188-187
Ephraim (tribe of Israil) connection to gems 319
Epitaph on the Marchioness of Winchester, An by John Milton 165
Epigrams by Claudian (on crystal) 181-182
Epiphanius, St. on Gems of the Breastplate 13, 17, 186,
        an author of *Legend of the Valley of Diamonds* 89, on jasper 133
Epistides 242
Erman, Dr. Adolph (translator) Egyptian papyri 140
Esmeraude, emeraude, and emeralde (emerald) 93
Essay About the Origine and Virtues of Gems, An by Robert Boyle 19
"Ethiopic hematite" 121
Etymologiarum sive Originum Libri XX (*Etymologies*) by Isidore 14, 301
Euchite (birthstone in *CCAG*) 320
Eumetis 242
Evax, king of Arabs (Damigeron?) 15
"Evening emerald" (peridot) 169
Exebonos 242
Eye-agate 26

F

Fairy Stone State Park (staurolite source) 205
"Fairy-stone" (staurolite) 205-206
Family Dictionary, The by Dr. W. Salmon 37
"Fancy jasper" 136
Fayalite 67, 169
Feldspar 147, 295, 298
Filaterius 242
"Fire agate" 30
"Fire obsidian" 149
"Fire opal" 154, 315
Fire Worshipers, The by Sir Thomas Moore 32, 34
Flint (in the *Bible*) 255
Flint glass "crystal" 173
"Flowering obsidian" 148
Fluor-spar 170, 323
Fongites 243
Forbes' Oriental Memoirs (origin of emeralds in India) 103
Forsterite (peridot) 67, 169
"Fortification agate" 23

379
Index

Foundation Stones of the New Jerusalem 44, 49, 68, 99, 112, 135, 170, 197,
200, 208, 231, 294, 299, 310, 314
French Blue diamond (origin of Hope diamond?) 88
Freya and the Necklace (amber legend) 40-41

G

Gabelchover, Wolfgang (author) on ruby 187-188, on sapphire 193
Gad (tribe of Israel) connection to gems 319
Gagates (agate) 24, (jet) 137
Galactides 243
Galen, Claudius (Roman physician) 133, 137
Gardens of the Hesperides, as source of amber 34
Gargantua and Pantagruel by Francois Rabelais 91-92
Gargates 243
Garnet 108-119, as carbuncle 56, similar to zircon 232, in literature 267,
as birthstone 310, 312-315, 317, 319, 321, 327
Garuda (India, eagle-king) 104
Gasidana 243
Gemini 317-322, 327
Gemmarum et Lapidum Historia by Anselmus Boëtius de Boodt 47, 51, 134
Gemological Institute of America 21, 292
Gems and Gemology 21
Gems of Beauty by Countess of Blessington, on opal 159
Gesner, Conrad author of *De Omne Rerum Fossilium* and *Lapidum et Gemmarum* 13, 17-18
Glaeser (amber) root word of glass 41
Glaesum (amber) 34
Glanvilla, Bartholomæus, (author) on amber 36
Gleadow, Rupert author of *Origin of the Zodiac* 321-323
Glosopetra 243
Goette, John author or *Jade Lore* 128-132
"Godstones" (rock-crystal) 175
Goethe, Wolfgang von (on carnelian) 59
Golden beryl, as beryl 48, indentified as beryl (Pliny) 51, as chrysoberyl 66
Gonelli, Josephi (Italian physician) on chalcedony 63-64, on diamonds 83
Goshenite 39, 257
Granate (garnet) 267
Granitus (garnet) 111
Granum (garnet) 111
Green jasper 136

380
Gems:

Green onyx 151
Green sapphire 95
Green spinel 204
Grossular garnet 118
Grossularite (garnet) 110
Guzman, Fuentes y (author) on obsidian 149
Gyges, Mystic Ring of (opal) 155, 158, 159
Gyoku (Japanese, jade) 124

H

Haemachates (red agate) 24
Haematite (hematite) 322
Hamalât 2
Hamid, Abdul II 88
Hamlet by Shakespeare 279, 285
Hamonis 243
Hartshorn 239
History and Mystery of Precious Stones, by William Jones , on emerald 100, on chrysolite 171, on zircon 232
He-goat (blood conquers diamond) 90-91
Hei-tiki (jade amulet) 131
Heliodor (beryl) 47, as birthstone 327
Heliotrope (bloodstone) 52, 53, as birthstone 273, 320
Hematite 119-121, as black pearl 119, in literature 274, as birthstone 317, 320
Henry IV by Shakespeare 279, 284
Henry V by Shakespeare 284, 285
Henry VI by Shakespeare 281, 282, 285
Hephoestite 273
Heraclean stone (lodestone) 142
Hermes Trismegetus (Greek for Egyptian Moon god) 104
Hermon (on emerald) 104
Hermonian (emerald, Pliny) 95
Hesoid (author) first reference, diamond 76
Hessonite (garnet) 110, named jacinth 121, as birthstone 327
Heywood, Thomas (on Queen Elizabeth and pearls) 164
Hexacontalite (opal?) 275
Hexaemeron by Ambrose (Greek, source for Medieval writers) 301
Hexagonal crystal system 8, 331
High-Priest's Breastplate, (see Breastplate of . . .) 296

Hilda, Saint (on jet) 138
Hildegarde, Saint 208, (on topaz) 81
Hill, John (translator of *Peri Lithon*) 11
Hind Horn and Maid Rimnald (English tale, diamonds) 83-84
Hidden light 331-332
Hippocrates (medical use of amber) 34
Historia Naturalis (Natural History) by Pliny the Elder 11-12
History of Pretious Stones, . . . The by Thomas Nichols 19, 18-19
"Hoematite" (hematite) in literature 274
Holland, Philemon, (first English translation of Pliny's Natural History) 12
Holme, Saxe author of *My Tourmaline* 213-214
Homer, author of *Odyssey*, (amber, first reference) 38-39
Hope Diamond 88
Hope, Henry Francis (owner of Hope diamond) 88
Hope, Henry Phillip Hope Pelham-Clinton (Hope diamond named for) 88
Hopi legends 227-228
Hopi and the Rescue From the Flood, legend of 227
Horta, Garcias ab (author) on diamond 81
Hu (Chinese, white jade tablet) 128
Huang (Chinese, black jade tablet) 128
Huysmans, Joris Karl author of *Against the Grain* 81
Hyacinth (zircon) 121, 230-232, in the Bible 252, in literature 263, 267,
    as birthstone 297, 308, 313, 315, 319, 320, 327
Hyacinthosontes (beryl, Pliny) 51
Hydrogrossular (garnet) 110
"Hydrophane" (opal) 156
Hyena stone 243

I

Iaspis (jasper) as birthstone 297
"Iaspisachates" (agate, Pliny) 24
Ibn-el-Beithar (Arabian botanist) on turquois 218
"Idem" in literature 288-290
Ijada (jade, first English name) 125
"Imperial jasper" 136
"Imperial topaz" 207
"Indicolite" (tourmaline) 211, as birthstone 322
Indra (India, King of Heavens) 79
Innini (Sumerian goddess) use of lapis-lazuli 139-140
Iolite, as birthstone 327

Gems:

Iona 211
Iron, as planetary ring 325
Ishtar (Innini) 139-140
Isidore, Bishop of Seville 14, on birthstones 301
Isometric crystal system <u>8</u>, <u>330</u>
Issachar (tribe of Israel) connection to gems 319

J

Jacinth (zircon) 121, 230-232, legendary source (like diamond) 89,
    in the *Bible* 252, 256, in literature 259,
    as birthstone 296-298, 300-301, 310, 313, 320-321, 324, 325
Jade 122-132, Chinese symbol for king <u>124</u>,, serpentine substituted for 201,
    in the *Bible* 252, as birthstone 297, 298, 317, 327
Jade Commandments, The (Chinese) 132
Jade Emperor (Yü Huang) 128
Jade Lore by John Goette 128-131
Jadeite 122-123, 125
James, Saint 200, 319
James the Lesser 319
"Jardin" (French, emerald inclusions) 93
Jargoon (zircon) 121, 230, as birthstone 324
Jasper 132-136, as pseudo emerald (Pliny) 12, 96,
    "bloody jasper" (bloodstone) 52, as chalcedony 62,
    as viewing mirror 94-95, in Egyptian jewelry 140
Jasper confused with tourmaline 212 turcica (turquois, the sixth jasper) 216
    in the *Bible* 253, 255, 257, in literature 258, 262, 264, 272,
    as birthstone 294-299, 301, 310, 312-313, 317, 319-321, 322
"Jaspis" (jasper) 133, 135, 294, 297
"Jaspis Aerizusa" (turquois) 216
Jerome, Saint (on sapphire) 193, on birthstones 294
Jet 136-138, in legend regarding turquois 228-229,
    in literature 269, 282, 287, 288
"Jewel of widows" (jet) 112
Jewelry Industry Council 311
Jinns 7
John the Apostle, St. 99, 165, 319, Foundation Stones in *Revelations* 301-309
John of Trevisa (author) on lodestone 144, on sapphire 193-194
Jones, William (author of *History and Mystery of Precious Stones*) on emer-
    ald 100, on chrysolite 171, on zircon 232-233
Josephus, Flavius 293-295, 298, 309, 318, 319

Josheph (tribe of Israel) connection with gems 319
Judah (tribe of Israel) connection with gems 186, 317
Judas, the Apostle 319
Julius Caesar by Shakespeare 189, 287
Juraté and Kastytis, Myth of (on amber) 39-40

K

"Keraunos" (red onyx) 320
Ketu (India, node of the moon) 326-327, 329
King John by Shakespeare 284, 286
King Lear by Shakespeare 281, 285
King of Benin (on coral) 74
Kircher, Athanasius (author) on birthstones 320
Koran 291-292
Krishna (India, god) 185
Kronos 120
"Krystallos" 173
Kshatriya (caste of gem, diamond or ruby) 185
Kunz, George Frederick author of *The Curious Lore of Precious Stones* 20, and *The Magic of Jewels and Charms* 20
Kunzite 317
Kyranide 248

L

"Labradorite" 315
"Lace agate" 23
Lalla Rookh by Sir Thomas Moore 101
"Landscape agate" 23
Lapidaire en Vers by Marbode 14, 186, 259-278
Lapidaire, Le by Chevalier Jean de Mandeville (see Mandeville)
Lapidario of Alfonso X 59
Lapides cancri (crab's eyes) 240
Lapidum et Gemmarum by Conrad Gesner 17
Lapis aquilaris (eagle-stone) 235
Lapis armenus (copper oxide) eye wash 6, regarding sapphire 192-193, regarding turquois 217
Lapis nephriticus (jade) 124-125
Lapis-crucifer (chiastolite, cross-stone) 65, (staurolite, cross-stone) 205

Lapis-lazuli 138-142, as blue agate 24, carved as signet 133,
    misnomer sapphire 191-192, like turquois 216, 218,
    in the *Bible* 255, 257, as birthstone 296, 298, 311, 314, 324
Law of Correspondence 321
Lazurite 139
Lead, as planetary ring 325
Legend of Chief Morning Green on turquois 227-228
Leonardus, Camillus author of *Speculum Lapidum* (*Mirror of Stones*) 16, 35,
    36, 44-45, 51, 60, 63, 69, 73, 113, 114, 135, 144, 146, 152,163, 174-
    175, 177, 196, 199, 200, 231, 238-239, 241-242, 248
"Leopard jasper" 136
Lesh´em (Hebrew, undefined stone) 253, 296, 297
"Leucachates" 24
Levi (tribe of Israel) connection to gems 319
Libellus de lapidibus (*Lapidaire en Vers*) by Marbode 14, 186, 259-278
Liber Memorobilium by Solinus (as source for Medieval writers) 15, 301
Lignite (jet) 137
Ligure (amber?) 35, in the *Bible* 253 as birthstone 294, 297
Lily Adair by Thomas Holly Chivers (on diamond) 75
Limoniatis (beryl, Pliny) 96
"Lion agate" 24-25
Lodestone or loadstone 142-144, as birthstone 317
Louis, Saint (on diamond) 82
Lover's Complaint by Shakespeare 92, 287-288, 290
Love's Labour's Lost by Shakespeare 23, 279, 280, 283, 286
Loves of the Angels by Sir Thomas Moore 79, 162
Lucien, author of *Veri Historia* 45
Lucrece by Shakespeare 288-289
Lychnis (tourmaline?, Pliny) 211, 320

M

Macbeth by Shakespeare 189, 281, 285, 287
"Madeira topaz" (citrine) 70
Magic of Jewels and Charms, The by George Frederick Kunz 20
Magical Papyrus of Leyden (Egyptian) on bloodstone 53, on hematite 120
Magnes (legendary source of name "magnet") 142
Magnet, as birthstone in *CCAG* 320
Magnetite (gray diamond?) 78, 142, 321, 323
Magnus, Albertus author of *De Mineralibus* 15, 49, 68, 72-73, 99, 101, 141,
    155, 193, 219, 241

"Mahogany obsidian" 149
Malachite 145-146, as emerald (Pliny) 12, 95-96, healing stone, Egypt 140,
 in literature 277, as birthstone 298, 317, 322
"Malaya garnet" 116
Manasseh (tribe of Israel) connection with gems 319
Mandeville, Jean de, author of *Le Lapidaire* 27, 35, 45, 50, 60, 63, 68, 84,
 97-98, 100, 115-116, 117, 125-126, 133-134, 141, 151-152, 163, 166,
 170-171, 177, 186-187, 189, 194-195, 198-199, 209, 219, 232
Mangala (India, Mars) 327
Mani Málá by Surindro Mohun Tagore 85-86, 222, 336
Mani málá (India, a necklace, "chain of gems") 344
Manilus, Roman poet, author of *Astronomica* on diamond 80
Marbod, Marbodus, Marbode Bishop of Rennes 15, 42, 68, 74, 97, 122, 144,
 147, 155, 171, 176, 186, 193, 196, 200, 208, 236, 245, 258-278
Marco Polo, (author of *The Book of Marvels)* on diamond 81, 90, on jade 125,
 on pearl 166, on ruby 186,
Margarita (pearl) 163
Martin Luther's ring (ruby) 186
Mat, Egyptian Goddess of Truth 140
"Matrix turquois" 215
Matthew, Saint 208, 300, 319
Matthias, Saint 301
McLean, Edward B. (regarding Hope diamond) 88
McLean, Mrs Edward (regarding Hope diamond) 88
Measure for Measure by Shakepeare 286
Meckenbach, Johann (author) on "oil-of-amber" 37
Median (emerald, Pliny) 95
Meleager 34
Meleagrides 34
"Melitite hematite" 121
Merchant of Venice by Shakepeare 221, 280, 282, 287
"Merchants' stone" (citrine) 70
Merry Wives of Windsor by Shakepeare 280, 282, 283, 287
Meruân, Abû (Abenzoar) on emerald 103
Meteria Medica on use of bezoar 239
"Mexican lace agate" 29
"Mexican turquois" 215
Midsummer Night's Dream by Shakepeare 283, 284, 286, 287
Milk stone (white agate) 27
Milton, John *Epitaph on the Marchioness of Winchester, An* on pearls 165
Mirror of Stones (Speculum Lapidum) by Camillus Leonardus 16, 35, 44-45,
 63, 73, 135, 144, 146, 163, 196, 199, 238-239

386
Gems:

Mitchell-Hedges, F.A. (explorer) 179
Mitchell-Hedges Skull 179
Mizauld, Antoine (on moonstone) 147
"Mocha agate," "mocha stone" 23, 27
"Molochites" (malachite) 277
Monardes of Seville, Dr. (author) on jade 124-126
money stone 70
Monoclinic system 9
Moonstone 146-147, in literature 272, as birthstone 314, 315, 325, 327
Moore, Sir Thomas (poet) 32, 34, 79, 101, 162
Morganite (beryl) 48, 51
Morion (smoky quartz) 172
Mosaic Breastplate see Aaron's Breastplate
"Moss agate" 23, 29
Mouth jades (Chinese) 127
Much Ado About Nothing by Shakespeare 279, 283
Muhammed Ibn Mansur (author) on turquois 219
Musaeum Metallicum in Libros IV Distributum by Ulyssis Aldrovandi 18
My Tourmaline by Saxe Holme 213-214
Mythical gems 234-249

N

Naharare (Hindu physician) on topaz 209
Naphtali (tribe of Israel) connection to gems 319
National Association of Goldsmiths of Great Britain 311
National Association of Jewelers 25, 310-311
Natural History, see Pliny the Elder 12
Natural History of Gems by C.W. King, on amber 37, on lapis-lazuli 141
Navajo Origin Legend (turquois) 228
Navaratna (natal stones) 327
Nephrite (jade) 122-123
Nephtali (tribe of Israel) connection to gems 319
Nero, Emperor (use of emeralds) 94
Nicander (author of earliest Greek medical text) on jet 137
Nicias, Athenian general (source of amber) 33
"Night stones" (opal) 158
Ningyo and the Origin of Pearls (Japanese legend) 168
Nophek (Hebrew, ruby? or garnet?) 186, 296-297
Nordenskjöld, mineralogist, named alexandrite 31

O

Obsidian 148-150
"Occidental topaz" 207
"Oculis cancri" (crab's eye) 240
Odem (Hebrew, carnelian or sard) 58, 295, 297
Odyssey by Homer (amber, first reference) 38-39
Oil-of-amber 35, 36-37
Old Dream Book 187
Oleum Succini of the Pharmacopeia (oil-of-amber) 36
Olivine (peridot) 169, 322
Onomacritis (Greek priest) on rock-crystal 173, 174
Onychinus (Latin), onychion (Greek) birthstone 297
Onyx 151-153, as chalcedony 62, component in sardonyx 200,
    in the *Bible* 265, in literature 215-216
    as birthstone 294-298, 312-313, 317, 319, 321-323
"Onyx obsidian" 148
Opal 153-159, in literature 275, 277, 282, 290,
    as birthstone 296, 311, 313-315, 317, 322
Ophite (birthstone in *CCAG*) 320
"Orange spinel" 204
d'Orchamps, Baron (on emerald) 100
"Oriental ruby" 185
"Oriental topaz" 207
"Oriental-chrysolite" (peridot) 169
"Orientisher Turckise, ein" (turquois) 216
Origin of the Zodiac by Rupert Gleadow 321
Orpheus (poet) 24, 72, 144, 154
Orthorhombic crystal system 9
Ortus Sanitatis, by Johannis de Cuba 4, 38, 55, 113, 167, 195, 233, 235, 237,
    241, 247, 249
Othello by Shakespeare 280, 285
"Ox-blood coral" 71

P

Padmarâga (India, ruby grade) 185, 186
"Palmyra topaz" (citrine) 70
"Pantheros" (opal) 277
Papyrus of King Snefru (*Baufra's Tale*) turquois 217
Paracelsus the Great, on zircon 234

Gems:

Paschali, Michaele (on emerald) 103
Passionate Pilgrim by Shakespeare 287-289
"Patterned agate" 29-30
Paul, Saint 197, 299
Peanita 244
Pearl 160-167, 239, in the *Bible* 254, 256, 257,
　　　in literature 277, 281, 283-285, 287-290, 291-292
　　　as birthstone 311-315, 321, 3235, 327
Pearl Age (late Renaissance Europe) 164
Pearl's orient 160
Pearls, searching for (using agate) 26
"Pearly gates," entrance to New Jerusalem 165
Pepper, George H. (archaeologist) on turquois 225
Peri Lithon (*Of Stones*) by Theophrastus 11
Pericles by Shakespeare 282
Peridot 168-171, named chrysolite 67, as topaz 208,
　　　as birthstone 295, 298, 311, 314, 315, 321, 322, 327
Peridotite 169
Peripatetic School (Theophrastus) 11
"Persian emerald" 95
"Persian turquois" 215
Peter, Saint 135, 299, 319
Petrie, author of Egyptian Tales (turquois) 219
Petrified wood 62, 172
Phaeton, Myth of (amber) 39
Philip, Saint 300, 319
Phillpotts, Eden (on magic) 11
Physiologus (Greek, Christian writer) 26, 63, 80-81
Pi (Chinese, green jade tablet) 128
"Picture jasper" 136
Piezoelectric (quartz)172
Pinctada (pearl oyster) 160
"Pink topaz" 315
"Pink Tourmaline" 315
Pinkerton, John (coined word gemology) on jade 125
Pitdah (Hebrew, topaz or peridot) 297
Pitt, Dr. Robert, author of *The Craft and Frauds of Physick Expos'd* 19-20
Planetary signs 321-329
Plasma (gem) 62
Plato 11, 142, 155, 244

Pliny the Elder 3, 11-13, 17, 24, 25, 34-35, 43, 48, 51, 56, 72, 78, 80-81, 90-91, 93-97, 112-113, 121, 133, 137, 142-143, 147, 149, 154, 161, 173, 184, 207, 211, 216, 217, 230, 232, 296
Poems of Shakespeare 287-290
Popes and gems
    Pope Adrian (amulet of pearls) 165-166,
    Pope Clement VI (healing topaz) 210,
    Pope Gregory II (healing topaz) 210,
    Pope Innocent II (sapphire ring) 196,
    Pope Leo X (moonstone) 147
Porta, Baptista (author) on gathering toadstones 247-248
Porta, Giovanni (on opal) 156
"Prairie Agate" 30
"Prase" (chrysoprase) 62, 68
Precious cat's-eye (cat's-eye chrysoberyl) 60-62, 66
Pribill, Rennaisance physician (quartz) 173
Psellus (author) on diamond 81, on emerald 103, on jacinth 230
Pseudo-smaragdus (emerald, Pliny) 96
Puabi, Sumerian queen 139
Pyrite, in lapis-lazuli 139, in literature 276, 278, as birthstone 320
Pyroelectric 173, 211-212
Pyrope (garnet) 109, 115, 118

Q

Qabalah (mystic Rabbinical writings) 321
Quartz, also see rock-crystal 171-182, as diamond 78, as birthstone 327
"Quartz topaz" (citrine) 172
Queen Victoria (wearing gems) 138, 157-158, 204
Quetzalcoatl, Aztec god 224
Quicksilver (mercury) as planetary ring 325
Quirinus or quirus 244
Qur'an 291-292

R

R. Jehudah of Mesopotamia 91
Rabelais, Francois (author) *The Doors of the Temple of Bacchus* 91-92
Ragiel, Rabbi author of *Book of Wings* 59-60, 196, 209
Rahu (node of the Moon) 326, 327, 329
"Rainbow jasper" 136

Gems:

"Rainbow obsidian" 149-150
Raleigh, Sir Walter (on jade) 126
Ratnapaiska (India, gem texts) 78
Raven stone 244
"Red jasper" 136
Republic by Plato (story of Gyges) opal 155
Reynard the Fox from *Roman de Renard* by William Caxton
    (tourmaline) 212-213
Rhodolite (garnet) 110, 118-119
Richard II by Shakespeare 286
Richard III by Shakespeare 284-285
Ring of Polycrates, The (on emerald) 105-106
Rock Crystal (quartz) 171-182, in literature 286, 289-290,
    as birthstone 314-315, 320-321, 323, 325
Romeo and Juliet by Shakespeare 279, 286
Romount of the Rose by Geofrey Chaucer 116-117
Roosevelt, President Theodore (staurolite amulet) 206
Rosalie Lee by Thomas Holly Chivers (on chrysoprase) 67
"Rose chalcedony" (whetstone) 64
"Rose quartz" 172, 178-179, as birthstone 322
Rosicrucians (on emerald) 102
"Rubellite" (tourmaline) 211
Rubicelle (as caste of ruby) 185, as spinel 203
Ruby 182-189, rank with diamond 81-82, garnet misnomer 115,
    spinel confused for 203, like tourmaline 212,
    in the Bible 252, 254, 255,
    in literature 267, 279, 286-287, 289-290, 292
    as birthstone 296-298, 311-315, 317, 319, 321, 324-325, 327
Rueben (tribe of Israel) connection to gems 319
Rueus, Fanciscus author of *De Gemmis Aliquot* 17, 102
Rumphius (author) on carbuncle 117
Ruskin, John, (on opal) 157
"Rutilated quartz" 179

S

Sadi (poet) on pearl 162
Sahagun, Bernardino de , author of *Booke of Thinges That are Brought from
    the West Indies* 53
Saint John's Island (source of peridot) 169-170

Sang de beouf (reddish-brown citrine) 70
Sapphire 189-197, green as emerald (Pliny) 12 and (Gesner) 17,
   lapis-lazuli misnomer (Pliny) 141, zircon misnomer 230,
   in the *Bible* 251-252, 254-257,
   in literature 258, 262, 279, 283, 287, 290
   as birthstone 294-299, 301, 303, 310-315, 317, 319-322, 324-325,
   327
"Sapphirine" (blue spinel) 203
"Sarcicon emerald" (Pliny) 95
Sard 197-199, as chalcedony 62, like jasper 133, in literature 258, 265, 270,
   as birthstone 305-306, 320
"Sardine emerald" 320
Sardonyx 199-200, as chalcedony 62, form of onyx 151, in the *Bible* 257,
   in literature 258, 265, as birthstone 294-295, 300-301, 305, 310-314,
   317, 319, 320-321
Saxo, Arnoldus (author) on turquois 216
"Saxon topaz" (citrine) 70
"Saxony-chrysolite" (peridot) 169
Saxony, Duke of (pearls) law regarding use of 164
Schist with hematite 121
Scholasticism (Albertus Magnus, founder of school) 15
"Schorl" (tourmaline) 211
"Schwindelstein" (quartz) 177
Scot, Reginald author of *Discoverie of Witchcraft* 49, 73
Scott, Sir Walter (author of *Bridal of the Trierman*) on diamond 84,
   on pearl 163, (author of Anne of Geirstein) on opal 157
Scrying (crystal reading) 178
Scythian emerald (emerald, Pliny) 95, in literature 264, as birthstone 320
Second Breastplate 319
Secundus, Gaius Plinius, see also Pliny the Elder 11
Sehagun (author) on Aztec dwellings 224
Selinite in literature 272, as birthstone in *CCAG* 320,
   Chaldaid planetary stone 324
Septuagint (Greek Christian writings) 296
Serpent stone 244-245
Serpentine 201-202, as birthstone 295, 298
"777" by Albert Crowley 321
Shakespeare, William 114, 189, 221, 248, 278-290
Shamir (rabbinical, see Hebrew yahalom, diamond?) 87
Shani (India, Saturn) 327
Shipley, Robert M. (founder Gemological Institute of America and The American Gem Society) 21

"Sheen obsidian" 148, 150
Shevoh' (Hebrew, precious stone, agate?) 296-297
Shoham (Hebrew, onyx) 296-297
Shukra (India, Venus) 327
"Siberite" (tourmaline) 211, as birthstone 320
Sidereal science (Eastern astrology) 322, 325-329
Sidereal signs for the days of the week 316
"Silente" (jade? Mandeville) 126
Silver, as planetary ring 325
Simeon (tribe of Israel) connection to gems 319
Simon, Fray (story teller) *Child of the Son* 106-107
Simon Zelotes, Saint 301, 319
Sinbad the Sailor, in *The Legend of the Valley of Diamonds* 90
"Smaragdachates" (green agate, Pliny) 24
Smaragdus (see emerald)
Smith, Captain John (use of staurolite) 205
Smithsonian Institution (owner of Hope diamond) 88
Smithsonite (emerald misnomer, Attican, Pliny) 12, 95
"Smoky Quartz" 172, as birthstone 315, topaz substitute 207
"Smoky topaz" (citrine) 207
"Snakeskin agate" 30
"Snakestone" 321
"Snowflake obsidian" 148, 150
Solinus (author of *Liber Memorobilium*) 15, 301
Solomon and the Shamir, legend of (on diamond) 87
Sonnet CXXX by Shakespeare 288
Sonnet XLVI by Shakespeare 290
Sonnet XXXIV by Shakespeare 289
Sophocles, on tears of the Meleagrides (amber) 34
Sorel, Agn'es 82
Sotacus (source for Pliny) 121
"Spanish topaz" (citrine) 70
Speculum Lapidum (*Mirror of Stones*) by Camillus Leonardus 16, 35, 44-45, 63, 73, 135, 144, 146, 163, 196, 199, 238-239
Spessartite garnet 111, 119
"Spiderweb turquois" 215
Spinel 202-204, as birthstone 213, 327, confused with ruby 185
Spring, gems of 315
"Star ruby" 183, 315
"Star sapphire" 315, 322
Staurolite 204-206
Stones for the days of the week 315

Stones for the hours of the day 317
Stones of the seasons 314-315
Story of Balarama, The (origin of jade) 129-130
Story of the Naming, The (jade) 130
Succinite, succinum (amber) 33
Sudra, caste of diamond 77-78, caste of ruby 185, spinel 203
Summer, gems of 315
"Sunstone" 315
Surahs ('chapters of Koran) 291
Surya (India, Sun god) 185, 327
Swallow stone 245-246
Syrices (sapphire) 50

T

Table of Solomon (emerald) 105
Tabula Smaragdina (Greek translation, Eqyptian writings, on emerald) 104
Tactitus (author, on amber) 33
Tagore, Rajah Sir Surinda Mohun, author of *Mani Málá* 85-86, 222
Taming of the Shrew by Shakespeare 279, 280, 284
Tanzanite (as birthstone) 327
T'ao Hung Ching (Chinese author) on jade 128
Tarshish (Hebrew, chrysolite) 296-297
Tavernier (on agate) 27
Teifahi, Ahmed (gem dealer) on emerald 101-102
Teifascite, Ahmed (author) on turquois 218-219
Tempest, The by Shakespeare 280, 283
Ternary of Paradoxes, A, by Joh. Bapt. Von Helmont 196
Tetragonal crystal system 9
Thales of Miletus (first record of electricity) on amber 34
Theocritus (on adamas) 76
Theophilus (author of *The Book of Various Acts*) 173, 178
Theophrastus (author of *Peri Lithon, (of Stones)*) 11, 24, 34, 58, 93, 133, 212, 271, 296
Theosophical Society 321
Tofte, Richard (author) on amber 36
Thorndike, Lynn (author of A History ot the Magical Experimental Sciences, on Marbode) 13-15
Thousand and One Nights (*Legend of the Valley of Diamonds*) 90
"Tiger-eye" 172
Tiki carvings (jade) 131

394
Gems:

Timberlake (author of *Discourse of the Travels of Two English Pilgrims to Jerusalem, Gaza, etc.*) 244-245
Timon of Athens by Shakespeare 281
"Timur ruby" (spinel) 203
Tin, as planetary ring 325
Titus Andronicus by Shakespeare 114, 282, 285
Toad stone 246-248
"Tokay Lux-sapphires" (obsidian) 148
"Top crystal" (grade of ruby) 186
Topaz 206-210, named chrysolite 67, misnomer connected to peridot 169,
    in the Bible 252, 254, 255, 257, in literature 258, 266
    as birthstone 294, 295, 297, 298, 300, 301, 307, 310-315, 317, 319-322, 327
"Topaz quartz" (citrine) 70
Tortoise stone 248
"Tourmalinated quartz" 179
Tourmaline 210-214, misnomer for emerald 17, with quartz 179,
    as birthstone 296, 311, 314, 315, 317, 321, 322, 327
Trallianos, Alexander (author) on engraved gems 133,
    on green jasper (turquois) 217-218
"Transvaal jade" (hydrogrossular garnet) 110
Travels by Mandeville 117
"Tree Agate" 30
"Tree-stone" (agate) 27
Tribes of Israel 319
Triclinic crystal system 9
Trigonal (hexagonal) crystal system 9
Troilus and Cressida by Shakespeare 285
Tsavorite (garnet) 110, as birthstone 327
Ts'ung (Chinese, yellow cube of jade) 128
Turcica, turcicus, turcois (turquois) 216
Turquois 214-229, emerald misnomer (Pliny) 12, 95,
    beryl misnomer (tanos, Pliny) 96, with hematite in jewelry 120,
    in the Bible 257, in literature 287,
    as birthstone 311, 313-315, 317, 322, 325
Twelfth Night by Shakespeare 282, 284
"Twin stone" (staurolite) 205
"Two Brothers of Heavenly Love" or the "jade twins" (jade carvings) 127
Two Gentlemen of Verona by Shakespeare 283, 286

U

Unio (freshwater pearl mussels) 160
Uparatna stones (India, other stones as birthstones) 325
"Uralian emerald" (demantoid garnet) 115
Uvarovite garnet 111, 119

V

Vaidurya stones (India, most revered gems, cat's-eye) 62
Vaisya (India, ruby grade, spinel) 185, 203
Vala (India, demon god, source of many gems) 55-56, 61-62, 74, 79, 103-104, 118, 166, 174, 185, 192
Valentine, Saint (on amethyst) 44
Valley of the Diamonds, legend of *The* 89-90
"Variegated jasper" 136
Varuna 79
Vasari (author) on bloodstone 54
Vasuki (India, celestial serpent) 74, 103-104
Vedic texts (India, mythology) 55-56, 61-62, 74, 79, 103-104, 118, 166, 174, 185, 192, and natal stones 325-329
Vega, Garcilaso de la (author, on emerald) 96
Venus and Adonis by Shakespeare 288-290
Veri Historia by Lucien (on an altar of gems) 45
Voiture, Vincent (French author, on jade) 126
"Violet spinel" 204
Visukamma (legendary carver of the Emerald Buddha) 106
Volömer (Scandinavian god, Vulcan) 155
Von Helmont, Joh. Bapt. author of *A Ternary of Paradoxes* 196
Von Linné, Carl (pyro-electricity of tourmaline) 212
Vulture stone 248

W

"Water opal" 154
"Water sapphire" (beryl) 50
"Watermelon tourmalines" 211
Weiner (author, first use of word zircon) 230
Webster, Robert (gemologist, on zircon) 230
"Whetstone" (rose chalcedony) 64

Gems:

Whitby Abbey (source area, jet) 138
"Whitby jet" (from Whitby Abbey area) 138
"White agate" 30
"White carnelian (agate) 27
"White opal" 153
"White sapphire" (colorless) 190, 315
"White spinel" (colorless) 204
Winston, Harry (regarding Hope diamond) 88
Winter, gems of 315
Wittich, Johann (physician, on coral) 73
Wyckoff, Dorthy, (on Saint Epiphanius and Albertus Magnus) 13, 15

Y

Yahalom (Hebrew, hard stone, translated shamir and later diamond 76, 296-297
Yama (India, god of death) 79
Yamani (agate) 26
Yashepheh´ (Hebrew, jasper or jade) 296-297
"Yellow corundum" 207
"Yellow jasper" 136
"Yellow spinel" 204
Yu (Chinese, jade) 124
Yü Huang (Jade Emporer) 128
Yusuf Ali, Abdullah (translator of Koran) 291
Yü-t'ien District, China (translation, field of jade) 130

Z

Zacutus (author) on diamond 82-83
Zebulun (tribe of Israel) connection to gems 317
Zer (Egyptian queen) 215
Zircon 229-233, named jacinth 127,
    as birthstone 310, 313-315, 320, 321, 324, 325, 327
Zodiac, gems of the 294, 309, 313, 317-323, 327

Jewelers Press

Parachute, Colorado

Notes:

About the Author

Jeweler, gemologist, and educator Bruce G. Knuth holds a B.S. degree in fine arts and an M.A. in jewelry and metals from Western Michigan University. He has also studied at the University of Kansas, Colorado State University, The Gemological Institute of America and Colorado Mountain College. He is retired as a fine arts specialist and instructor from Adams School District 12 in Thornton, Colorado.

Knuth now resides in Parachute, Colorado where he continues work as a custom jeweler and author. His *Jeweler's Resource*, first published in 1994 and now available in a revised edition, has become a standard reference for jewelers and jewelry students.

His popular reference *Gems in Myth, Legend and Lore* is offered here in a revised and reformatted edition. The reduced format size and elimination of color photographs aids in making this new edition more affordable and comprehensive in its coverage.